Seaweed and microalgae as alternative sources of protein

It is widely recognised that agriculture is a significant contributor to global warming and climate change. Agriculture needs to reduce its environmental impact and adapt to current climate change whilst still feeding a growing population, i.e. become more 'climate-smart'. Burleigh Dodds Science Publishing is playing its part in achieving this by bringing together key research on making the production of the world's most important crops and livestock products more sustainable.

Based on extensive research, our publications specifically target the challenge of climate-smart agriculture. In this way we are using 'smart publishing' to help achieve climate-smart agriculture.

Burleigh Dodds Science Publishing is an independent and innovative publisher delivering high quality customer-focused agricultural science content in both print and online formats for the academic and research communities. Our aim is to build a foundation of knowledge on which researchers can build to meet the challenge of climate-smart agriculture.

For more information about Burleigh Dodds Science Publishing simply call us on +44 (0) 1223 839365, email info@bdspublishing.com or alternatively please visit our website at www.bdspublishing.com.

Related titles:

Achieving sustainable production of poultry meat Volume 2: Breeding and nutrition
Print (ISBN 978-1-78676-068-5); Online (ISBN 978-1-78676-070-8, 978-1-78676-071-5)

Achieving sustainable production of pig meat Volume 2: Animal breeding and nutrition
Print (ISBN 978-1-78676-092-0); Online (ISBN 978-1-78676-094-4, 978-1-78676-095-1)

Improving gut health in poultry
Print (ISBN 978-1-78676-304-4); Online (ISBN 978-1-78676-306-8, 978-1-78676-307-5)

Improving rumen function
Print (ISBN 978-1-78676-332-7); Online (ISBN 978-1-78676-334-1, 978-1-78676-335-8)

Reducing greenhouse gas emissions from livestock production
Print (ISBN 978-1-78676-439-3); Online (ISBN 978-1-78676-441-6, 978-1-78676-442-3)

Developing animal feed products
Print (ISBN 978-1-78676-463-8); Online (ISBN 978-1-78676-465-2, 978-1-78676-466-9)

Chapters are available individually from our online bookshop:
https://shop.bdspublishing.com

climate•SMART•publishing
IN AGRICULTURAL SCIENCE

BURLEIGH DODDS SERIES IN AGRICULTURAL SCIENCE
NUMBER 107

Seaweed and microalgae as alternative sources of protein

Edited by Professor Xin Gen Lei, Cornell University, USA

burleigh dodds
SCIENCE PUBLISHING

Published by Burleigh Dodds Science Publishing Limited
82 High Street, Sawston, Cambridge CB22 3HJ, UK
www.bdspublishing.com

Burleigh Dodds Science Publishing, 1518 Walnut Street, Suite 900, Philadelphia, PA 19102-3406, USA

First published 2021 by Burleigh Dodds Science Publishing Limited

Library of Congress Control Number: 2021940795

British Library Cataloguing in Publication Data
A catalogue record for this book is available from the British Library

ISBN 978-1-78676-620-5 (Print)
ISBN 978-1-78676-623-6 (PDF)
ISBN 978-1-78676-622-9 (ePub)
ISSN 2059-6936 (print)
ISSN 2059-6944 (online)

DOI 10.19103/AS.2021.0091

Typeset by Deanta Global Publishing Services, Dublin, Ireland

Contents

Series list

Title	Series number
Achieving sustainable cultivation of maize - Vol 1 From improved varieties to local applications Edited by: Dr Dave Watson, CGIAR Maize Research Program Manager, CIMMYT, Mexico	001
Achieving sustainable cultivation of maize - Vol 2 Cultivation techniques, pest and disease control Edited by: Dr Dave Watson, CGIAR Maize Research Program Manager, CIMMYT, Mexico	002
Achieving sustainable cultivation of rice - Vol 1 Breeding for higher yield and quality Edited by: Prof. Takuji Sasaki, Tokyo University of Agriculture, Japan	003
Achieving sustainable cultivation of rice - Vol 2 Cultivation, pest and disease management Edited by: Prof. Takuji Sasaki, Tokyo University of Agriculture, Japan	004
Achieving sustainable cultivation of wheat - Vol 1 Breeding, quality traits, pests and diseases Edited by: Prof. Peter Langridge, The University of Adelaide, Australia	005
Achieving sustainable cultivation of wheat - Vol 2 Cultivation techniques Edited by: Prof. Peter Langridge, The University of Adelaide, Australia	006
Achieving sustainable cultivation of tomatoes Edited by: Dr Autar Mattoo, USDA-ARS, USA; and Prof. Avtar Handa, Purdue University, USA	007
Achieving sustainable production of milk - Vol 1 Milk composition, genetics and breeding Edited by: Dr Nico van Belzen, International Dairy Federation (IDF), Belgium	008
Achieving sustainable production of milk - Vol 2 Safety, quality and sustainability Edited by: Dr Nico van Belzen, International Dairy Federation (IDF), Belgium	009
Achieving sustainable production of milk - Vol 3 Dairy herd management and welfare Edited by: Prof. John Webster, University of Bristol, UK	010
Ensuring safety and quality in the production of beef - Vol 1 Safety Edited by: Prof. Gary Acuff, Texas A&M University, USA; and Prof. James Dickson, Iowa State University, USA	011
Ensuring safety and quality in the production of beef - Vol 2 Quality Edited by: Prof. Michael Dikeman, Kansas State University, USA	012
Achieving sustainable production of poultry meat - Vol 1 Safety, quality and sustainability Edited by: Prof. Steven C. Ricke, University of Arkansas, USA	013
Achieving sustainable production of poultry meat - Vol 2 Breeding and nutrition Edited by: Prof. Todd Applegate, University of Georgia, USA	014

Acknowledgements

We wish to acknowledge the following for their help in reviewing particular chapters:

- Chapter 2: Dr Jianping Yu, National Renewable Energy Laboratory, USA
- Chapter 3: Dr Schonna R. Manning, University of Texas at Austin, USA
- Chapter 9: Dr Yanhong Liu, University of California-Davis, USA
- Chapter 10: Dr Robert Elkin, Penn State University, USA; and Dr Mike Azain, University of Georgia, USA

Introduction

With traditional sources of protein considered as major contributors to climate change, there is growing interest in alternative, more 'climate-smart' sources of protein. This collection assesses the viability of using macroalgae (seaweed) and microalgae as sustainable protein sources.

Part 1 of the book discusses the different types of macroalgae and microalgae and how they can be used as alternative sources of protein in the food animal sectors. Chapters focus on areas such as using algae in animal feeds, the use of cyanobacteria and eukaryotic microalgae for solar energy conversion, and the extraction of proteins and other functional components from red seaweed. Part 2 focuses on the cultivation and processing of algae, specifically developments in commercial scale farming, developments in algal processing, bioprocessing of microalgal protein and the environmental impacts of seaweed cultivation. Part 3 of the book highlights the use of algae in ruminant, pig, poultry and fish diets.

Part 1 Types of macroalgae and microalgae

The first chapter of the book discusses algae as a potential protein supplement in animal feeds. Chapter 1 begins by highlighting novel protein sources for animal feed and the key factors for selecting alternative protein sources for both human and animal nutrition. The chapter also examines the production of algae for foods and feeds, then goes on to discuss the nutrient composition of different algae. A section on algae as protein supplements in feed is also included. The chapter then provides a summary that emphasises the potential importance of using algae as an alternative in providing nutrients to animals.

The next chapter focuses on solar energy conversion, oxygen evolution and carbon assimilation in cyanobacteria and eukaryotic microalgae. Chapter 2 starts by discussing the evolution and diversity of the photoautotrophic cyanobacteria and microalgae. It then moves on to examine photosynthesis and conversion of solar energy, focusing on areas such as light-harvesting antenna and pigments, the importance of chlorophyll and how phycobilins can fill the gap of chlorophylls. The chapter also discusses carotenoids and how they can facilitate excitation energy transfer and photoprotection. A section on the functions of photosystems in converting light energy into chemical energy is also provided, which is then followed by an analysis of photosynthetic CO_2 assimilation and the potential of cyanobacteria and microalgae.

The final chapter of Part 1 examines the extraction of proteins and other functional components from red seaweed (*Rhodophyta*). Red seaweeds are a major industrial source of agar and carrageenan. They also contain high amounts of other polysaccharides and higher amounts of proteins than brown seaweeds. Some species of red seaweed also contain polyphenols with antioxidant activities as well as being a source of carotenoids and phycobiliproteins. However, there are currently no commercial-scale methods for extracting those bioactive compounds. Chapter 3 provides an overview of various methods for extracting all these functional compounds from red seaweeds. Whereas the main emphasis is on the extractions of proteins, extractions of polysaccharides (carrageenan and agar) and antioxidants (phenolic compounds) are covered. Multi-extraction approaches for extractions of bioactive compounds from red seaweeds are also described.

Part 2 Cultivation and processing

Part 2 opens with a chapter that discusses the developments in commercial scale farming of microalgae and seaweeds. Chapter 4 highlights how expanding markets for microalgae and macroalgae products have led to increased development of commercial farming operations. While microalgae and macroalgae, or seaweed, have historically been harvested in many parts of the globe, more recent developments seek to improve productivity, decrease production costs, increase scale, and mitigate environmental impacts of cultivation. The chapter includes some of those recent developments and identifies future focus areas for research and development.

Chapter 5 considers the developments in algal processing. The chapter outlines various developments in processing technologies used for the treatment of algal biomass with discussion of scalability, cost, time, and efficacy. Topics cover methods for harvesting and dewatering algae, drying algal biomass, biomass disruption, and nutrient recovery, including case studies with lessons learned. The final section presents biomass applications and product considerations. While there is no universally adopted approach for processing algal biomass, these studies provide the foundation for making informed decisions, considering the unique properties of the algae and the integrity of the desired end products.

The subject of Chapter 6 is bioprocessing of microalgal protein and their applications in the cosmetic, nutraceutical and food industries. Microalgae have long been recognised for their nutritional value and high protein contents. They are highly productive and can be grown without competing for arable land or freshwater resources. Most microalgal protein contains all essential amino acids and is hence more suitable as a meat replacement than any other plant protein. Microalgal products are also readily suitable for cosmetic and nutraceutical applications due to their significant health benefits. The chapter provides an overview of the use of various protein-rich microalgae in food, feed, nutraceuticals, and cosmetics. The latest technical advances in protein extraction from microalgae are presented, together with an overview of current knowledge on bioavailability and digestibility of microalgal proteins.

Chapter 7 addresses the environmental impacts of seaweed cultivation, focusing specifically on kelp farming and preservation. The chapter starts with a short overview of the life cycle assessment (LCA) methodology, and how it can be used to quantify the environmental impacts of seaweed supply chains. After a discussion of the overall environmental impacts of the preserved seaweed supply chain, the chapter focuses on specific life cycle stages: spore preparation and seeding of juvenile seaweed onto string in the hatchery, seaweed cultivation, harvesting preservation and storage of harvested seaweed. The chapter ends with a summary and discussion of future trends in the subject.

Part 3 Applications

The first chapter of Part 3 examines the nutritional and anti-methanogenic potentials of macroalgae for ruminants. Chapter 8 begins by discussing the nutritional value of macroalgae, then goes on to examine its digestibility as a feed or feed ingredients. The livestock production sector is facing challenges to find alternative feed resources and nutritional strategies to mitigate enteric methane (CH_2) emissions from ruminants. Recently, marine macroalgae have emerged as potential anti-methanogenic feed ingredients due to their ability to suppress enteric CH_4 production in ruminants. The anti-methanogenic properties of macroalgae have been ascribed to the contents of secondary metabolites, such as halogenated compounds e.g. bromoform in red species, and polyphenols or isoprenoids in brown species. The chapter also analyses the anti-methanogenic properties of macroalgae. A section on the processing and seasonal effects of these properties is also provided, before concluding with an analysis of potential future research trends and a section that emphasises the importance of using macroalgae as alternative ruminant feeds.

The next chapter addresses developing macroalgae as feed for pigs. Macroalgae are a promising source of nutritional ingredients including proteins, polysaccharides and minerals. The need to increase animal and feed production has increased interest in macroalgae as underutilised resources with promising applications as alternative animal feeds. Chapter 9 summarises the nutritional attributes of macroalgae in terms of macro and micronutrients as a source of protein and other compounds in pig nutrition. The benefits of macroalgae or macroalgal derived extracts in feed are discussed together with future trends and challenges in the development of effective feed formulations.

Moving on from Chapter 9, Chapter 10 examines microalgae and its importance as a source of dietary protein, lipids and other compounds for poultry. The chapter reviews effects of supplemental full- or de-fatted microalgal biomass in diets for broiler chickens, laying hens, and other types of poultry on their production performance, meat and egg qualities, nutrient metabolism, and molecular responses. Different sources of microalgal biomass have shown excellent potential to replace a good amount of soybean meal and/or corn without adverse effects. Meanwhile, microalgae are used to enrich chicken meat and eggs with N-3 polyunsaturated fatty acids and bioactive phytochemicals for adding human health-promoting values to these animal products. The dual application of microalgae for producing biofuels and replacing conventional feed protein stands as propitious mediator to reshape the junction between intensifying animal agriculture and meeting the global needs for energy, food, and environmental sustainability.

Chapter 11 focuses on developing macroalgae and microalgae as feed for fish. The rapid expansion of aquaculture industry is severely restricted by the shortage of key feed ingredients such as fishmeal and fish oil. Application of marine plants (macro and microalgae) as dietary ingredients could potentially overcome the limitation of key feed ingredients used in aquafeed. The chapter summarises the current knowledge of the use of macro and microalgae in aquafeed and their effects on overall fish performance.

Part 1

Types of macroalgae and microalgae

Chapter 1

Seaweed as a potential protein supplement in animal feeds

Sung Woo Kim, North Carolina State University, USA

1 Introduction

Global meat demand has been steadily increasing due to population and economic growth, putting pressure on feed supply to support animal production (FAO, 2003). Protein supplements are typically the most expensive components in animal feeds (Kim et al., 2019). The price of protein supplements varies largely depending on their source. Traditional protein supplements include oilseeds, animal coproducts, and distillers grains. Soybean meal is currently the most widely used protein supplement in animal feeds. Two hundred and thirty-eight million metric tons (MT) of soybean meal were used in animal feeds globally in 2020, representing almost 70% of all oilseed meal consumed (ASA, 2020; FAS, 2021). The use of soybean meal in animal feeding increased from 165 million MT in 2010 to 238 million MT in 2020, representing an annual increase of 3.8% (Table 1). The use of dried distillers grains (DDGs) in animal feeds has also become popular since 2000, increasing rapidly to 39 million MT in the US (ERS, 2020a).

Improvements in yield have been a major factor enabling crop production to meet this demand. In the United States, for example, soybean yield improved from 2.89 MT/ha in 2010 to 3.56 MT/ha in 2018 (NASS, 2018). However, there has also been a need to expand agricultural land use to support crop production to meet the demand for foods and feeds (NASS, 2018; ERS, 2020b; FAS, 2021). An

http://dx.doi.org/10.19103/AS.2021.0091.01

Table 1 Global production of soybean meal (million metric ton)[1]

Year	Argentina	Brazil	Canada	China	EU	India	Mexico	Paraguay	USA	Other	Total
09/10	26.6	26.1	1.0	38.6	10.0	6.2	2.9	1.2	37.8	14.7	165.2
10/11	29.3	28.2	1.1	43.6	9.7	7.5	2.9	1.3	35.6	15.9	175.0
11/12	27.9	29.5	1.1	48.3	9.2	8.2	2.9	0.7	37.2	15.8	180.9
12/13	26.1	27.3	1.2	51.5	10.0	8.6	2.9	2.3	36.2	16.1	182.2
13/14	27.9	28.5	1.2	54.6	10.3	7.0	3.2	2.8	36.9	18.1	190.4
14/15	30.9	31.3	1.3	59.0	11.4	6.2	3.3	2.9	40.9	21.3	208.5
15/16	33.2	30.8	1.5	64.5	11.8	4.4	3.5	3.0	40.5	22.8	216.0
16/17	33.3	31.3	1.4	69.7	11.4	7.2	3.6	2.9	40.6	24.5	225.9
17/18	28.4	34.3	1.5	71.3	11.8	6.2	4.2	3.0	44.7	27.4	232.7
18/19	31.2	33.4	1.6	67.3	12.8	7.7	4.4	3.1	44.3	28.7	234.4
19/20	33.8	34.0	1.5	66.9	12.6	6.4	4.7	3.0	44.9	30.1	237.8

[1] Data were adapted from FAS (2021).

estimated 11% of the land is used globally for crop production, covering about 40% of the land potentially available for agriculture (FAO, 2003). However, there are clear limitations to further expansion of agriculture land due partly to the availability of water (McDaniel et al., 2017; Sloat et al., 2018). About 10% of land currently used for agriculture could face water shortages (Fitton et al., 2019). Limitations in available agriculture land and water will potentially limit further expansion of animal production and could even threaten the existing level of production (Sloat et al., 2018). There is, therefore, a need to consider alternative nutrient sources to support growing animal production.

It is important to note that there are a number of challenges in using alternative protein sources. Protein sources from plants contain compounds that interfere with nutrient digestion or impairing animal intestinal health. These compounds are collectively called antinutritional factors and include trypsin inhibitor, lectin, glycinin, β-conglycinin, raffinose family oligosaccharide, and β-galactomannan (Kim et al., 2003; Hong et al., 2004; Taliercio et al., 2014). While the digestive tract of mature animals can cope with most of these antinutritional factors, this is not true for young animals which suffer inflammatory response and oxidative damage in the intestinal epithelium, leading to impaired nutrient digestion and absorption and ultimately reduced growth (Zhao et al., 2014; Yin et al., 2017; Chen et al., 2017; Sun and Kim, 2017).

To avoid problems with antinutritional factors, animal-based protein supplements, mostly rendered animal coproducts, have been used extensively in feeding young animals. These coproducts include meat meal, meat and bone meal, poultry coproduct meal, fish meal, blood coproducts (blood meal, blood cells, and blood plasma), and milk coproducts (dried skim milk, whey protein concentrates, and casein) which are mostly free of typical antinutritional factors. However, the use of rendered animal coproducts has been limited by restrictions in supply. Global production of fish meal, as an example, has fallen from 7 million MT in 1995 to 5 million MT in 2019 (Tacon et al., 2011; Rabobank, 2019). Some rendered animal by-products have also been banned for their potential associations with zoonotic diseases and are less acceptable to consumers (Kim et al., 2019).

2 Novel protein sources for animal feeds

With these limitations in the amount and types of protein supplements available, animal nutritionists have been seeking alternative sources to replace conventional protein supplements. Key factors for selecting alternative sources include safety for both animals and consumers, nutritional value and palatability to animals, affordability and availability in large quantities, consistency in composition, and environmental sustainability (Hard, 2004; Jędrejek et al., 2016; Dee et al., 2018).

Emerging candidates for alternative protein supplements include:

- insect protein,
- single cell-based protein, and
- seaweed.

Larvae of selected insects such as black soldier flies and yellow mealworms can accumulate proteins (35%–55%) and lipids (10%–40%) and have been extensively reviewed (Rumpold and Schlüter, 2013; Makkar et al., 2014; DiGiacomo and Leury, 2019). Single-cell protein is derived from microorganisms (cells), including bacteria, fungi, microalgae, and yeast, with the latter being most developed (Spark et al., 2005; Shen et al., 2009; Espinosa et al., 2020). Reviews of the use of single-cell proteins include Kim et al. (2019) and Jones et al. (2020). The focus of this chapter is on the use of seaweed in animal feeds.

3 Seaweed production for foods and feeds

Seaweed is the common name for a wide range of multicellular macroalgae growing in the marine environment. Seaweed is often classified into three groups:

- Chlorophyta (also known as green macroalgae),
- Phaeophyta (also known as brown macroalgae), and
- Rhodophyta (also known as red macroalgae).

Seaweed has a crucial role in maintaining marine ecology as a source of nutrients and oxygen (Krause-Jensen and Duarte, 2016).

There are nearly 6000 species of seaweeds with a limited number of species used as food and feed sources. Currently, almost 30 million MT/year of seaweed is produced globally, mainly for biomass and food production, with only 1% used for animal feeds (Rajauria, 2015). Seaweed has long been cultured and used as food in Asian countries in particular (FAO, 2018). Production of cultured seaweed increased twofold from 14.3 MT in 2006 to 29.4 MT in 2015. China and Indonesia are major producers accounting for 47% and 38% of global production, respectively, followed by the Philippines (5.3%) and the Republic of Korea (4.1%), as shown in Fig. 1 (FAO, 2018).

Cultured seaweeds from China and Korea are mainly used for foods, whereas those from Indonesia and the Philippines are mainly for the extraction of carrageenan and agar used as an ingredient in food processing and other applications (FAO, 2018). The major cultured seaweed in China is *Saccharina japonica* (a brown seaweed), whereas Korea mainly produces *Pyropia tenara* (a red seaweed) (Abbot, 1988). Major cultured seaweeds from Indonesia and

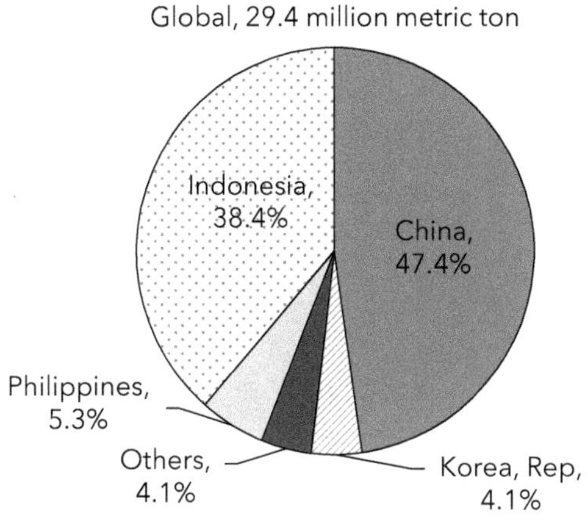

Figure 1 Global production of cultured seaweeds (FAO, 2018).

the Philippines are *Eucheuma* and *Kappaphycus* (Hurtado et al., 2017; Lucas and Southgate, 2012). In the northern Atlantic regions, *Ascophyllum nodosum* (a brown seaweed) is not cultured but extensively harvested for biomass and animal feeds (Anderson et al., 2006; Braden et al., 2007).

4 Nutrient composition of seaweeds

The nutrient composition of seaweeds varies significantly depending on species. On a dry matter basis, seaweeds are mainly composed of carbohydrates (30%–55%), minerals (20%–30%, mainly ash), and protein (7%–30%) with minor amounts of fat (1%–9%) and phenolic compounds (MacArtain et al., 2007; Kumar et al., 2015; Makkar et al., 2016; Ganesan et al., 2020). In general, green and red seaweeds have higher protein content than brown seaweeds (Cabrita et al., 2016). Nutrient composition within a species can also vary depending on season and environmental conditions. Protein and carbohydrate contents are generally highest in young seaweed at the end of winter, whereas fat and ash contents are highest in the spring (Dawes, 1987; Kumar et al., 2015).

Excluding water, carbohydrates are the major component of seaweed, comprising alginates, carrageenans, and agar (MacArtain et al., 2007; Rioux et al., 2007). Seaweed also contains β-glucans (Bobadilla et al., 2013; Nakashima et al., 2018). These are indigestible polysaccharides with a potential role as prebiotics (O'Sullivan et al., 2010; Charoensiddhi et al., 2020; Lopez-Santamarina et al., 2020). Seaweeds contain high levels of minerals such as K (1.5%–4.5%), Na (0.5%–4%), and Ca (0.4%–2%), while the content of other minerals varies depending on the type of seaweeds (Holdt and Kraan, 2011). According to Cabrita et al. (2016), Lorenzo et al. (2017), and Circuncisão et al. (2018), green seaweeds are generally high in Mg (about 2% in green seaweeds

Table 2 Amino acid profiles (relative to lysine) in major protein supplements and seaweeds[1]

%	Protein supplement[2]			Red seaweed[2]				Green seaweed[2]				Brown seaweed[2]							
Lys	SBM	FM	PM	AS	GE	PP	Avg	UL	UF	EF	Avg	PG	LD	AN	SS	FS	SL	MP	Avg
Ala	70	86	105	65	106	157	109	154	87	160	134	86	151	125	118	21	168	73	106
Arg	117	84	112	24	27	134	61	101	56	200	119	69	73	106	45	43	99	73	73
Asp	183	119	143	86	125	796	335	498	116	313	309	152	136	176	336	168	232	163	195
Cys	24	13	22	-	-	-		-	-	-		-	-	-	-	-	-	-	
Glu	289	173	221	144	193	1169	502	632	139	338	369	266	166	304	546	264	232	163	277
Gly	67	103	145	71	149	-	110	-	100	209	155	77	-	-	-	-	-	-	77
His	43	32	42	81	102	28	70	40	27	36	34	61	41	34	23	18	54	35	38
Ile	72	56	63	70	194	92	119	72	77	124	91	77	66	79	76	49	77	67	70
Leu	122	98	116	93	111	170	125	139	131	240	170	124	112	139	143	84	125	114	120
Lys	100	100	100	100	100	100	100	100	100	100	100	100	100	100	100	100	100	100	100
Met	22	38	29	31	65	64	53	58	24	50	44	48	34	40	19	11	49	33	33
Phe	81	54	66	104	235	120	153	115	81	166	121	153	73	89	83	47	94	82	89
Pro	85	63	106	34	38	-	36	-	11	187	99	91	-	-	-	-	-	-	91
Ser	80	53	92	-	-	-		-	-	-		-	-	-	-	-	-	-	
Thr	63	57	64	99	94	89	94	104	74	212	130	93	83	96	109	41	106	78	87
Trp	22	14	16	-	-	28	28	24	-	-	24	-	39	-	-	16	-	-	28
Tyr	54	41	46	88	136	81	102	49	64	109	74	258	39	43	66	30	-	42	80

Val	75	67	77	85	88	158	110	97	89	198	128	122	122	100	108	61	101	89	100
CP, % DM	-	-	-	-	-	22	22	16	18	-	16	-	9.4	7.4	10	12.6	7.6	8.3	9
Lys, % CP	6.20	7.21	6.17	5.26	3.24	0.65	3.05	1.29	5.24	2.02	2.85	2.64	4.69	6.45	3.57	6.03	5.33	7.99	5.24

[1] Data for protein supplements are adapted from NRC (2012) and for seaweeds from Duis et al. (1995), Galland-Irmouli et al. (1999), Cruz-Suárez et al. (2009), Kolb et al. (2004), Ortiz et al. (2006), Cruz-Suarez et al. (2009), Holdt and Kraan (2011), Makkar et al. (2016), Saravana et al. (2016), Tibbetts et al. (2016), Lorenzo et al. (2017), Catarino et al. (2018), and Gaillard et al. (2018).

[2] SBM, soybean meal (dehulled, solvent extracted, IFN: 5-04-612); FM, fish meal (combine, IFN: 5-01-977); PM, poultry meal (IFN: 5-03-798), AS, *Acanthophora spicifera*; GE, *Gracilaria edulis*; PP, *Palmaria palmate*; UL, *Ulva lactuca*; UF, *Ulva fasciata*; EF, *Enteromorpha flexuosa*; PG, *Padina gymnospora*; LD, *Laminaria digitate*; AN, *Ascophillum nodosum*; SS, *Sagassum* spp.; FS, *Fucus* spp.; SL, *Saccharina latissimi*; MP, *Marocystis pyrifera*.

compared with less than 1% in others) and Fe (about 0.2% compared to less than 0.05% in others). Brown and red seaweeds are generally high in Mn (about 200 mg/kg compared to 50 mg/kg in green seaweed), Zn (about 50 mg/kg compared to <20 mg/kg), and I (about 500 mg/kg compared to <100 mg/kg).

Protein contents in seaweeds vary depending on the source. They are mostly lower than those in other major protein sources but are higher than those in cereal grains. Red and green seaweeds contain 16%–22% crude protein (CP), whereas brown seaweeds contain lower CP at 7%–10% (Table 2). Lysine content in seaweeds (0.5%–0.8%) is significantly lower than that in other protein sources (3%–6%) due to low CP content.

Amino acid profiles are similar across seaweeds. Amino acid profiles are shown in Table 2. Arginine (relative to lysine) is about 80% lower in green seaweeds than red and brown seaweeds. Methionine and isoleucine (relative to lysine) are about 50% lower in red seaweeds than green and brown seaweeds. Other amino acids (relative to lysine) are at similar levels across seaweeds. Compared to other major sources of protein supplementation, threonine and tryptophan levels in seaweeds are high, whereas arginine levels are low.

5 Seaweeds as protein supplements in feeds

Considering significant protein content and unique amino acid profiles, some types of seaweed have potential as protein supplements in animal feeds. Beharrell (1942) was among the first to review seaweed as a potential feed supplement. There have been a number of studies evaluating the effects of adding extracts of *Laminaria* spp. (brown seaweed) to swine feeds at ranges of 0.01%–0.05% showing enhanced growth of nursery pigs (Gahan et al., 2009; McAlpine et al., 2012). Dierick et al. (2009) supplemented 1–2% of *Ascophyllum nodosum* (brown seaweed) in swine feeds with similar enhancement in weight gain of nursery pigs. Turner et al. (2002) supplemented 1–2% of *A. nodosum* extract and observed enhanced weight gain in nursery pigs. Feeding A. nodosum to beef cattle improved carcass quality without affecting growth performance (Anderson et al., 2006).

However, there is currently limited understanding of how well nutrients in seaweed are digested and utilized by animals. A limited number of in vitro digestibility studies indicate that protein digestibility varies, depending on the types of seaweed as well as the methods used. Based on in vitro digestibility tests, protein in *Ulva lactuca* is 75–90% digestible (Wong and Cheung, 2001; Bikker et al., 2020), whereas the protein in *Palmaria palmata* is about 57–82% digestible (Galland-Irmouli et al., 1999; Marrion et al., 2003; Bikker et al., 2020). Future investigation of evaluating seaweeds as a protein supplement should also consider the digestibility of amino acids and protein in various seaweeds compared with traditional protein supplements.

6 Conclusion

Considering the limited availability of traditional feedstuffs, seaweed is an attractive alternative in providing nutrients to animals. Protein and mineral contents are generally high, with unique amino acid profiles of potential benefit to animal nutrition. High indigestible polysaccharide content suggests potential applications as prebiotics in feeds. However, there is a need to better understand the nutrient digestibility of seaweed. Large variation in nutrient compositions of different seaweeds also remains a challenge.

7 Where to look for further information

Data on historical changes of seaweed production can be found in the publication from the FAO (2018):

- http://www.fao.org/3/CA1121EN/ca1121en.pdf.

Basic information about the classification, production, processing, composition, and use of seaweed can be found in a book called 'Seaweed Sustainability':

- Tiwari, B. K. and Troy, D. J. 2015. Seaweed Sustainability. Academic Press. pp. 472. Cambridge, MA, USA.

There are recent papers thoroughly reviewing the current challenges and future development of protein supplements for animal production:

- Jędrejek, D., Levic, J., Wallace, J. and Oleszek, W. 2016. Animal by-products for feed: characteristics, European regulatory framework, and potential impacts on human and animal health and the environment. *J. Anim. Feed Sci.* 25(3):189–202. doi: 10.22358/jafs/65548/2016.
- Jones, S. W., Karpol, A., Friedman, S., Maru, B. T. and Tracy, B. P. 2020. Recent advances in single cell protein use as a feed ingredient in aquaculture. *Curr. Opin. Biotechnol.* 61:189–197. doi: 10.1016/j.copbio.2019.12.026.
- Kim, S. W., Less, J. F., Wang, L., Yan, T., Kiron, V., Kaushik, S. J. and Lei, X. G. 2019. Meeting global feed protein demand: challenge, opportunity, and strategy. *Annu. Rev. Anim. Biosci.* Meeting global feed protein demand: challenge, opportunity, and strategy 7:221–243. doi: 10.1146/annurev-animal-030117-014838.

8 References

Abbot, I. A. 1988. Food and food products from seaweeds. In: Lembi, C. A. and Waaland, J. R. (Eds). *Algae and Human Affairs*. Cambridge: Cambridge University Press. p. 141. ISBN 978-0-521-32115-0.

Anderson, M. J., Blanton Jr., J. R., Gleghorn, J., Kim, S. W. and Johnson, J. W. 2006. Ascophyllum nodosum supplementation strategies that improve overall carcass merit of implanted English crossbred cattle. *Asian Australas. J. Anim. Sci.* 19(10):1514–1518.

ASA. 2020. Soy stats 2020: a reference guide to important soybean facts and figures. Available at: https://soygrowers.com/news-media/soy-stats/. St. Louis, MO: American Soybean Association. 36 pp.

Beharrell, J. 1942. Seaweed as a food for livestock. *Nature* 149(3776):306–307.

Bikker, P., Stokvis, L., van Krimpen, M. M., van Wikselaar, P. G. and Cone, J. W. 2020. Evaluation of seaweeds from marine waters in Northwestern Europe for application in animal nutrition. *Anim. Feed Sci. Technol.* 263:114460.

Bobadilla, F., Rodriguez-Tirado, C., Imarai, M., Galotto, M. J. and Andersson, R. 2013. Soluble β-1,3/1,6-glucan in seaweed from the southern hemisphere and its immunomodulatory effect. *Carbohydr. Polym.* 92(1):241–248. doi: 10.1016/j.carbpol.2012.09.071.

Braden, K. W., Blanton, J. R. Jr., Montgomery, J. L., van Santen, E., Allen, V. G. and Miller, M. F. 2007. Tasco supplementation: effects on carcass characteristics, sensory attributes, and retail display shelf-life. *J. Anim. Sci.* 85(3):754–768. doi: 10.2527/jas.2006-294.

Cabrita, A. R. J., Maia, M. R. G., Oliveira, H. M., Sousa-Pinto, I., Almeida, A. A., Pinto, E. and Fonseca, A. J. M. 2016. Tracing seaweeds as mineral sources for farm-animals. *J. Appl. Phycol.* 28(5):3135–3150.

Catarino, M. D., Silva, A. M. S. and Cardoso, S. M. 2018. Phycochemical constituents and biological activities of Fucus spp. *Mar. Drugs* 16(8):249.

Charoensiddhi, S., Abraham, R. E., Su, P. and Zhang, W. 2020. Seaweed and seaweed-derived metabolites as prebiotics. *Adv. Food Nutr. Res.* 91:97–156. doi: 10.1016/bs.afnr.2019.10.001.

Chen, H., Zhang, S., Park, I. and Kim, S. W. 2017. Effects of supplemental protease on growth performance, nutrient digestibility, and gut health in nursery pigs fed diets with corn or sorghum. *Anim. Nutr.* 3(4):359–365. doi: 10.1016/j.aninu.2017.09.005.

Circuncisão, A. R., Catarino, M. D., Cardoso, S. M. and Silva, A. M. S. 2018. Minerals from macroalgae origin: health benefits and risks for consumers. *Mar. Drugs* 16(11):400. doi: 10.3390/md16110400.

Cruz-Suarez, L. E., Tapia-Salazar, M., Nieto-López, M. G. and Guajardo, C. 2009. Comparison of Ulva clathrata and the kelps Macrocystis pyrifera and Ascophyllum nodosum as ingredients in shrimp feeds. *Aquac Nutr.* 15:421–430.

Dawes, C. J. 1987. Physiological ecology of two species of *Sargassum* (Fucales, Phaeophyta) on the west coast of Florida. *Bull. Mar. Sci.* 40:198–209.

Dee, S. A., Bauermann, F. V., Niederwerder, M. C., Singrey, A., Clement, T., de Lima, M., Long, C., Patterson, G., Sheahan, M. A., Stoian, A. M. M., Petrovan, V., Jones, C. K., Jong, J., Ji, J., Spronk, G. D., Minion, L., Christopher-Hennings, J., Zimmerman, J. J., Rowland, R. R. R., Nelson, E., Sundberg, P. and Diel, D. G. 2018. Survival of viral pathogens in animal feed ingredients under transboundary shipping models. *Plos One* 13:e019.

Dierick, N., Ovyn, A. and De Smet, S. 2009. Effect of feeding intact brown seaweed Ascophyllum nodosum on some digestive parameters and on iodine content in edible tissues in pigs. *J. Sci. Food Agric.* 89(4):584–594.

DiGiacomo, K. and Leury, B. J. 2019. Review: insect meal: a future source of protein feed for pigs? *Animal* 13(12):3022–3030. doi: 10.1017/S1751731119001873.

Duis, K., Inglis, V., Beveridge, M. C. M. and Hammer, C. 1995. Leaching of four different antibacterials from oil- and alginate-coated fish-feed pellets. *Aquacult. Res.* 26(8):549–556.

ERS. 2020a. U.S. Bioenergy statistics. USDA. Available at: https://www.ers.usda.gov/data-products/us-bioenergy-statistics/. Economic Research Service.

ERS. 2020b. Oil crops sector at a glance. USDA. Available at: https://www.ers.usda.gov/topics/crops/soybeans-oil-crops/oil-crops-sector-at-a-glance/. Economic Research Service.

Espinosa, C. D., Lagos, L. V. and Stein, H. H. 2020. Effect of torula yeast on growth performance, diarrhea incidence, and blood characteristics in weanling pigs. *J. Anim. Sci.* 98(10). doi: 10.1093/jas/skaa307.

FAO. 2003. *World Agriculture: Towards 2015/2030. An FAO Perspective*. FAO Earthscan Publications Ltd. p. 432. ISBN 9251048355.

FAO. 2018. *The Global Status of Seaweed Production, Trade and Utilization. Globefish Research Program*, Vol 124. Rome: Food and Agriculture Organization. 120 pp.

FAS. 2021. *Oilseeds: World Markets and Trade*. United States Department of Agriculture Foreign Agricultural Service. Available at: https://www.fas.usda.gov/data.

Fitton, N., Alexander, P., Arnell, N., Bajzelj, B., Calvin, K., Doelman, J., Gerber, J. S., Havlik, P., Hasegawa, T., Herrero, M., Krisztin, T., van Meijl, H., Powell, T., Sands, R., Stehfest, E., West, P. C. and Smith, P. 2019. The vulnerabilities of agricultural land and food production to future water scarcity. *Glob. Environ. Change* 58:101944.

Gahan, D. A., Lynch, M. B., Callan, J. J., O'Sullivan, J. T. and O'Doherty, J. V. 2009. Performance of weanling piglets offered low-, medium- or high-lactose diets supplemented with a seaweed extract from Laminaria spp. *Animal* 3(1):24–31.

Gaillard, C., Bhatti, H. S., Novoa-Garrido, M., Lind, V., Roleda, M. Y. and Weisbjerg, M. R. 2018. Amino acid profiles of nine seaweed species and their in situ degradability in dairy cows. *Anim. Feed Sci. Technol.* 241:210–222.

Galland-Irmouli, A. V., Fleurence, J., Lamghari, R., Luçon, M., Rouxel, C., Barbaroux, O., Bronowicki, J. P., Villaume, C. and Guéant, J. L. 1999. Nutritional value of proteins from edible seaweed *Palmaria palmata* (Dulse). *J. Nutr. Biochem.* 10(6):353–359.

Ganesan, A. R., Subramani, K., Shanmugam, M., Seedevi, P., Park, S., Alfarhan, A. H., Rajagopal, R. and Balasubramanian, B. 2020. A comparison of nutritional value of underexploited edible seaweeds with recommended dietary allowances. *J. King Saud Univ. Sci.* 32(1):1206–1211. doi: 10.1016/j.jksus.2019.11.009.

Hard, D. L. 2004. Innovative developments in the production and delivery of alternative protein sources. In: *Protein Sources for the Animal Feed Industry. Expert Consultation and Workshop Bangkok*. Rome: Food and Agriculture Organization. pp. 125–139.

Holdt, S. L. and Kraan, S. 2011. Bioactive compounds in seaweed: functional food applications and legislation. *J. Appl. Phycol.* 23(3):543–597.

Hong, K. J., Lee, C. H. and Kim, S. W. 2004. Aspargillus Oryzae GB-107 fermentation improves nutritional quality of food soybeans and feed soybean meals. *J. Med. Food* 7(4):430–435.

Hurtado, A. Q., Critchley, A. T. and Neish, I. C. 2017. *Tropical Seaweed Farming Trends, Problems and Opportunities: Focus on Kappaphycus and Eucheuma of Commerce (Volume 9 of Developments in Applied Phycology)*. Springer. ISBN 978-3319634982.

Jędrejek, D., Levic, J., Wallace, J. and Oleszek, W. 2016. Animal by-products for feed: characteristics, European regulatory framework, and potential impacts on human and animal health and the environment. *J. Anim. Feed Sci.* 25(3):189–202. doi: 10.22358/jafs/65548/2016.

Jones, S. W., Karpol, A., Friedman, S., Maru, B. T. and Tracy, B. P. 2020. Recent advances in single cell protein use as a feed ingredient in aquaculture. *Curr. Opin. Biotechnol.* 61:189–197. doi: 10.1016/j.copbio.2019.12.026.

Kim, S. W., Knabe, D. A., Hong, K. J. and Easter, R. A. 2003. Use of carbohydrases in corn-soybean meal-based nursery diets. *J. Anim. Sci.* 81(10):2496–2504. doi: 10.2527/2003.81102496x.

Kim, S. W., Less, J. F., Wang, L., Yan, T., Kiron, V., Kaushik, S. J. and Lei, X. G. 2019. Meeting global feed protein demand: challenge, opportunity, and strategy. *Annu. Rev. Anim. Biosci*. Meeting global feed protein demand: challenge, opportunity, and strategy 7:221–243. doi: 10.1146/annurev-animal-030117-014838.

Kolb, N., Vallorani, L., Milanovi, N. and Stocchi, V. 2004. Evaluation of marine AlgaeWakame (Undaria pinnatifida) and Kombu (Laminaria digitata japonica) as food supplements. *Food Technol. Biotechnol.* 42:57–61.

Krause-Jensen, D. and Duarte, C. M. 2016. Substantial role of macroalgae in marine carbon sequestration. *Nat. Geosci.* 9(10):737–742. doi: 10.1038/ngeo2790.

Kumar, S., Sahoo, D. and Levine, I. 2015. Assessment of nutritional value in a brown seaweed Sargassum wightii and their seasonal variations. *Algal Res.* 9:117–125.

Lopez-Santamarina, A., Miranda, J. M., Mondragon, A. D. C., Lamas, A., Cardelle-Cobas, A., Franco, C. M. and Cepeda, A. 2020. Potential use of marine seaweeds as prebiotics: a review. *Molecules* 25(4):1004. doi: 10.3390/molecules25041004.

Lorenzo, J. M., Agregán, R., Munekata, P. E. S., Franco, D., Carballo, J., Sahin, S., Lacomba, R. and Barba, F. J. 2017. Proximate composition and nutritional value of three macroalgae: *Ascophyllum nodosum*, *Fucus vesiculosus* and *Bifurcaria bifurcate*. *Mar. Drugs* 15(11):360.

Lucas, J. S. and Southgate, P. C. 2012. *Aquaculture: Farming Aquatic Animals and Plants* (2nd edn.). Chichester, West Sussex: Blackwell Publishing. p. 276. ISBN 978-1-4443-4710-4. OCLC 778436274.

MacArtain, P., Gill, C. I. R., Brooks, M., Campbell, R. and Rowland, I. R. 2007. Nutritional value of edible seaweeds. *Nutr. Rev.* 65(12 Pt 1):535–543.

Makkar, H. P. S., Tran, G., Heuzé, V. and Ankers, P. 2014. State-of-the-art on use of insects as animal feed. *Anim. Feed Sci. Technol.* 197:1–33.

Makkar, H. P. S., Tran, G., Heuze, V., Giger-Reverdin, S., Lessire, M., Lebas, F. and Ankers, P. 2016. Seaweeds for livestock diets: a review. *Anim. Feed Sci. Technol.* 212:1–17.

Marrion, O., Schwertz, A., Fleurence, J., Gueant, J. L. and Villaume, C. 2003. Improvement of the digestibility of the proteins of the red alga Palmaria palmate by physical processes and fermentation. *Mol. Nutr. Food Res.* 47:339–344.

McAlpine, P., O'Shea, C. J., Varley, P. F., Flynn, B. and O'Doherty, J. V. 2012. The effect of seaweed extract as an alternative to zinc oxide diets on growth performance, nutrient digestibility, and fecal score of weaned piglets. *J. Anim. Sci.* 90 (Suppl. 4):224–226.

McDaniel, R. L., Munster, C. and Nielsen-Gammin, J. 2017. Crop and location specific agricultural drought quantification: part III. Forecasting water stress and yield trends. *Trans. ASABE* 60(3):741–752.

Nakashima, A., Yamada, K., Iwata, O., Sugimoto, R., Atsuji, K., Ogawa, T., Ishibashi-Ohgo, N. and Suzuki, K. 2018. beta-glucan in foods and its physiological functions. *J. Nutr. Sci. Vitaminol. (Tokyo)* 64(1):8–17. doi: 10.3177/jnsv.64.8.

NASS. 2018. Crop production. United States Department of Agriculture National Agricultural Statistics Service. Available at: https://www.nass.usda.gov/Publications/Todays_Reports/reports/crop1118.pdf.

NRC. 2012. *Nutrient Requirements of Swine* (11th rev. edn.). Washington, DC: The National Academies Press.

Ortiz, J., Romero, N., Robert, P., Araya, J., Lopez-Hernández, J., Bozzo, C., Navarrete, E., Osorio, A. and Rios, A. 2006. Dietary fiber, amino acid, fatty acid and tocopherol contents of the edible seaweeds Ulva Lactuca and Durvillaea antarctica. *Food Chem.* 99(1):98–104.

O'Sullivan, L., Murphy, B., McLoughlin, P., Duggan, P., Lawlor, P. G., Hughes, H. and Gardiner, G. E. 2010. Prebiotics from marine macroalgae for human and animal health applications. *Mar. Drugs* 8(7):2038–2064. doi: 10.3390/md8072038.

Rabobank. 2019. *Global Animal Protein Outlook 2020*. Utrecht, The Netherlands: RoboResearch Food & Agribusiness, Robobank. 28 pp.

Rajauria, G. 2015. Chapter 15: Seaweeds: a sustainable feed source for livestock and aquaculture. In: Tiwari, B. K. and Troy, D. J. (Eds). *Seaweed Sustainability*. Cambridge, MA: Academic Press. pp. 389–420.

Rioux, L. E., Turgeon, S. L. and Beaulieu, M. 2007. Characterization of polysaccharides extracted from brown seaweeds. *Carb. Poly* 69:530–537.

Rumpold, B. A. and Schlüter, O. K. 2013. Potential and challenges of insects as an innovative source for food and feed production. *Innov. Food Sci. Emerg. Technol.* 17:1–11.

Saravana, P. S., Choi, J. H., Park, Y. B., Woo, H. C. and Chun, B. S. 2016. Evaluation of the chemical composition of brown seaweed (Saccharina japonica) hydrolysate by pressurized hot water extraction. *Algal Res.* 13:246–254.

Shen, Y. B., Piao, X. S., Kim, S. W., Wang, L., Liu, P., Yoon, I. and Zhen, Y. G. 2009. Effects of yeast culture supplementation on growth performance, intestinal health, and immune response of nursery pigs. *J. Anim. Sci.* 87(8):2614–2624. doi: 10.2527/jas.2008-1512.

Sloat, L. L., Gerber, J. S., Samberg, L. H., Smith, W. K., Herrero, M., Ferreira, L. G., Godde, C. M. and West, P. C. 2018. Increasing importance of precipitation variability on global livestock grazing lands. *Nat. Clim. Chang.* 8(3):214–218.

Spark, M., Paschertz, H. and Kamphues, J. 2005. Yeast (different sources and levels) as protein source in diets of reared piglets: effects on protein digestibility and N-metabolism. *J. Anim. Physiol. Anim. Nutr. (Berl)* 89(3–6):184–188. doi: 10.1111/j.1439-0396.2005.00552.x.

Sun, Y. and Kim, S. W. 2017. Intestinal challenge with enterotoxigenic Escherichia coli in pigs, and nutritional intervention to prevent postweaning diarrhea. *Anim. Nutr.* 3(4):322–330. doi: 10.1016/j.aninu.2017.10.001.

Tacon, A. G. J., Hasan, M. R. and Metian, M. 2011. *Demand and Supply of Feed Ingredients for Farmed Fish and Crustaceans: Trends and Prospects*. FAO Fisheries and Aquaculture Technical Paper No. 564. Food and Agriculture Organization. 87 pp.

Taliercio, E., Loveless, T. M., Turano, M. J. and Kim, S. W. 2014. Identification of epitopes of the β-subunit of soybean β-conglycinin that are antigenic in four animal species. *J. Sci. Food Agric.* 94(11):2289–2294.

Tibbetts, S. M., Milley, J. E. and Lall, S. P. 2016. Nutritional quality of some wild and cultivated seaweeds: nutrient composition, total phenolic content and in vitro digestibility. *J. Appl. Phycol.* 28(6):3575–3585.

Turner, J. L., Dritz, S. S., Higgins, J. J. and Minton, J. E. 2002. Effects of Ascophyllum nodosum extract on growth performance and immune function of young pigs challenged with Salmonella typhimurium. *J. Anim. Sci.* 80(7):1947–1953. doi: 10.2527/2002.8071947x.

Wong, K. H. and Cheung, P. C. K. 2001. Nutritional evaluation of some subtropical red and green seaweeds. Part II. In vitro protein digestibility and amino acid profiles of protein concentrates. *Food Chem.* 72(1):11–17.

Yin, J., Conlon, M. and Kim, S. W. 2017. Nutrients and inflammatory diseases. *Mediators Inflamm.* 2017:6134909. doi: 10.1155/2017/6134909.

Zhao, Y., Weaver, A. C., Fellner, V. and Kim, S. W. 2014. Amino acid fortified diets for weanling pigs replacing the use of fish meal and whey protein concentrate: effects on growth, immune status, and gut health. *J. Anim. Sci. Biotechnol.* 5:57.

Chapter 2

Solar energy conversion, oxygen evolution and carbon assimilation in cyanobacteria and eukaryotic microalgae

Gaozhong Shen, The Pennsylvania State University, USA

1 Introduction

Oxygenic photosynthesis in cyanobacteria, eukaryotic algae and plants is a central biological process that provides for us all of our food to live, oxygen to breathe and, ultimately, most of the energy resources used by mankind. Using sunlight as the energy source, CO_2 of the atmosphere, and water as chemicals, natural photosynthesis in cyanobacteria, algae and plants carries out important light and dark reactions: decomposition of water into molecular oxygen, accompanied by reduction of CO_2 to carbohydrates and other carbon-rich products. Eukaryotic algae are classified into many divisions of the kingdom Plantae. Their sizes range from the single cells of picophytoplankton microalgae to multicellular macroalgae seaweeds. In general, microalgae are defined as a polyphyletic group of eukaryotic, unicellular and photosynthetic organisms that can be found in a variety of sizes, structures and forms. Among the most ancient and diverse organisms on the planet, microalgae include species of diatoms, dinoflagellates and green flagellates. They are distributed

http://dx.doi.org/10.19103/AS.2021.0091.02

worldwide in the sea, in freshwater and in most conditions on land. Because of their high photosynthetic efficiency and the lack of heterotrophic tissues, single-celled microalgae are among the most productive photoautotrophic organisms in nature. Although the term 'blue-green algae' for cyanobacteria was used early on from a bioenergetic point of view regarding their typically plant-type photosynthetic machinery (Fritsch, 1945) and is still widely used in the academic literature, now it has proven more reflective to adapt the classification of the oxygenic photosynthetic prokaryotes as (cyano-)bacteria for purely systematic reasons (Rippka et al., 1979). Cyanobacteria (with a single class, Cyanophyceae) are simply classified as photosynthetic bacteria that have revealed their prokaryotic nature: a single circular chromosome and plasmid(s), lack of membrane-bound organelles and gram-negative cell walls (Waterbury, 2006). Cyanobacterial diversification occurs through ecological adaptation, facilitated by asexual reproduction, homologous recombination and horizontal gene transfer, alongside selection pressures on ecotypes leading to speciation (Willis and Woodhouse, 2020). Genome plasticity allows cyanobacteria to rapidly adapt and be resilient to environmental change, illustrating the means of their persistence and is an important aspect of their biology. Compared to the plastid-containing eukaryotic algae, cyanobacteria are oxygenic photosynthetic prokaryotes.

The storage of solar energy in energetically rich organic compounds and the accompanying evolution of oxygen represents the basis of life on Earth. In addition, oxygenic photosynthesis is extremely important for the maintenance of the delicate equilibrium between oxygen production and CO_2 consumption, which has a severe impact on the world climate and the CO_2 content of the oceans alongside the pH value of oceans, the latter which influences the abundance of other organisms. In the photoautotrophic cyanobacteria and microalgae, the light-driven photosynthetic reactions occur in two distinct stages: (i) light harvesting and (ii) a series of electron-transfer reactions that constitute the 'light-induced reactions'. Water oxidation to molecular oxygen is an important process in the photosystem II (PS II) reaction centers (RCs). The electrons are taken through a series of uphill and downhill steps to generate energy-rich intermediates in the part known as photosystem I (PS I) RCs, as illustrated in Fig. 1. Subsequently, a series of light-independent ('dark') reactions occur in which the products of the light-induced reactions (ATP and NADPH) are used to form C-C covalent bonds of carbohydrates. Therefore, they play a significant role in the global carbon cycle, having substantial implications in the maintenance of the biosphere. The natural diversity and global distribution of different species of cyanobacteria and microalgae make them invaluable resources, such as natural products for food, nutrition, biomedicals, fuel and more. This chapter will focus on the solar energy conversion in light-harvesting and light-driven electron transport for the

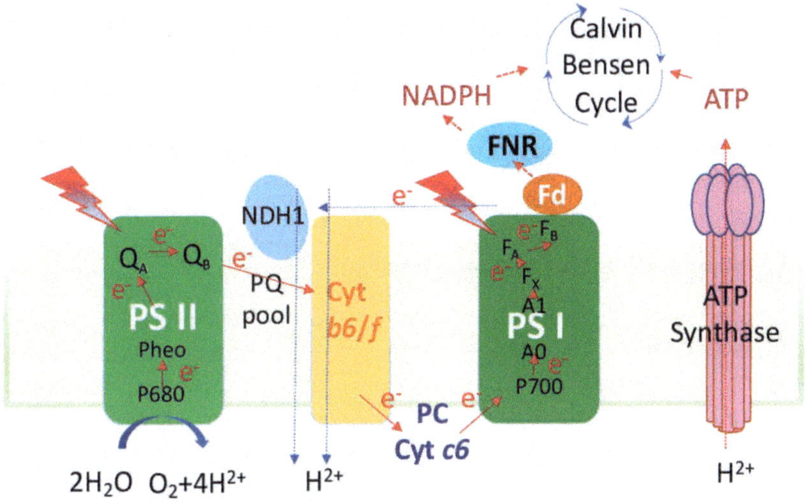

Figure 1 Schematic illustration of the thylakoid membrane showing the major photosynthetic complexes (PS I, PS II, Cyt $b_6 f$ and ATP synthase) structurally and functionally correlated in the light-driven electron transport pathways. Together, PS II generates the most positive redox potential while PS I generates the reducing power to reduce the ferredoxin (Fd), which is a strong reductant, transfers electrons to FNR to generate the powerful naturally occurring reductant NADPH. The electron transfer reactions are coupled with the formation of an electrochemical gradient across the thylakoid membranes, which is essential for driving ATPase. Eventually, NADPH and ATP are utilized in CO_2 assimilation. Major light-harvesting complexes are not shown.

production of reducing power for CO_2 fixation in prokaryotic cyanobacteria and eukaryotic microalgae.

2 Evolution and diversity of the photoautotrophic cyanobacteria and microalgae

2.1 Cyanobacteria and evolution of oxygenic photosynthesis

Most of us take our richly oxygenated world for granted and expect to find O_2 everywhere – after all, it makes up 21% of the modern-day atmosphere. Mostly, oxygenic photosynthesis is the only significant source of free oxygen on Earth's surface (Olson, 2006; Schirrmeister et al., 2013). Oxygenic photosynthesis provides an abundant source of energy and essentially all of the oxygen on the planet. This reaction was the prerequisite for the development of all higher life on Earth, which led to a dramatic change for life on Earth from anaerobic to aerobic conditions (Schirrmeister et al., 2013). Cyanobacteria are a specialized group of photosynthetic organisms that appeared approximately 3.5 billion years ago and began to drive the rise of oxygen in Earth's atmosphere, setting the stage for the evolution of new forms of more complex, multicellular organisms. As such, cyanobacteria have played a significant role in Earth's history as primary producers and the ultimate source of atmospheric oxygen. Cyanobacteria account for 20–30% of Earth's primary photosynthetic productivity and convert solar energy into biomass-stored chemical energy at the rate of ~450 TW (Waterbury et al., 1979).

2.2 Systemic diversity of cyanobacteria

The systematics of cyanobacteria is generally considered on classification schemes based on cell or colony shape (Rippka et al., 1979). Among prokaryotic organisms, cyanobacteria form a morphologically and developmentally diverse group of oxygenic prokaryotic phototrophs, consisting of unicellular, colonial and filamentous strains, some with differentiated cell types (Stanier and Cohen-Bazire, 1977; Waterbury, 2006; Cohen and Gurevitz, 2006). Cyanobacteria differentiation occurs as single cells, simple filaments or the most complex form of branching filaments. The capacity of nitrogen fixation has been reported in many cyanobacteria. Since the nitrogen-fixing enzyme complex, nitrogenase, is extremely oxygen-sensitive, oxygen-evolving cyanobacteria have had to adopt strategies to achieve an anaerobic environment for this enzyme. Many filamentous cyanobacteria solve this issue through cell differentiation. For heterocystous cyanobacteria, oxygenic photosynthesis and nitrogen fixation are separated spatially, with photosynthesis carried out in vegetative cells, whereas nitrogen fixation takes place in specialized cells called heterocysts that have low oxygen permeability (Wolk et al., 1994). Under aerobic growth conditions, heterocystous cyanobacteria's vegetative cells perform oxygenic photosynthesis for O_2 evolution and CO_2 fixation, and then they differentiate heterocystous cells for harboring nitrogenase for nitrogen assimilation with degradation of the O_2 evolving PS II complexes. Heterocysts are specialized cells that provide an oxygen-free environment for nitrogen fixation: the energy-expensive enzymatic process for reducing atmospheric nitrogen gas (N_2) to ammonia (NH_3^-) accompanying the formation of molecular hydrogen. N_2 fixation and H_2 formation are closely linked processes. Hydrogenase recycles the H_2 produced in N_2 fixation for minimizing the loss of energy during nitrogenase catalysis. Other non-heterocystous cyanobacteria, such as filamentous *Trichodesmium*, and unicellular *Gloeocapsa* and *Cyanothece* species (Gallon, 2004), solve this problem by separating photosynthesis and nitrogen fixation by time: photosynthesis during the light period and nitrogen fixation in the dark period. Therefore, at night, nitrogenase is not exposed to the photosynthetically produced O_2 and respiration might then utilize most of the O_2 in the air to provide for anaerobic conditions, especially in dense cultures or in biofilms.

Through adaptation and evolution, cyanobacteria have expanded into a wide range of aquatic and terrestrial environments, including such extreme habitats as hot springs, deserts and polar regions (Whitton and Potts, 2000; Whitton, 2012; Willis and Woodhouse, 2020). They are found in highly diverse and extreme environments, concerning temperature (Antarctica and hot springs) (Nowack et al., 2015), salinity (marine and freshwater environments), pH levels (from acidic hot springs to highly alkaline lakes) and water availability (tropical/Antarctic deserts). However, many cyanobacteria species remain yet to be discovered, across many alternate environments. The known species are distributed among 323 genera (Rippka et al., 1979). It's estimated 8000 species of cyanobacteria, with 4700 remaining to be described. It is suggested that large numbers of cyanobacteria species are yet to be discovered, considering the global biodiversity of cyanobacteria and, especially, of the oceanic cyanobacteria *Prochlorococcus* and *Synechococcus* that are considered to be the most numerous organisms on Earth (Flombaum et al., 2013).

2.3 *Marine picocyanobacteria:* Prochlorococcus *and* Synechococcus

Marine picocyanobacteria of the genera *Prochlorococcus* and *Synechococcus* are globally abundant and ubiquitously distributed marine phytoplankton that dominate the photoautotrophic picoplankton over vast tracts of the world's oceans, where they occupy a key position at the base of the marine food web and contribute significantly to global primary productivity. Together, *Prochlorococcus* and *Synechococcus* contribute to ~25% of marine net primary production (Waterbury et al., 1986; Partensky et al., 1999).

Although they are closely related and often coexist in the same oceanic habitat, marine *Synechococcus* and *Prochlorococcus* possess very different photosynthetic light-harvesting antennas. Whereas *Synechococcus* and the majority of cyanobacteria use phycobilisomes (PBSs), *Prochlorococcus* has evolved to use a chlorophyll a_2/b_2 light-harvesting complex (LHC) although it still possesses genes (*cpeB* and *cpeA*) encoding phycoerythrin (PE). It appears that differences in the absorption properties and cellular costs between chlorophyll a_2/b_2 and PBS antennas in extant *Prochlorococcus* and *Synechococcus* play a role in differentiating their ecological niches in the ocean environment. Therefore, although they often co-exist in the ocean, marine *Synechococcus* and *Prochlorococcus* exhibit different spatial distributions on a worldwide and local scale. Whereas the oceanic *Synechococcus* has a wider habitat range from coastal to open ocean habitats that extend deep into temperate high-latitudes and are generally more abundant in nutrient-rich surface waters, more than 20 clades of marine *Synechococcus* cyanobacteria have adapted distinct strategies for growth and survival across a range of marine conditions and display distinct geographical or seasonal distributions (Farrant et al., 2016). *Prochlorococcus* is largely confined to a 40°N–40°S latitudinal band, being generally absent from brackish or well-mixed waters (Moore et al., 1998). *Prochlorococcus* is often in low abundance or entirely absent from coastal waters where *Synechococcus* thrives and generally extends deeper in the water column than *Synechococcus*. *Prochlorococcus* is most abundant at ocean temperatures above 20°C.

In the oceans, the wavelengths of solar radiation available for fueling photosynthesis differ from those of terrestrial ecosystems. Water rapidly attenuates longer wavelengths of light, so red light does not penetrate below ~15 m. In the open ocean, near-surface waters (i.e. above 30 m) are illuminated primarily with blue and green wavelengths and deeper waters (100–200 m) are enriched with blue light (400–500 nm) (Kirk, 1994). Therefore, the ability to absorb and utilize efficiently the specific wavelengths enriching a particular region and depth of ocean waters would provide a photosynthetic organism with a competitive advantage in that environment. *Prochlorococcus*'s advantage in absorbing blue light wavelengths is consistent with its dominance of deep waters at depths of 100–200 m. Although not able to absorb blue light, as well as *Prochlorococcus* can, *Synechococcus* is capable of absorbing green wavelengths because of containing PBSs. *Synechococcus* also consists of numerous genetically distinct clades, of which the best characterized is the motile clade, whose members are all capable of a unique form of swimming motility. Another genetically distinct clade consists of isolates that lack the chromophore, PUB, which is present in members of all other clades (Rocap et al., 2002). The no-PUB ecotype might be the dominant PE-containing *Synechococcus* in coastal waters, whereas the *Synechococcus* clades that possess PUB to dominate in the open ocean. Related to their distribution in different depth and light conditions in the ocean, the high- and low-light-adapted *Prochlorococcus* ecotypes differ in the number of genes encoding the major chlorophyll-binding, light-harvesting antenna proteins (Pcb) (Moore et al., 1998). Most of the low-light-adapted ecotypes are found in significant abundance only in deeper waters, while the high-light-adapted cells typically dominate surface waters. Genomic sequencing suggests that the presence of multiple *pcb* genes in low-light-adapted ecotypes (seven in SS120) compared with high-light-adapted ecotypes (one in MED4) could be an important factor in acclimation to low irradiance levels (Flombaum et al., 2013).

2.4 Endosymbiosis and evolution of plastids in eukaryotic algae

Photosynthetic organelles (plastids or chloroplasts) were primarily derived from the endosymbiotic uptake of a cyanobacterium (Martin and Kowallik, 1999; Gould et al., 2008; Hohmann-Marriott and Blankenship, 2011; Shih,

2015). The study of plastid multi-gene phylogeny using molecular clock analyses placed the origin of the first alga before 1558 MYA (Yoon et al., 2004; Parfrey et al., 2011). The primary endosymbiosis evolution gave rise to the plastids in three unicellular eukaryotic microalgae. Cyanobacterium was retained as primary plastid over time in three distinct evolutionary lineages: red algae, green algae and glaucophytes. It has been widely accepted that all well-recognized types of primary plastid-containing organisms fall into three groups, each clearly monophyletic: the green algae (including, of course, land plants), red algae and glaucophytes, and all three of these lineages 'probably' trace back to the same cyanobacterial endosymbiosis; that is, primary plastids arose only once. Subsequent to this key event, the primary plastids of red and green algae were laterally transferred to other eukaryotic clades by secondary and tertiary endosymbiosis, spawning some of the most abundant and ecologically important aquatic photosynthesizers on Earth (McFadden, 2001; Palmer, 2003; Parfrey et al., 2011; Gould, 2012; Curtis et al., 2012). Moreover, evidence of horizontal gene transfer from bacteria and archaea to algae to help adapt to extreme environments is also emerging (Flores et al., 2008; Schönknecht et al., 2013), as demonstrated in the genome of *G. sulphuraria* with notable contributions from a pan-domain gene pool, which suggests that its environmental adaptation was facilitated by horizontal gene transfer from various bacteria and archaea.

Eukaryotic microalgae have evolved to adapt to a wide range of environments and consequently have proven to be a rich source of genetic and chemical diversity. For example, it has been estimated that the Rhodophyta (red algae) are currently classified into six classes with more than 6000 species, about 30 marine species in the small phylum of Glaucophyta (glaucophytes), and more than 8000 species for the Chlorophyta (green algae) with 12 classes sorted into marine vs. freshwater/subaerial classes. The great phylogenetic diversity of photosynthetic microalgae corresponds with a wide range of interesting and useful bioproducts for human society.

3 Photosynthesis and conversion of solar energy

3.1 Light-harvesting antenna and pigments

Oxygenic photosynthesis begins with the capture of light energy by pigment-protein antenna complexes, which surround the RCs to increase their absorption cross-section for the incoming sunlight. The solar spectrum provides the source of electromagnetic energy to drive photosynthesis. Harvesting light in the utilization of solar energy is the first step of photosynthesis in cyanobacteria, algae and also higher plants. To adapt to the extraordinary diversity of light environments, photosynthetic organisms have evolved a wide range of light-harvesting antenna systems and adapted different mechanisms of light energy harvesting and transferring for optimizing the utilization of solar energy. The different chromophores are used to absorb in a spectral region in which radiant energy is available and to transfer the energy to the RC where electron transfer processes store a fraction of the energy in the photon as oxidized and reduce species. The spectrum of light absorbed by these light-harvesting antenna complexes is controlled by the properties of the pigments and the interactions of the pigments with one another and with the protein environment in which they are located.

Antenna systems are found in every known photosynthetic organism and typically consist of pigments specifically bound to membrane-associated proteins (Green and Parson, 2003; Perrine et al., 2012). The nature and size of the antenna systems vary widely among different types of photosynthetic organisms. The antennae can absorb solar photons of many different colors, substantially increasing the effective absorption cross-section of the photosynthetic RCs. Light-harvesting antenna systems are characterized by a well-defined structure, that is, the spatial arrangement of the chromophores (embedding the light-harvesting pigments in the protein matrix) for maximizing the potential for light absorption and energy transfer. Pigments are arranged with a high absorption cross-section to increase the number of photons that can be absorbed at any time. A greater range of absorption across the solar spectrum is achieved by using complexes with a variety of pigments and also protein conformation adjustments in binding chromophores. The main pigments used in natural light harvesting are substituted tetrapyrroles. The absorption properties of these pigments vary in the extent of the conjugation and the number and nature of substitutions, creating a 'palette' that covers the visible and far-red regions of the solar spectrum. In cyanobacteria and microalgae, the light-harvesting antenna systems are mostly composed of three categories of chromophores: chlorophylls, phycobilins and carotenoids.

3.2 Chlorophylls in light absorption and excitation energy transfer

Chlorophyll (Chl) is an essential component of the photosynthetic apparatus. Chls are a fundamental class of tetrapyrroles and function as the central RC, accessory and photoprotective pigments in photosynthesis. Their unique individual photochemical properties are a consequence of the tetrapyrrole macrocycle, the structural chemistry and coordination behavior of the phytochlorin system, and specific substituent pattern. The Chl absorption properties are highly tunable, as the saturation of chemical bonds and the use of different substituents on the pyrrole rings strongly influence energy levels. Five chemically distinct Chls have been identified in light-harvesting complexes of oxygenic phototrophs: Chls *a*, *b*, *c*, *d* and *f*. The chemical difference of Chls is the fact that in Chl *b*, Chl *c*, Chl *d* and Chl *f*, in contrast to Chl *a*, a methyl or vinylic group is substituted by a formyl group, found in oxygenic photosynthetic organisms. They achieve their full potential in solar energy conversion by working in concert in highly complex, supramolecular structures such as the RCs and light-harvesting complexes of photobiology. Embedded into Chl-binding proteins, Chl molecules play a central role in light harvesting and charge separation within the photosystems. It is critical for the photosynthetic cell to not only ensure the synthesis of a sufficient amount of new Chl-binding proteins but also avoid any imbalance between apoprotein synthesis and the formation of potentially phototoxic Chl molecules. The biochemical function of these structures depends on the controlled interplay of structural and functional principles of the apoprotein and pigment cofactors.

All oxygenic phototrophs have Chl *a*. Chl *a* serves a dual role in oxygenic photosynthesis: in light harvesting as well as in converting energy of absorbed photons to chemical energy. Chl *a*, depending on its protein environment, functions either as a light harvester or as a redox participant in electronic excitation trapping (primary charge separation) and electron transporting events in the PS II and PS I RCs of oxygenic organisms through photochemical charge separation to create strong oxidants and reductants, sending low-potential electrons through the electron transport chain. Synthesized from Chl *a* by chlorophyll-*a*-oxygenase, Chl *b* is less widely distributed than Chl *a*. The Chl *b* molecules, found in prochlorophyte, green algae, Characean algae and plants, serve as accessory pigments in the light-harvesting complex I (LHC I) and II (LHC II). Chl *c*, which occurs mainly as Chl c_1, c_2 or c_3, distinguished by the residues at the porphyrin ring system, is characterized by the lack of a phytol ester. Chl *c* and derivatives serve as accessory pigments with functions in light-harvesting and energy transfer. Most, but by far not all, LHC complexes of the 'red lineage' microalgae bind Chl *c* in addition to Chl *a*. Chl *d* was first discovered as a minor component in red algae (Manning and Strain, 1943), but in the cyanobacterial group of *Acaryochloris marina*, Chl *d* can replace nearly all of the functions of Chl *a* (Miyashita et al., 1996). The structure of Chl *d* is identical to Chl *a*, apart from a substitution of the C3 vinyl group in ring I of Chl *a* for a formyl group, which leads to a red shift in its absorbance properties for the advantages in absorbing light filtered through Chl *a*-containing organisms.

Chl *f* molecules are the most red-shifted chlorophylls (Chen and Blankenship, 2011), that are found in cyanobacteria capable of far-red light photoacclimation (FaRLiP) (Gan et al., 2014a,b). In oxygenic photosynthesis, the spectral range that can be utilized effectively for photosynthesis is primarily limited to the visible range, approximately from 400 to 700 nm, which makes up only about half of the incident solar spectrum. This spectral range is called photosynthetically active radiation (PAR), which on an energy basis is about 43% of the full solar spectrum on the surface of the Earth. However, various photoacclimation mechanisms have evolved that allow cyanobacteria and microalgae to utilize longer wavelength light in the far-red spectral range. Far-red light (FRL) is particularly enriched in shaded conditions. When a stratification exists in a forest canopy, water column in an ocean or lake, microbial mat or other biological community, FRL tends to penetrate the first few layers of phototrophs. These photoacclimations provide distinct advantages to the different cyanobacteria in adaptation to the different light environments for effective utilization of the light source. In some cyanobacterial species, FaRLiP includes the remodeling of PBS antennas and PS I and PS II RCs (Gan et al., 2014a,b) and the production of FRL-absorbing Chls *d* and *f* (Ho et al., 2016; Shen et al., 2019). FaRLiP is regulated using a red/far-red knotless phytochrome RfpA, as well as RfpB and RfpC, which are involved in a signaling cascade that induces the expression of the FaRLiP gene cluster (Zhao et al., 2015a). With the synthesis of Chl *d* and Chl *f*, cyanobacteria capable of FaRLiP express a set of FaRLiP genes: six genes (*psaA2, psaB2, psaF2, psaJ2, psaL2 and psaI2*) for assembly of FRL-PS I, and five genes (*psbA3, psbD3, psbB2, psbC2 and psbH2*) for assembly of FRL-PS II for harboring far-red light-absorbing Chl *f* and Chl *d*. Chl *f* is located in both FRL-PS I and FRL-PS II, but only in FRL-PS II for Chl *d*. As characterized structurally and spectroscopically, Chl *f* molecules in FRL-PS I function as accessory pigments for extension of the light-harvesting region (Gisriel et al., 2020a,b; Cherepanov et al., 2020).

Microalgae adapt a very different strategy for utilization of FRL in comparison to cyanobacteria. Some microalgae, including many diatoms and some Eustigmatophytes (Wolf and Blankenship, 2019), can absorb FRL by express specialized light-harvesting pigment-protein complexes that use interactions between pigments and their protein environments to spectrally tune the absorption of Chl *a* with shifting the absorption spectrum of Chl *a* in their LHC-type antennas instead of utilizing a chemically distinct pigment.

3.3 Phycobilins filling the 'green gap' of chlorophylls

The absorption spectra of Chls exhibit very small absorbance in the region of 500–600 nm, known as the 'green gap'. Cyanobacteria and some eukaryotic algae use special phycobilins and carotenoids to fill the absorbance gap to complement the absorption of Chl *a*. Phycobilins, open tetrapyrroles that absorb between 550 and 680 nm, are the chromophores of light-harvesting PBS in cyanobacteria and red algae. Phycobiliproteins (PBPs) with linear tetrapyrrole prosthetic groups (bilins), in their functional state, are covalently linked to specific cysteine residues of the proteins (Glazer and Cohen-Bazire, 1971; Bryant et al., 1979). PBPs are abundant in cells and serve as a major light-harvesting complex for photosynthesis. However, they also serve as a nitrogen reservoir, especially in dealing with nitrogen starvation. All PBPs contain one or two of either the phycocyanobilin or phycoerythrobilin chromophores, and may also contain one of the three minor bilins. Each bilin has unique spectral characteristics that may be further modified by the interactions of the protein subunits and of the chromophore with the apoprotein (Glazer and Cohen-Bazire, 1971; Bryant et al., 1979; Schluchter et al., 2010; Adir et al., 2020). Based on absorption wavelength, the PBPs in cyanobacteria and red algae are generally classified into four groups known as (1) PE; λ_{max} = 490–570 nm), (2) phycoerythrocyanin (PEC; λ_{max} = 560–600 nm), (3) phycocyanin (PC; λ_{max} = 610–625 nm) and (4) allophycocyanin (AP; λ_{max} = 650–660 nm), though two classes of PBPs exist in crypotomobads exist: PE (λ_{max} = 540–570 nm) and PC (λ_{max} = 610–650 nm). Also, PCs are subdivided into C-PC and R-PC. There are two types of PE, namely R-PE and B-PE, which were named after the taxa of the photosynthetic organisms in which they were isolated. R-PE was isolated first from Rhodophyta and B-PE from the Bangiales order (red seaweed) of Rhodophyta (Sidler, 1994).

PBSs appear to be restricted to the cyanobacteria, glaucophytes, red algae and the secondary endosymbiotic ancestors of the red algae, such as the cryptophytes. Structurally, PBS is an extra-membrane supramolecular complex composed of many chromophore (bilin)-binding proteins and linker proteins, which generally are colorless. It has a central core, which is made of two to five AP cylinders, surrounded by six to eight rods that are made from stacked discs of the PC, which primarily has the pigment phycocyanobilin covalently bound (Sidler, 1994; Adir et al., 2020). Both rods and cores are cylindrical structures composed of a series of discs of PBPs, which consist of chromophorylated hetero-subunits α and β that are together called a 'monomer'. Two stacks of three monomers form a disc. The linker proteins hold the discs together, keep the rods connected to the core, and facilitate unidirectional energy transfer from the outer portions of the rods into the photosystems. In addition, specialized linker proteins keep the core, and thus the PBS itself, associated with the thylakoid membrane. While different PBSs may be composed of rods with various types of PBPs, the core-distal discs always have absorption wavelength maxima that are shorter than or equivalent to the core-proximal ones to ensure unidirectional energy transfer into the RC. As demonstrated in the PBS structures from red algae *G. pacifica* (Zhang et al., 2017), representing the largest PBS complex with a molecular weight of over 16.8 MDa, the PBS structure contains 862 unique protein chains and over 2000 chromophore molecules. Seventy-two linker proteins were modeled within the core cylinders and rods.

The PBSs are found attached to the stromal side of thylakoid membranes, filling up most of the gap between individual thylakoids. PBSs collect light energy of a wide range of wavelengths, funnel it to the central core, and then transfer it to the photosynthetic RCs. Energy transfer from shorter wavelength-absorbing PBPs of peripheral rods to longer wavelength-absorbing PBPs of the AP cores and then to Chls is efficient due to spatiotemporal fine-tuning of each bilin chromophore, which is achieved by chromophore-chromophore interactions and chromophore-protein interactions. Energy absorbed by the PBSs is usually funneled to the PS II RCs, and also can be directed to either PS II or PS I RCs in adaptation to alternate light conditions (Liu et al., 2013). The heterogeneity of spectral properties of core PBPs likely enables the core PBS to collect excitation energy from the peripheral rods and to funnel it to the energy sink, which is usually PS II. Energy is transferred from the peripheral rods to PS II very efficiently. However, upon exposure to light that excites PS II more than PS I, energy transfer switches from PS II to PS I. The different light absorption properties of the two photosystems and their associated light-harvesting proteins require a frequent balancing of the input excitation energy between the two RCs. Such adjustments can act on a very short timescale (seconds to minutes) and these are called State Transitions (Mullineaux, 2008; Dong et al., 2009), which involves rapid changes in the specific associations of PS I, PS II and PBSs in cells. Preferential excitation of PS II leads to State 2, in which excess energy is channeled to PS I, while over excitation of PS I leads to State 1, during which energy is redistributed to PS II, which is induced by changes in light color/wavelength, and involves relative movements of PBSs and/or photosystems that modify the contacts between PBSs and the two photosystems as well as between PS II and PS I.

In addition to light irradiance variation, cyanobacteria experience differences in light color in their natural environment. When exposed to different light colors, many species are able to adjust the composition of their PBSs through a process known as complementary chromatic acclimation or adaptation (CCA) (Gutu and Kehoe, 2012; Ho et al., 2017), which refers to the light wavelength (color)-dependent alteration of PBPs (PE and PC) that occur in the peripheral rods of PBSs. However, CCA can only occur in cyanobacterial species that make both PC and PE, although not all such species are capable of this process. It is noteworthy that although red macroalgae also use PBSs for light harvesting, none are known to undergo CCA. In adaptation to growth with the utilization of FRL, cyanobacteria capable of FaRLiP express a special set of AP genes (*apcB2*, *apcD2*, *apcD3*, *apcE2* and *apcD5*) for assembly of FRL-PBS core complexes (Gan et al., 2014a,b; Ho et al., 2017; Bryant et al., 2020; Soulier et al., 2020) to extend the light-harvesting region to FRL (λ = 700–800 nm). Therefore, when grown under FRL conditions, these cyanobacteria produce specialized, paralogous APs that comprise the cores of FRL-PBS or smaller, bicylindrical core complexes, which enable cells to absorb FRL. Also, as observed in the mutants unable to synthesize the FRL-specific APs associated with growth in FRL, mutant cells were found to be strongly depleted or even devoid of Chl *d*, suggesting that the thiol-rich FRL-AP cores may play a role in producing the small amount of Chl *d* required for growth in FRL (Bryant et al., 2020). In some low-light ecotypes of the Type A thermophilic *Synechococcus* species, a phenomenon termed low-light photoacclimation (LoLiP) had evolved to improve light harvesting at low light irradiance (Nowack et al., 2015; Becraft et al., 2015). The genomes of LoLiP *Synechococcus* strains contain an operon with a pair of AP genes (*apcD4* and *apcB3*) and a putative Chl-binding protein gene, *isiX*, for cells to perform LoLiP with the extension of light absorption beyond 700 nm (Soulier et al., 2020), for improving growth at low irradiance and far-red fluorescence emission that was absent in high light ecotypes.

3.4 Carotenoids in facilitating excitation energy transfer and photoprotection

In photosynthetic organisms, carotenoids are widespread pigments with important structural and functional roles. Carotenoids are a group of structurally and functionally diverse isoprenoid pigments naturally produced in cyanobacteria, and in eukaryotic algae and plants. A double bond in its structure forms a chromophore group, which gives them different colorations: yellow, orange or red. Most carotenoids share a common C40 backbone structure of isoprene units (termed terpenoid). Carotenoids can be classified as carotenes, having a non-oxygenated hydrocarbon structure, and under their oxygenated derivatives, the xanthophylls. The hydrophobic character of carotenoids leads to their preferential presence in the lipid membrane environment. The majority of carotenoids, especially xanthophylls, are located in the outer, cytoplasmic and thylakoid membranes. In cyanobacteria, the most abundant carotenoids are β-carotene and various xanthophylls, such as synechoxanthin, canthaxanthin, caloxanthin, echinenone, myxoxanthophyll, nostoxanthin and zeaxanthin (Graham and Bryant, 2008; Zhu et al., 2010). They do not synthesize β-rings and therefore contain β-, but not α- or δ-, carotenes and their oxygenated forms. Glycosylated carotenoids are also very common in cyanobacteria (Hirschberg and Chamovitz, 1994). Microalgae contain some special carotenoids, such as the acetylenic fucoxanthin in diatom *Chaetoceros gracilis* and 9′-*cis* neoxanthin in green algae. Evolutionarily and functionally, carotenoid pigments afford valuable clues about the chemistry and biology of both modern and ancient aquatic environments (Cui et al., 2020). Carotenoids might be evolutionarily restructured in eukaryotic algae during the endosymbiosis of cyanobacteria (Frommolt et al., 2008).

In cyanobacteria and microalgae, most carotenoids are located in the thylakoid membranes and are bound with the photosynthetic complexes.

Structurally, carotenoids play crucial roles in the assembly and stabilization of photosynthetic complexes and function in transferring the captured energy to Chls and initiating electron transport. As mentioned previously, carotenoids can function as accessory light-harvesting pigment for filling the 'green gap'. The carotenoids function to absorb light and quench excess energy in photosynthetic metabolism. Some primary carotenoids serve as accessory pigments that can transfer absorbed energy to Chls. Noticeably, the excited-state lifetime of carotenoids is very short, typically on the order of 10 ps, which means that excitations are almost immediately lost if left unchecked. Therefore, carotenoids are always found in close contact with Chls, allowing excitation energy transferred to Chl within 1 ps, after which excitations are stabilized for several nanoseconds (Croce and van Amerongen, 2014). One good example for carotenoid fucoxanthin in light harvesting is that the light-harvesting fucoxanthin-binding proteins of brown algae are discovered to be associated with PS I or PS II. These supramolecular structures are characterized by a very high pigment content, since they bind several fucoxanthin, violaxanthin, Chl *c* and Chl *a* molecules. In these aggregates, fucoxanthin has a light-harvesting function, transferring its excitation energy to Chl *a* (Katoh et al., 1989). Mostly, those carotenoid compounds perform important functions in the primary processes of photosynthesis. Among carotenoids, β-carotene and several xanthophylls serve as constituents of functional multiprotein photosynthetic complexes, such as PS I and PS II, Cyt b_6f complexes, and the light-harvesting complexes. These pigments can be essential for the assembly of the protein complexes (especially the light-harvesting complexes and photosystems) and for maintaining membrane integrity.

In light harvesting and excitation energy transfer, carotenoids also provide photoprotection. Chl excitations can lead to dangerous triplets, especially when organisms are exposed to excessive light, which causes the formation of reactive oxygen species (ROS), which can be lethal for cells (Hirschberg and Chamovitz, 1994; Zhu et al., 2010). ROS are very harmful to the cells, since they can damage pigments, proteins and lipids. Carotenoids are highly efficient quenchers of Chl triplets when they are in direct contact with the Chls, and the overlap of their wave functions allows the exchange of electrons and the quenching of the triplet excited state of Chls and direct scavenging of singlet oxygen. Structurally, most Chls in the antenna are in direct contact with a carotenoid. For example, hydroxy-echinenone or echinenone serves as an activator switch in the orange carotenoid protein, which is responsible for non-photochemical quenching in cyanobacteria (Kirilovsky and Kerfeld, 2016) and the protection of cells from oxygen radicals. Due to their anti-oxidant properties, carotenoids can protect cells from reactive radicals, prevent lipid peroxidation and promote the stability and functionality of the photosynthetic apparatus.

4 Functions of photosystems in converting light energy into chemical energy

The light reactions of oxygenic photosynthesis are carried out in specialized membranes called thylakoids. The thylakoid membranes of cyanobacteria and microalgae are a complex intracytoplasmic membrane system that houses two photosystems with the photosynthetic electron transport chains, and also most of the respiratory electron transport complexes. Thylakoid membrane pairs separate a lumenal space from the cytoplasm: electron transport results in proton translocation across the membrane and into the lumen, and the resultant proton motive force is used to power ATP synthesis. As illustrated in Fig. 1, the core enzymatic machinery of light reactions is made up of four large membrane complexes, two photosystems PS II and PS I, the Cyt b_6f complex and ATP synthase (F-ATPase, proton-motive-force-driven ATP synthase). The photosynthetic RC is an integral membrane pigment-protein and functions by carrying out a series of photo-induced electron transfer reactions. Following photon absorption, antenna pigments channel excitation energy via exciton transfer to a closely arranged pair of Chls in the photosynthetic RCs, PS II and PS I. This excited special pair of Chls ($P680^+$ in PS II and $P700^+$ in PS I) transfers an electron to an acceptor molecule, resulting in a charge-separated state between the reduced acceptor and the oxidized donor molecule. After the initial electron transfer event, a series of electron transfer reactions take place that eventually stabilize the stored energy in reduction equivalents and ATP.

4.1 PS II in generation of oxidative power for water splitting and oxygen evolution

Cyanobacteria and microalgae are an important group of photosynthetic organisms that have been credited with the creation of an oxygenic environment on Earth. It has been estimated that the abundance of cyanobacteria in both marine and fresh water is the reason that more than 25% of all oxygen that we breathe is still produced by their photosynthetic activities. While there exist a wide variety of photosynthetic organisms from cyanobacteria to eukaryotic algae and plants, the cellular machinery responsible for water splitting is unique and uses the same mechanism across all oxygenic photosynthetic species. The environmentally and energetically most important photosynthetic light reaction is the light-powered water-splitting reaction at the donor side of PS II, which generates oxygen. The photosynthetic electron transfer is initiated at PS II, which catalyzes the unique light-driven oxidation of water, which is catalyzed by an inorganic Mn_4O_5Ca cluster (Ferreira et al., 2004). PS II generates an oxidant with a redox potential high enough to enable it to oxidize H_2O, a substrate so abundant that it assures a practically unlimited electron source for life on Earth.

Structurally, although the low-molecular-weight and extrinsic subunits of PS II vary among diverse oxygenic phototrophs, the RC core is remarkably conserved in PS II, which is composed of D1 (PsbA), D2 (PsbD), CP43 (PsbC), CP47 (PsbB) and Cyt b_{559} (α- and β-subunits). Each monomer of the dimeric PS II protein complex consists of up to 20 protein subunits, 35 Chl *a* molecules, 20–25 lipids, 12 β-carotenes, 2–3 plastoquinone, 2 pheophytins, 2 hemes, the water-oxidizing complex (Mn_4O_5Ca), 4 Ca^{2+} ions, 3 Cl^- ions and non-heme iron. Most redox centers are coordinated by the central transmembrane subunits D1 and D2, whereas light harvesting is mainly mediated by Chl molecules that are bound to the intrinsic antenna proteins CP43 and CP47. PS II from cyanobacteria is composed of 17 transmembrane subunits, three peripheral proteins and a number of cofactors, with a total molecular weight of 350 kDa (Umena et al., 2011; Shen, 2015). Eukaryotic PS II in microalgae possesses additional subunits not found in cyanobacterial PS II. As resolved in the structure from a centric diatom *Chaetoceros gracilis* (Pi et al., 2019), the diatom PS II core is composed of 19 intrinsic subunits and five extrinsic subunits that are associated with the oxygen-evolving complex. Also, the diatom PS II core is associated with fucoxanthin–Chl *a/c* binding proteins (FCPs). The PS II-FCP II supercomplex contains 35 protein subunits. 11 FCP subunits are associated with each PS II core monomer, and 22 FCP subunits constitute a PS II dimer.

PS II contains the photoreaction center and water-oxidizing complex, which includes the water-splitting cluster [Mn_4O_5Ca] and a nearby tyrosine residue Y_z. The light-driven photochemistry of PS II leads to water splitting and oxygen evolution. The light-induced oxidation of water is catalyzed by a Mn_4O_5Ca cluster that cycles through several different redox states (S_i, $i = 0$–4) on the extraction of each electron by the PS II RC, P_{680}. When four electrons and four protons are extracted from two molecules of water, one molecule of dioxygen is formed (Kok et al., 1970; Joliot, 2003). Electron transfer from H_2O to plastoquinone (PQ) proceeds as the following chain of redox reactions is initiated by the light-induced separation of charges in the RC of PS II, as shown in Fig. 1. Two electrons extracted from the water molecule are used to reduce PQ to plastohydroquinone (PQH_2). Thus, PS II acts as water:plastoquinone oxidoreductase, which results in the products: molecular (triplet) oxygen ($^3O_2$), protons and reduced PQH_2. The light-induced process in PS II is very efficient, with a quantum yield of over 90% and an energy efficiency of about 20%.

4.2 PS I in generation of reducing power for production of strong reductants

In oxygenic photosynthesis, the PS I and PS II RCs provide the energy transducing machinery to drive electrons against a thermodynamic gradient at the expense of light. PS I generates the most negative redox potential in nature and largely determines the global amount of enthalpy in living systems. The PS I RC is the largest known membrane protein complex that is involved in solar energy conversion that oxidizes plastocyanin/Cyt *c*6 and reduces Fd, a strong power reductant. Although PS II is unique in its ability to extract electrons from water, PS I is arguably the most efficient photoelectric apparatus in nature, exhibiting a quantum efficiency of almost 100% in its utilization of light for electron transport (Nelson and Yocum, 2006).

PS I consist of the PS I core and the peripheral antenna system, as well as different types of light-harvesting pigments, to adapt to various light environments. Although it is generally accepted that PS II mainly forms a dimeric structure in all oxyphototrophs, PS I exists in different oligomeric states in different organisms, mostly trimers or even tetramers in cyanobacteria. However, the PS I complexes are organized mostly as monomers in eukaryotic algae and plant, which evolutionarily resulted from the need for binding various numbers of trans-membrane light-harvesting antenna around the PS I RC core. The core complex of PS I, which is a heterodimer (PsaA/PsaB), contains the RC Chls (P_{700}) and three other electron transfer components operating sequentially. The first PS I structure from the thermophilic cyanobacterium *Synechococcus elongatus* displayed a complex with 12 protein subunits and 127 cofactors (96 Chls, 22 carotenoids, 2 phylloquinones, 3 Fe_4S_4 clusters and 4 lipids), providing the very first detailed molecular architecture of PS I (Jordan et al., 2001; Grotjohann and Fromme, 2005).

While the core structure of PS I RCs remains highly conserved during 2.5 billion years of evolution from cyanobacteria to eukaryotic algae and higher plant with only changes in small subunit composition, such as the loss of PsaM and PsaX and gain of PsaG, PsaH, PsaN and PsaO in eukaryotic algae and plant. Also, different from prokaryotic PS I complexes, membrane-spanning LHCs are usually associated with the eukaryotic PS I RC, forming PS I-LHC supercomplexes, as determined in the PS I-LHC structures from red algae (Pi et al., 2018) and green algae (Suga et al., 2019). The red algal PS I core showed features of both cyanobacterial and higher plant PS I, suggesting an intermediate type during evolution from prokaryotes to eukaryotes. Two forms of PS I-LHCs from the red alga *Cyanidioschyzon merolae* have been determined, one with three Lhcr subunits attached to the side, similar to that of higher plants, and the other with two additional Lhcr subunits attached to the opposite side, indicating an ancient form of PS I-LHC I (Pi et al., 2018). In the structure of the PS I-LHC I supercomplex of green algal *Chlamydomonas reinhardtii*, PS I cores are monomers and associate with LHC I. The structure reveals that eight Lhca proteins form two tetrameric LHC I belts attached to the PsaF side while the other two Lhca proteins form an additional Lhca2/Lhca9 heterodimer attached to the opposite side (Suga et al., 2019).

In light-driven photosynthetic reactions, light energy absorbed by antenna systems is transferred to the PS I RC, where excitation energy is converted into redox energy by a series of redox components. First, light-induced charge separation oxidizes the primary electron donor P_{700}, a Chl *a/a'* heterodimer, and reduces the primary electron acceptor A_0, a Chl *a* monomer. The electron is transferred to A_1, a phylloquinone, then to F_X, an interpolypeptide (4Fe-4S) cluster, and, finally, to F_A and F_B, which are 4Fe-4S clusters bound to the extrinsic subunit PsaC. The RC of PS I consists of six Chls organized into two almost identical branches in which each branch is bound by either PsaA (A-branch) or PsaB (B-branch). The existence of the two almost identical branches raised the question if both sides are active for electron transport to F_x (Redding and van der Est, 2006). Although no consensus in the literature exists regarding a contribution of A and B branches into electron flow through PS I, there is experimental evidence that in cyanobacterial PS I (at least under certain experimental conditions), electron transport is asymmetrical, with the majority of electron transfer taking place through the A-branch of cofactors (Jagannathan et al., 2012). The central Chl *a/a'* pair (P_{700}) is oriented almost perpendicular to the membrane plane. To complicate matters, another pair of Chl *a* monomers is located approximately halfway between P_{700} and A_0 that is assigned as accessory Chls that may participate in excitation and/or electron transfer. From there, each branch contains two Chl *a* (accessory Chl and the A_0, or $ec2_{A/B}$ and $ec3_{A/B}$) and a phylloquinone (A_1, PhQ_A or PhQ_B). The two branches join again at the 4Fe–4S-cluster F_X, which is followed by two additional 4Fe–4S-clusters, F_A and F_B, which are bound to PsaC.

The reducing side of the PS I RC relies heavily on Fe-S clusters. The oxygen build-up on Earth played an essential role as the driving force in evolving the reducing side of PS I. Evolutionarily, the increase in binding interactions, achieved through the development of a network of oppositely charged amino acids in the proteins harboring F_X and F_A/F_B Fe-S clusters, was critical to the protection of the clusters by the formation of an oxygen-impenetrable interface between them. Fastening of PsaC to the PsaA/PsaB heterodimer was further ensured by the acquisition of PsaD and PsaE subunits. These proteins flank PsaC on either side of the stromal ridge of PS I and provide additional solvent shielding to F_A/F_B Fe-S clusters (Jagannathan et al., 2012).

Fd is a soluble protein with one 2Fe-2S cluster that accepts electrons from the F_B cluster on the stromal side of the PS I complex. Fd not only functions as the electron acceptor but also plays a significant role in determining the enthalpy on Earth. Fd reduces $NADP^+$ via Fd-NADP+ oxidoreductase (FNR) for various reductive biosynthetic pathways, thioredoxin via Fd-thioredoxin reductase for redox regulation, and also enzymes such as acyl carrier protein desaturase, nitrite reductase and glutamic acid synthase. Fd also reduces the Cyt b_6f complex or plastoquinone in a cyclic electron transfer pathway around

PS I that generates a proton gradient and, hence, ATP synthesis. Cyanobacteria grown under iron-depleted conditions synthesize flavodoxin that replaces Fd functionally in most, but not all, reactions (Karlusich and Carrillo, 2017). PS I is involved not only in linear electron transfer but also in cyclic electron transfer, which is required for the production of ATP without the generation of NADPH. The resulting NADPH and ATP are used to fix CO_2 into carbohydrates. The reduced Fd is subsequently used for NADPH production, which provides the reducing power for the conversion of carbon dioxide into organic molecules.

4.3 Mediator of the electron transport and energy transduction – the Cyt b_6f complex

The Cyt b_6f complex is one of the three hetero-oligomeric integral membrane protein complexes responsible for electron transport and energy transduction in oxygenic photosynthetic membranes and functions as a plastoquinol: Cyt c_6/plastocyanin oxidoreductase. It mediates the electron transport between the RC complexes of the oxygenic photosynthetic electron transport chain. As shown in Fig. 1, Cyt b_6f occupies an electrochemically central position in the noncyclic or "linear" electron-transport chain. Cyt b_6f is a dimeric integral membrane protein complex of about 220 kDa, with each monomer consisting of four major subunits Cyt b_6, subunit-IV, Cyt *f* and the Rieske iron-sulfur protein), as well as four minor subunits (PetG, L, M and N). The three-dimensional crystal structure of the b_6f complex has been determined from the cyanobacterium *Mastigocladus laminosus* (Kurisu et al., 2003) and also from the eukaryotic green algal *Chlamydomonas reinhardtii* (Stroebel et al., 2003). Interestingly, the structure of the photosynthetic Cyt b_6f complex bears similarities to the respiratory cytochrome bc_1 complex, but also exhibits some unique features, such as binding one β-carotene and one Chl *a* molecule, and an unexpected heme sharing a quinone site.

In the linear electron transport, the Cyt b_6f complex functionally interconnects with two photosystems, in which electron transfer through the complex is coupled with proton transfer from the electrochemically negative (*n*) to the positive (*p*) side of the complex through receiving electrons from PS II by plastoquinol and passing them to PS I by reducing plastocyanin or Cyt c_6. This process results in proton uptake from the stroma and generates a proton electrochemical gradient across the membrane, powering the Q-cycle and boosting ATP synthesis at the expense of reducing equivalents (Cramer et al., 2006). Furthermore, The Cyt b_6f complex can switch from linear electron transfer between both photosystems to a cyclic mode of electron transfer around PS I. Therefore, Cyt b_6f is critical in controlling the relative efficiency of linear and cyclic flow, either indirectly (by setting the rate of plastoquinol oxidation) or directly, for example, via the formation of a biochemically stable

PS I-Cyt b_6f complex as observed in unicellular green algae (Minagawa, 2016). The cyclic electron flow around PS I provides extra ATP with balancing the ATP/NADPH production ratio and for protecting both photosystems from damage caused by stromal over reduction.

5 Photosynthetic CO_2 assimilation

Biological processes on land and in the oceans strongly affect the global carbon cycle across all time scales. In both components of the biosphere, oxygenic photosynthesis is responsible for virtually all of the biochemical production of organic matter. Photosynthetic primary production is critical. Total global net primary production has been estimated to be 104.9 petagrams per year, with roughly equal amounts arising from terrestrial and marine photosynthesis. The cyanobacteria and other phytoplankton, chiefly in the oceans, contribute nearly 50% of global annual primary productivity (Field et al., 1998). Probably as much as 25% of annual global productivity is achieved by deep ocean cyanobacteria (Liu et al., 1997). Cyanobacteria and microalgae contribute considerably to the worldwide generation of light-powered energy as they are among the first components of the food chain in all water habitats.

5.1 CO_2 concentrating mechanisms and carbon fixation

Carbon dioxide is arguably one of the most important gases on Earth, even though it is present in only relatively low concentrations in the atmosphere. CO_2 is an essential resource for oxygenic photosynthesis in organic carbon synthesis and often limits substrate for photosynthetic carbon assimilation in microalgae and cyanobacteria and other photosynthetic organisms. Carbon fixation through the reductive pentose phosphate pathway (Calvin-Benson cycle) is the most biologically abundant and economically relevant pathway (Beardall and Raven, 2020), in which CO_2 combines with a five-carbon sugar, RuBP, to yield two molecules of a three-carbon compound glycerate 3-phosphate, also known as 3-phosphoglycerate. All oxygenic photosynthetic organisms utilize the ribulose-1,5-bisphosphate carboxylase/oxygenase (RuBisCO) enzyme for CO_2 fixation – the net carbon assimilation for the conversion of CO_2 to organic carbon through carboxylation. Cyanobacteria and microalgae evolved a range of different forms of RuBisCO with differences in their kinetic properties and selectivity to O_2 and CO_2.

However, the concentrations of CO_2 and O_2 in the environment are major factors determining the rate of photosynthesis and respiration in the oxygenic phototrophs. The atmosphere's relatively low CO_2 concentration represents stress frequently confronted by photosynthetic organisms, largely because of the low CO_2 affinity of RuBisCO. When RuBisCO first evolved in cyanobacteria some 3.5 billion years ago, it did so in an environment under which CO_2 was many folds higher than present levels and O_2 levels were very low. Under these ancient conditions, RuBisCO was saturated with CO_2 and seems to have been efficient, possessing a relatively high carboxylation rate. Through the mechanism of RuBisCO, O_2 is a competitive substrate with respect to CO_2. As O_2 levels increased and CO_2 decreased in the atmosphere, this CO_2/O_2 ratio declined. In an evolutionary response to the changed atmospheric conditions, a diversity of CO_2 concentrating mechanisms (CCMs) were evolved in cyanobacteria and eukaryotic algae (Badger and Price, 2003; Spalding, 2008) to enhance uptake in low-CO_2 conditions to saturate RuBisCO with CO_2. This adaptation strategy evolutionarily led to a highly efficient CCM that can accumulate CO_2 around RuBisCO by a factor of 1000-fold above ambient levels to improve photosynthetic performance, particularly under CO_2-limiting conditions (Badger and Price, 2003; Price, 2011). After being concentrated, CO_2 was then transported into the biosolar cell factory to contribute to organic carbon skeleton synthesis through CO_2 sequestration and assimilation. In cyanobacteria and eukaryotic algae, subcellular compartments are used to concentrate carbon-fixing machinery (e.g. carboxysomes and pyrenoids).

5.2 Carboxysomes: cyanobacteria - an essential part of CCM

RuBisCO in cyanobacteria is encapsulated in unique proteinaceous microcompartments known as carboxysomes, which act as the site of CO_2 elevation. Carboxysomes are thought to have evolved as a consequence of the progressive increases in oxygen concentration in the ancient atmosphere: this is because O_2 is a competing substrate to CO_2 in the RuBisCO reaction (Badger and Price, 2003). The average unicellular cyanobacterial cells would normally possess 5–15 carboxysomes per cell depending on species and growth conditions. Carboxysomes range from 90 nm to 250 nm in diameter, depending on the cyanobacterial species type, and are proteinaceous, polyhedral bodies containing tightly packed RuBisCO molecules (L_8S_8) bounded by a thin protein shell of about 3 nm in width (Kerfeld and Melnicki, 2016). The carboxysome has an outer shell composed of a few thousand protein subunits, which encapsulates a CO_2-producing enzyme (carbonic anhydrase (CA)) and the carbon-fixing enzyme, RuBisCO. The two carboxysome types, alpha and beta, differ fundamentally in their components (including the form of RuBisCO they enclose), assembly, and species distribution (Rae et al., 2013). The carboxysome is the core component of the cyanobacterial CCM that also involves the active uptake of inorganic carbon into the cell and its intracellular accumulation, primarily as bicarbonate (HCO_3^-). Carboxysomes contain most, if not all, of the RuBisCO content of the cell. The CCM functions to actively transport inorganic carbon species (Ci, HCO_3^- and CO_2) resulting in conditional accumulation of HCO_3^- within the cell. HCO_3^- diffuses through the carboxysome shell; within the lumen, an encapsulated CA converts it into CO_2, promoting the carboxylation reaction of RuBisCO.

5.3 *Pyrenoids in eukaryotic algae*

The pyrenoid, a microcompartment within the chloroplasts, is found in many eukaryotic algae. With analogous structures to the carboxysomes of cyanobacteria, pyrenoids are crystal-like proteinaceous, which may play important roles in carbon concentrating mechanisms (Badger et al., 1998). Pyrenoids can have very diverse morphologies, from simple aggregates of RuBisCO to more complicated structures like thylakoid membrane spanning through the pyrenoid body. In the marine diatom *P. tricornutum*, the structure of the pyrenoid has been reported as a lens-shaped body residing in the central part of the stroma with two-layered thylakoid membranes that penetrate the center of the pyrenoid (Giordano et al., 2005). Pyrenoids, where present, contain the majority of the cellular RuBisCO, with much smaller amounts in the stroma. In the unicellular green alga *C. reinhardtii*, most RuBisCO proteins are located in the pyrenoid. Recently, two plastidic fructose-1,6-bisphosphatases have been located to reside in the pyrenoid of the diatom *P. tricornutum* (Allen et al., 2011), indicating that diatom pyrenoids also play an important role in controlling carbon dioxide assimilation. The thylakoid membranes that traverse the pyrenoid are thought to supply a high concentration of CO_2 to RuBisCO in the matrix, increasing the rate of CO_2 fixation. This concentration is mediated by the coordinated action of HCO_3^- transporters and CAs, which catalyze the conversion of CO_2 to HCO_3^-. A CA in the periplasm of the algal cell facilitates the conversion of CO_2 to HCO_3^-. Noteworthily, all pyrenoid-containing algae have CCMs, but not all eukaryotic algae with CCMs possess pyrenoids (Badger et al., 1998). In species lacking pyrenoids, RuBisCO is found entirely in the stroma.

6 Potential of cyanobacteria and microalgae

Cyanobacteria and microalgae are rich sources of numerous nutrients (proteins and carbohydrates) and health-beneficial components, including vitamins, minerals, proteins with essential amino acids, polyunsaturated fatty acids, antioxidants and dietary fiber. The photosynthetic ability of cyanobacteria and microalgae to use sunlight and CO_2 as energy and carbon sources allow them to stand out from other organisms in scientific research and industrial application. These photoautotrophic organisms are interesting hosts for the production of bioproducts and chemicals due to their abilities to harness light energy and fix CO_2. These oxygenic phototrophs are especially attractive in an era during which sustainable biotechnological processes are of increasing importance. They have the advantage of fast growth rates, low nutritional requirements, and potential for large-scale cultivation in bioreactors and even open ponds. Also, the geographic diversity of cyanobacteria and microalgae provides these hosts an added advantage as their biosynthetic machineries can be robust, sustainable, and, in turn, commercially attractive. Moreover, advances in genomic sequencing of microalgae and cyanobacteria have been greatly facilitating fundamental studies in the exploration around the molecular mechanisms of light harvesting and conversion of solar energy into organic bioproducts and valuable metabolites, further expanding perceived potential in a biotechnological and commercial application (Angermayr et al., 2015; Dittmann et al., 2015). Advances in molecular strategies and genetic engineering tools have facilitated the manipulation of various aspects of oxygenic organisms for improving photosynthetic properties with enhanced light-harvesting capacities and efficiency, optimizing the light-driven electron transport and improving carbon fixation rates, which in turn result in their higher productivity.

Compared to algae and plants, cyanobacteria have advantages in genetic manipulation, considering their small genomes with simple genetic systems, natural DNA uptake, fast growth and so on. On the other hand, many of the standard biological parts available for synthetic biology applications in *Escherichia coli* and other prokaryotes can also be used in cyanobacteria. In comparison with eukaryotic algae and plants, cyanobacteria are more genetically tractable. Many model cyanobacterial strains are naturally transformable, due to the existence of an efficient mechanism of recombination (RecA; Murphy et al., 1987), which makes integration of DNA into the genome by homologous recombination possible. As most of the model cyanobacteria strains are naturally transformable, designed genetic materials can be transferred relatively easily into target hosts directly. Many cyanobacteria contain multiple copies of their chromosomes (e.g. about 10 copies per cell in the unicellular *Synechococcus* sp. PCC 7002 (Xu et al., 2011) and about 12 genomic copies per cell in *Synechocystis* sp. PCC 6803 (Labarre et al., 1989). Therefore, only several rounds of colony screening are necessary for the segregation of the transformants. In recent years, the development of synthetic biology tools has provided new approaches for genetic manipulation of microalgae and cyanobacteria in gene expression and metabolic engineering (Nogales et al., 2012). For example, a few gene expression systems have been designed specifically to work in diverse strains: *Synechocystis* sp. PCC 6803 (Vermaas, 1996), *Synechococcus* sp. PCC 7002 (Xu et al., 2011), *Synechococcus elongatus* PCC 7942 (Kim et al., 2017), *Nostoc* sp. PCC 7120 (Wolk et al., 1984) and more.

One stand-out cyanobacterium is the marine cyanobacterium *Synechococcus* sp. PCC 7002 (hereafter, *Synechococcus* 7002). The unicellular *Synechococcus*

7002 is a fast-growing photoautotroph (Shen and Bryant, 1995), which can tolerate various light intensities, exposures to oxidative stress, changes in nutrient supply, temperature, and salinity (Sakamoto and Bryant, 1997; Nomura et al., 2006; Xu et al., 2011; Ludwig and Bryant, 2012). Due to its salt tolerance, *Synechococcus* 7002 can utilize unprocessed seawater and is capable of surviving salt concentrations as high as 1.7 M (Batterton and Baalen, 1971). These properties make it an ideal candidate for experiments that would be difficult to have accomplished using less robust cyanobacteria. Mostly, the growth of *Synechococcus* 7002 can achieve high biomass concentrations and unprecedented photoautotrophic biomass productivities. Furthermore, the photoautotrophic *Synechococcus* 7002 can also grow mixotrophically or heterotrophically and tolerates a wide range of temperature. Along with the establishment of genetic manipulation methods, synthetic biological tools and gene expression platforms (Ruffing et al., 2016; Markley et al., 2015; Hendry et al., 2016; Pérez et al., 2016; Ruffing et al., 2016; Zess et al., 2016; Gordon et al., 2016; Vogel et al., 2017), *Synechococcus* 7002 has been successfully used as an ideal host for the production of bioproducts, including of proteins and carotenoids (Xu et al., 2011; Hasunuma et al., 2019), free fatty acids (Ruffing et al., 2016), limonene and bisabolene (Davies et al., 2014), lauric acid (Work et al., 2015), cellulose (Zhao et al., 2015b), ethanol (Kopka et al., 2017), mannitol (Madsen et al., 2018) and other valuable chemicals (Vu et al., 2013). It has demonstrated that with the expression of the chlorophyll *f* synthase gene *chlF*, Chl *f* molecules synthesized in *Synechcococcus* 7002 were inserted into trimeric PS I complexes (Shen et al., 2019) and functionally connected to the RC of the PS I complex for maintaining efficient energy transfer within the PS I complex (Tros et al., 2020). Therefore, the introduction of Chl *f* molecules extends the photosynthetically active radiation of the new hybrid PS I complexes up to 750 nm for the enhancement of light utilization capacity.

Photoautotrophic CO_2 fixation underlies nearly all biological processes on Earth. Carbon fixation through the reductive pentose phosphate pathway is the most biologically abundant and economically relevant pathway. Through three distinct phases (carboxylation, reduction and regeneration) of photosynthesis, several key enzymes directly influence the rate and efficiency of photosynthesis, especially the low activity of RuBisCO. Although reengineering RuBisCO has been a primary objective for enhancing the photosynthetic process, most successful engineering approaches to increasing rates of carbon fixation in autotrophs have not involved modification of RuBisCO itself, but rather sought to improve CCMs and concentrating carbon strategies and modifying the neighboring pathways and enzymes (Badger and Price, 2003; Price et al., 2013). The development of synthetic biological tools has provided novel strategies for designing and creating efficient carbon fixation enzymes and modules that are based on the diversity of natural enzymes and pathways in cyanobacteria and microalgae.

7 Conclusion and future trends

Photosynthetic O_2 production and CO_2 assimilation have established the composition of the biosphere and provides all life forms with essential food and fuel. Nature has evolved oxygenic photosynthesis first in cyanobacteria and then in eukaryotic microalgae and plant species with the use of two photochemical systems connected in series to generate the electric potential difference required to split water and produce ATP and NADPH. Oxygenic photosynthesis, the principal converter of sunlight into chemical energy on Earth, is catalyzed by four multi-subunit membrane-protein complexes: PS I, PS II, the Cyt b_6f complex and F-ATPase. PS II generates an oxidant that has a redox potential high enough to enable it to oxidize H_2O, a substrate so abundant that it assures a practically unlimited electron source for life on Earth. PS I promotes the most negative redox potential in nature and largely determines the global amount of enthalpy in living systems. Turning to photosynthesis, cyanobacteria and microalgae adapted one straightforward strategy for improving the efficiency limit by tuning the light-absorbing pigments (chlorophylls, phycobilins and carotenoids) to extend the range of solar light absorption in optimizing their capacities to produce a diverse array of complex organic compounds through light-driven CO_2 assimilation. For adaptation to a wide range of environments and growth conditions, cyanobacteria and microalgae expanded globally their diversity and variety, which provided promising new sources of green biomass that may complement agricultural production to meet increasing global demand for food, nutrition, biofuels and valuable products.

Photosynthetic capacity and feasibility for genetic engineering have made cyanobacteria and microalgae more attractive in recent years, mostly because these oxygenic phototrophs have simple growth requirements, and use light, CO_2 and other inorganic nutrients efficiently. New molecular approaches have been developed for genetically engineering strains with improved properties. Especially in recent years, great progress has been made in the 'omics' of different cyanobacteria and eukaryotic microalgae, which give a comprehensive knowledge and explanation for all molecular mechanisms and biochemical conversions taking place within a living cell of different photosynthetic organisms. Synthetic biological approaches permit heterologous gene expression and rational construction of novel metabolic pathways that are based on the combination of multiple enzymatic activities of different origins. In addition, the genetic manipulation of whole genome and metabolic networks by CRISPR-based and sRNA-based technologies with multiple parallel targets will further expand and improve the efficiency of genome editing in cyanobacteria and microalgae. Those developments lead to the engineering of photosynthetic hosts in order to optimize genetic and regulatory processes in cells to increase photosynthetic efficiency, optimize cell growth, and facilitate cellular reprogramming and refinement of genetic circuits used to express the heterologous genes as green cell factories.

8 Where to look for further information

To learn more about classification and collections of different photosynthetic microorganisms or to obtain any strain of the photosynthetic microorganism for research studies, metabolic engineering and biotechnogical applications, it is highly recommended to check out the following websites for more information on classification and collections of different strains of cyanobacteria, microalgae and macroalgae (and even more):

- The Pasture Culture Collection of Cyanobacteria (PCC): https://research.pasteur.fr/en/team/collection-of-cyanobacteria/.
- American Type Culture Collection (ATCC): https://www.atcc.org.
- UTEX Culture Collection of Algae: https://utex.org/.
- Roscoff Culture Collection (RCC): http://roscoff-culture-collection.org/.

9 References

Adir, N., Bar-Zvi, S. and Harris, D. (2020). The amazing phycobilisome. *Biochim. Biophys. Acta Bioenerg.* 1861(4): 148047.

Allen, A. E., Moustafa, A., Montsant, A., Eckert, A., Kroth, P. G. and Bowler, C. (2011). Evolution and functional diversification of fructose bisphosphate aldolase genes in photosynthetic marine diatoms. *Mol. Biol. Evol.* 29(1): 367–379.

Angermayr, S. A., Rovira, A. G. and Hellingwerf, K. J. (2015). Metabolic engineering of cyanobacteria for the synthesis of commodity products. *Trends Biotechnol.* 33(6): 352-361.

Badger, M. R. and Price, G. D. (2003). CO_2 concentrating mechanisms in cyanobacteria: molecular components, their diversity and evolution. *J. Exp. Bot.* 54(383): 609–622.

Badger, M. R., Andrews, T. J., Whitney, S. M., Ludwig, M., Yellowlees, D. C., Leggat, W. and Price, G. D. (1998). The diversity and coevolution of Rubiscos, plastids, pyrenoids and chloroplast-based CO_2-concentrating mechanisms in algae. *Can. J. Bot.* 76(6): 1052–1071.

Batterton, J. C. Jr. and Baalen, C. (1971). Growth responses of blue-green algae to sodium chloride concentration. *Arch. Mikrobiol.* 76(2): 151–165.

Beardall, J. and Raven, J. A. (2020). Acquisition of inorganic carbon by microalgae and cyanobacteria. In: Wang, Q. (Ed.) *Microbial Photosynthesis*. Springer, Singapore, pp. 151–168.

Becraft, E. D., Woodm, J. M., Rusch, D. B., Kuhl, M., Jensen, S. I., Bryant, D. A., Roberts, D. W., Cohen, F. M. and Ward, D. M. (2015). The molecular dimension of microbial species: 1: Ecological distinctions among, and homogeneity within, putative ecotypes of *Synechococcus* inhabiting the cyanobacterial mat of Mushroom Spring, Yellowstone National Park. *Front. Microbiol.* 6: 590.

Bryant, D. A., Guglielmi, G., Tandeau de Marsacm, N. T., Castets, A. M. and Cohen-Bazire, G. (1979). The structure of cyanobacterial phycobilisomes: A model. *Arch. Microbiol.* 123(2): 113–127.

Bryant, D. A., Shen, G., Turner, G. M., Soulier, N., Laremore, T. N. and Ho, M.-Y. (2020). Far-red light allophycocyanin subunits play a role in chlorophyll *d* accumulation in far-red light. *Photosynth. Res.* 143(1): 81–95.

Chen, M. and Blankenship, R. E. (2011). Expanding the solar spectrum used by photosynthesis. *Trends Plant Sci.* 16(8): 427–431.

Cherepanov, D. A., Shelaev, I. V., Gostev, F. E., Aybush, A. V., Mamedov, M. D., Shen, G., Nadtochenko, V. A., Bryant, D. A., Semenov, A. Y. and Golbeck, J. H. (2020). Evidence that chlorophyll *f* functions solely as an antenna pigment in far-red-light photosystem I from *Fischerella thermalis* PCC 7521. *Biochim. Biophys. Acta Bioenerg.* 1861(5–6): 148184.

Cohen, Y. and Gurevitz, M. (2006). The cyanobacteria-ecology, physiology and molecular genetics. In: Dworkin, M., Falkow,, S., Rosenberg, E., Schleifer, K. H. and Stackebrandt, E. (Eds) *The Prokaryotes*. Springer, New York, pp. 1074–1098.

Cramer, W. A., Zhang, H., Yan, J., Kurisu, G. and Smith, J. L. (2006). Transmembrane traffic in the cytochrome b_6f complex. *Annu. Rev. Biochem.* 75: 769– 790.

Croce, R. and van Amerongen, H. (2014). Natural strategies for photosynthetic light harvesting. *Nat. Chem. Biol.* 10(7): 492–501.

Cui, X., Liu, X. L., Shen, G., Ma, J., Husain, F., Rocher, D., Zumberge, J. E., Bryant, D. A. and Summons, R. E. (2020). Niche expansion for phototrophic sulfur bacteria at the Proterozoic-Phanerozoic transition. *Proc. Natl. Acad. Sci. U. S. A.* 117(30): 17599–17606.

Curtis, B. A., Tanifuji, G., Burki, F., Gruber, A., Irimia, M., Maruyama, S., Arias, M. C., Ball, S. G., Gile, G. H., Hirakawa, Y., Hopkins, J. F., Kuo, A., Rensing, S. A., Schmutz, J., Symeonidi, A., Elias, M., Eveleigh, R. J., Herman, E. K., Klute, M. J., Nakayama, T., Oborník, M., Reyes-Prieto, A., Armbrust, E. V., Aves, S. J., Beiko, R. G., Coutinho, P., Dacks, J. B., Durnford, D. G., Fast, N. M., Green, B. R., Grisdale, C. J., Hempel, F., Henrissat, B., Höppner, M. P., Ishida, K., Kim, E., Kořený, L., Kroth, P. G., Liu, Y., Malik, S. B., Maier, U. G., McRose, D., Mock, T., Neilson, J. A., Onodera, N. T., Poole, A. M., Pritham, E. J., Richards, T. A., Rocap, G., Roy, S. W., Sarai, C., Schaack, S., Shirato, S., Slamovits, C. H., Spencer, D. F., Suzuki, S., Worden, A. Z., Zauner, S., Barry, K., Bell, C., Bharti, A. K., Crow, J. A., Grimwood, J., Kramer, R., Lindquist, E., Lucas, S., Salamov, A., McFadden, G. I., Lane, C. E., Keeling, P. J., Gray, M. W., Grigoriev, I. V. and Archibald, J. M. (2012). Algal genomes reveal evolutionary mosaicism and the fate of nucleomorphs. *Nature* 492(7427): 59–65.

Davies, F. K., Work, V. H., Beliaev, A. S. and Posewitz, M. C. (2014). Engineering limonene and bisabolene production in wild type and a glycogen-deficient mutant of *Synechococcus* sp. PCC 7002. *Front. Bioeng. Biotechnol.* 2: 1–11.

Dittmann, E., Gugger, M., Sivonen, K. and Fewer, D. P. (2015). Natural product biosynthetic diversity and comparative genomics of the cyanobacteria. *Trends Microbiol.* 23(10): 642–652.

Dong, C., Tang, A., Zhao, J., Mullineaux, C. W., Shen, G. and Bryant, D. A. (2009). ApcD is necessary for efficient energy transfer from phycobilisomes to photosystem I and helps to prevent photoinhibition in the cyanobacterium *Synechococcus* sp. PCC 7002. *Biochim. Biophys. Acta* 1787(9): 1122–1128.

Farrant, G. K., Dore, H., Cornejo-Castillo, F. M., Partensky, F., Ratin, M., Ostrowski, M., Pitt, F. D., Wincker, P., Scanlan, D. J., Iudicone, D., Acinas, S. G. and Garczarek, L. (2016). Delineating ecologically significant taxonomic units from global patterns of marine picocyanobacteria. *Proc. Natl. Acad. Sci. U. S. A.* 113(24): E3365–E3374.

Ferreira, K. N., Iverson, T. M., Maghlaoui, K., Barber, J. and Iwata, S. (2004). Architecture of the photosynthetic oxygen-evolving center. *Science* 303(5665): 1831–1838.

Field, C. B., Behrenfeld, M. J., Randerson, J. T. and Falkowski, P. (1998). Primary production of the biosphere: Integrating terrestrial and oceanic components. *Science* 281(5374): 237-240.

Flombaum, P., Gallegos, J. L., Gordillo, R. A., Rincón, J., Zabala, L. L., Jiao, N., Karl, D. M., Li, W. K., Lomas, M. W., Veneziano, D., Vera, C. S., Vrugt, J. A. and Martiny, A. C. (2013). Present and future global distributions of the marine Cyanobacteria *Prochlorococcus* and *Synechococcus*. *Proc. Natl. Acad. Sci. U. S. A.* 110(24): 9824–9829.

Flores, E., Muro-Pastor, A. M. and Meeks, J. C. (2008). Gene transfer to cyanobacteria in the laboratory and in nature. In: Herrero, A. and Flores, E. (Eds) *The Cyanobacteria: Molecular Biology, Genomics and Evolution*. Caister Academic Press, Norfolk, UK, pp. 45–57.

Fritsch, F. E. (1945). *The Structure and Reproduction of the Algae* (vol. II). Cambridge University Press, Cambridge.

Frommolt, R., Werner, S., Paulsen, H., Goss, R., Wilhelm, C., Zauner, S., Maier, U. G., Grossman, A. R., Bhattacharya, D. and Lohr, M. (2008). Ancient recruitment by chromists of green algal genes encoding enzymes for carotenoid biosynthesis. *Mol. Biol. Evol.* 25(12): 2653–2667.

Gallon, J. R. (2004). N_2 fixation by non-heterocystous cyanobacteria. In: Klipp, W., Masepohl, B., Gallon, J. R. and Newton, W. E. (Eds) *Genetics and Regulation of Nitrogen Fixation in Free-Living Bacteria. Nitrogen Fixation: Origins, Applications, and Research Progress* (vol. 2). Springer, Dordrecht, pp. 111–139.

Gan, F., Shen, G. and Bryant, D. A. (2014a). Occurrence of far-red light photoacclimation (FaRLiP) in diverse cyanobacteria. *Life (Basel)* 5(1): 4–24.

Gan, F., Zhang, S., Rockwell, N. C., Martin, S. S., Lagarias, J. C. and Bryant, D. A. (2014b). Extensive remodeling of a cyanobacterial photosynthetic apparatus in far-red light. *Science* 345(6202): 1312–1317.

Giordano, M., Beardall, J. and Raven, J. A. (2005). CO_2 concentration mechanisms in algae: Mechanisms, environmental modulation and evolution. *Annu. Rev. Plant Biol.* 56: 99–131.

Gisriel, C. J., Shen, G., Kurashov, V., Ho, M. Y., Zhang, S., Williams, D., Golbeck, J. H., Fromme, P. and Bryant, D. A. (2020a). The structure of photosystem I acclimated to far-red light illuminates an ecologically important acclimation process in photosynthesis. *Sci. Adv.* 6(6): eaay6415.

Gisriel, C. J., Wang, J., Brudvig, G. W. and Bryant, D. A. (2020b). Opportunities and challenges for assigning cofactors in cryo-EM density maps of chlorophyll-containing proteins. *Commun. Biol.* 3(1): 408.

Glazer, A. N. and Cohen-Bazire, G. (1971). Subunit structure of the phycobiliproteins of blue-green algae. *Proc. Natl. Acad. Sci. U. S. A.* 68(7): 1398–1401.

Gordon, G. C., Korosh, T. C., Cameron, J. C., Markley, A. L., Begemann, M. B. and Pfleger, B. F. (2016). CRISPR interference as a titratable, trans-acting regulatory tool for metabolic engineering in the cyanobacterium *Synechococcus* sp. strain PCC 7002. *Metab. Eng.* 38: 170–179.

Gould, S. B. (2012). Evolutionary genomics: Algae's complex origins. *Nature* 492(7427): 46–48.

Gould, S. B., Waller, R. F. and McFadden, G. I. (2008). Plastid evolution. *Annu. Rev. Plant Biol.* 59: 491–517.

Graham, J. E. and Bryant, D. A. (2008). The biosynthetic pathway for synechoxanthin, an aromatic carotenoid synthesized by the euryhaline, unicellular cyanobacterium *Synechococcus* sp. strain PCC 7002. *J. Bacteriol.* 190(24): 7966–7974.

Green, B. R. and Parson, W. W. (2003). *Light-Harvesting Antennas*. Kluwer Academic Press, Dordrecht.

Grotjohann, I. and Fromme, P. (2005). Structure of cyanobacterial photosystem I. *Photosynth. Res.* 85(1): 51–72.

Gutu, A. and Kehoe, D. M. (2012). Emerging perspectives on the mechanisms, regulation, and distribution of light color acclimation in cyanobacteria. *Mol. Plant* 5(1): 1–13.

Hasunuma, T., Takaki, A., Matsuda, M., Kato, Y., Vavricka, C. J. and Kondo, A. (2019). Single-stage astaxanthin production enhances the nonmevalonate pathway and photosynthetic central metabolism in *Synechococcus* sp. PCC 7002. *ACS Synth. Biol.* 8(12): 2701–2709.

Hendry, J. I., Prasannan, C. B., Joshi, A., Dasgupta, S. and Wangikar, P. P. (2016). Metabolic model of *Synechococcus* sp. PCC 7002: Prediction of flux distribution and network modification for enhanced biofuel production. *Bioresour. Technol.* 213: 190–197.

Hirschberg, J. and Chamovitz, D. (1994). Carotenoids in cyanobacteria. In: Bryant, D. A. (Ed.) *The Molecular Biology of Cyanobacteria: Advances in Photosynthesis* (vol. 1). Springer, Dordrecht, pp. 559–579.

Ho, M.-Y., Shen, G., Canniffe, D. P., Zhao, C. and Bryant, D. A. (2016). Light-dependent chlorophyll *f* synthase is a highly divergent paralog of PsbA of photosystem II. *Science* 353 (6302), aaf9178.

Ho, M.-Y., Soulier, N. T., Canniffe, D. P., Shen, G. and Bryant, D. A. (2017). Light regulation of pigment and photosystem biosynthesis in cyanobacteria. *Curr. Opin. Plant Biol.* 37: 24–33.

Hohmann-Marriott, M. F. and Blankenship, R. E. (2011). Evolution of photosynthesis. *Annu. Rev. Plant Biol.* 62: 515–548.

Jagannathan, B., Shen, G. and Golbeck, J. H. (2012). The evolution of type I reaction centers: The response to oxygenic photosynthesis. In: Burnap, R. L. and Vermaas, W. F. J. (Eds) *Functional Genomics and Evolution of Photosynthetic Systems*. Springer, Dordrecht, The Netherlands, pp. 285–316.

Joliot, P. (2003). Period-four oscillations of the flash-induced oxygen formation in photosynthesis. *Photosynth. Res.* 76(1-3): 65–72.

Jordan, P., Fromme, P., Witt, H. T., Klukas, O., Saenger, W. and Krauß, N. (2001). Three-dimensional structure of cyanobacterial photosystem I at 2.5 Å resolution. *Nature* 411(6840): 909–917.

Karlusich, J. J. P. and Carrillo, N. (2017). Evolution of the acceptor side of photosystem I: Ferredoxin, flavodoxin, and ferredoxin-NADP+ oxidoreductase. *Photosynth. Res.* 134(3): 235–250.

Katoh, T., Mimuro, M. and Takaichi, S. (1989). Light-harvesting particles isolated from a brown alga, *Dictyota dichotoma*. *Biochim. Biophys. Acta* 976(2-3): 233–240.

Kerfeld, C. A. and Melnicki, M. R. (2016). Assembly, function and evolution of cyanobacterial carboxysomes. *Curr. Opin. Plant Biol.* 31: 66–75.

Kim, W. J., Lee, S. M., Um, Y., Sim, S. J. and Woo, H. M. (2017). Development of SyneBrick vectors as a synthetic biology platform for gene expression in *Synechococcus elongatus* PCC 7942. *Front. Plant Sci.* 8: 293.

Kirilovsky, D. and Kerfeld, C. A. (2016). Cyanobacterial photoprotection by the orange carotenoid protein. *Nat. Plants* 2(12): 16180.

Kirk, J. T. O. (1994). The nature of the underwater light field. In: Kirk, J. T. O. (Ed.) *Light and Photosynthesis in Aquatic Ecosystems*. Cambridge University Press, Cambridge; Melbourne, p. 509.

Kok, B., Forbush, B. and McGloin, M. (1970). Cooperation of charges in photosynthetic oxygen evolution. I. A linear four step mechanism. *Photochem. Photobiol.* 11(6): 457–475.

Kopka, J., Schmidt, S., Dethloff, F., Pade, N., Berendt, S., Schottkowski, M., Martin, N., Dühring, U., Kuchmina, E., Enke, H., Kramer, D., Wilde, A., Hagemann, M. and Friedrich, A. (2017). Systems analysis of ethanol production in the genetically engineered cyanobacterium *Synechococcus* sp. PCC 7002. *Biotechnol. Biofuels* 10: 56.

Kurisu, G., Zhang, H., Smith, J. L. and Cramer, W. A. (2003). Structure of the cytochrome b_6f complex of oxygenic photosynthesis: Tuning the cavity. *Science* 302(5647): 1009–1014.

Labarre, J., Chauvat, F. and Thuriaux, P. (1989). Insertional mutagenesis by random cloning of antibiotic resistance genes into the genome of the cyanobacterium *Synechocystis* strain PCC 6803. *J. Bacteriol.* 171(6): 3449–3457.

Liu, H., Zhang, H., Niedzwiedzki, D. M., Prado, M., He, G., Gross, M. L. and Blankenship, R. E. (2013). Phycobilisomes supply excitations to both photosystems in a megacomplex in cyanobacteria. *Science* 342(6162): 1104–1107.

Liu, H. B., Nolla, H. A. and Campbell, L. (1997). *Prochlorococcus* growth rate and contribution to primary production in the equatorial and subtropical North Pacific Ocean. *Aquat. Microb. Ecol.* 12: 39–47.

Ludwig, M. and Bryant, D. A. (2012). *Synechococcus* sp. strain PCC 7002 transcriptome: Acclimation to temperature, salinity, oxidative stress and mixotrophic growth conditions. *Front. Microbiol.* 3: 354.

Madsen, M. A., Semerdzhiev, S., Amtmann, A. and Tononm, T. (2018). Engineering mannitol biosynthesis in *Escherichia coli* and *Synechococcus* sp. PCC 7002 using a green algal fusion protein. *ACS Synth. Biol.* 7(12): 2833–2840.

Manning, W. M. and Strain, H. H. (1943). Chlorophyll *d*, a green pigment of red algae. *J. Biol. Chem.* 151(1): 1–19.

Markley, A. L., Begemann, M. B., Clarke, R. E., Gordon, G. C. and Pfleger, B. F. (2015). Synthetic biology toolbox for controlling gene expression in the cyanobacterium *Synechococcus* sp. strain PCC 7002. *ACS Synth. Biol.* 4(5): 595–603.

Martin, W. and Kowallik, K. V. (1999). Annotated English translation of Mereschkowsky's 1905 paper "Uber Natur und Ursprung der Chromatophoren im Pflanzenreiche". *Eur. J. Phycol.* 34: 287–295.

McFadden, G. I. (2001). Primary and secondary endosymbiosis and origin of plastid. *J. Phycol.* 37(6): 951–959.

Minagawa, J. (2016). A supercomplex of cytochrome b_6f and photosystem I for cyclic electron flow. In: Cramer, W. and Kallas, T. (Eds) *Cytochrome Complexes: Evolution, Structures, Energy Transduction, and Signaling. Advances in Photosynthesis*

and Respiration (Including Bioenergy and Related Processes) (vol. 41). Springer, Dordrecht, pp. 453-462.
Miyashita, H., Ikemoto, H., Kurano, N., Adachi, K., Chihara, M. and Miyachi, S. (1996). Chlorophyll *d* as a major pigment. *Nature* 383(6599): 402-402.
Moore, L. R., Rocap, G. and Chisholm, S. W. (1998). Physiology and molecular phylogeny of coexisting *Prochlorococcus* ecotypes. *Nature* 393(6684): 464-467.
Mullineaux, C. W. (2008). Phycobilisome-reaction centre interaction in cyanobacteria. *Photosynth. Res.* 95(2-3): 175-182.
Murphy, R. C., Bryant, D. A. and Porter, R. O. (1987). Molecular cloning and characterization of the *recA* gene from the cyanobacterium *Synechococcus* sp. PCC 7002. *J. Bacteriol.* 169: 2739-2747.
Nelson, N. and Yocum, C. F. (2006). Structure and function of photosystem I and II. *Annu. Rev. Plant Biol.* 57: 521-565.
Nogales, J., Gudmundsson, S., Knight, E. M., Palsson, B. O. and Thiele, I. (2012). Detailing the optimality of photosynthesis in cyanobacteria through systems biology analysis. *Proc. Natl. Acad. Sci. U. S. A.* 109(7): 2678-2683.
Nomura, C. T., Sakamoto, T. and Bryant, D. A. (2006). Roles for heme-copper oxidases in extreme high-light and oxidative stress response in the cyanobacterium *Synechococcus* sp. PCC 7002. *Arch. Microbiol.* 185(6): 471-479.
Nowack, S., Olsen, M. T., Schaible, G. A., Becraft, E. D., Shen, G., Klapper, I., Bryant, D. A. and Ward, D. M. (2015). The molecular dimension of microbial species: 2. *Synechococcus* strains representative of putative ecotypes inhabiting different depths in the Mushroom Spring microbial mat exhibit different adaptive and acclimative responses to light. *Front. Microbiol.* 6: 1-13.
Olson, J. M. (2006). Photosynthesis in the Archean era. *Photosynth. Res.* 88(2): 109-117.
Palmer, J. D. (2003). The symbiotic birth and spread of plastids: How many times and whodunnit? *J. Phycol.* 39(1): 4-12.
Parfrey, L. W., Lahr, D. J. G., Knoll, A. H. and Katz, L. A. (2011). Estimating the timing of early eukaryotic diversification with multigene molecular clocks. *Proc. Natl. Acad. Sci. U. S. A.* 108(33): 13624-13629.
Partensky, F., Blanchot, J. and Vaulot, D. (1999). Differential distribution and ecology of *Prochlorococus* and *Synechococcus* in oceanic waters: A review. In: Charpy, L. and Larkum, A. W. D. (Eds) *Marine Cyanobacteria*. Special No. 19. Bull. Inst. Oceanogr., Monaco. pp. 457-475.
Pérez, A. A., Rodionov, D. A. and Bryant, D. A. (2016). Identification and regulation of genes for cobalamin transport in the *Cyanobacterium Synechococcus* sp. strain PCC 7002. *J. Bacteriol.* 198(19): 2753-2761.
Perrine, Z., Negi, S. and Sayre, R. T. (2012). Optimization of photosynthetic light energy utilization by microalgae. *Algal Res.* 1(2): 134-142.
Pi, X., Tian, L., Dai, H. E., Qin, X., Cheng, L., Kuang, T.-Y., Sui, S. F. and Shen, J. R. (2018). Unique organization of photosystem I-light-harvesting supercomplex revealed by cryo-EM from a red alga. *Proc. Natl. Acad. Sci. U. S. A.* 115(17): 4423-4428.
Price, G. D. (2011). Inorganic carbon transporters of the cyanobacterial CO_2 concentrating mechanism. *Photosynth. Res.* 109(1-3): 47-57.
Price, G. D., Pengelly, J. J. L., Forster, B., Du, J., Whitney, S. M., von Caemmerer, S., Badger, M. R., Howitt, S. M. and Evans, J. R. (2013). The cyanobacterial CCM as a source of genes for improving photosynthetic CO2 fixation in crop species. *J. Exp. Bot.* 64(3): 753-768.

Rae, B. D., Long, B. M., Badger, M. R. and Price, G. D. (2013). Functions, compositions, and evolution of the two types of carboxysomes: Polyhedral microcompartments that facilitate CO_2 fixation in cyanobacteria and some proteobacteria. *Microbiol. Mol. Biol. Rev.* 77(3): 357–379.

Redding, K. and van der Est, A. (2006). The directionality of electron transport in photosystem I. In: Golbeck, J. H. (Ed.) *Photosystem I. Advances in Photosynthesis and Respiration*, (vol. 24). Springer, pp. 413–437.

Rippka, R., Deruelles, J., Waterbury, J. B., Herdman, M. and Stanier, R. (1979). Generic assignments, strain histories and properties of pure cultures of cyanobacteria. *J. Gen. Microbiol.* 111: 1–61.

Rocap, G., Distel, D. L., Waterbury, J. B. and Chisholm, S. W. (2002). Resolution of *Prochlorococcus* and *Synechococcus* ecotypes by using 16S–23S ribosomal DNA internal transcribed spacer sequences. *Appl. Environ. Microbiol.* 68(3): 1180–1191.

Ruffing, A. M., Jensen, T. J. and Strickland, L. M. (2016). Genetic tools for advancement of *Synechococcus* sp. PCC 7002 as a cyanobacterial chassis. *Microb. Cell Fact.* 15(1): 190.

Sakamoto, T. and Bryant, D. A. (1997). Growth at low temperature causes nitrogen limitation in the cyanobacterium *Synechococcus* sp. PCC 7002. *Arch. Microbiol.* 169(1): 10–19.

Schirrmeister, B. E., de Vos, J. M., Antonelli, A. and Bagheri, H. C. (2013). Evolution of multicellularity coincided with increased diversification of cyanobacteria and the Great Oxidation Event. *Proc. Natl. Acad. Sci. U. S. A.* 110(5): 1791–1796.

Schluchter, W. M., Shen, G., Alvey, R. M., Biswas, A., Saunée, N. A., Williams, S. R., Mille, C. A. and Bryant, D. A. (2010). Phycobiliprotein biosynthesis in cyanobacteria: Structure and function of enzymes involved in post-translational modification. In: Hallenbeck, P. (Ed.) Recent Advances in Phototrophic Prokaryotes. Advances in Experimental Medicine and Biology (vol. 675). Springer, New York, NY, pp. 211–228.

Schönknecht, G., Chen, W. H., Ternes, C. M., Barbier, G. G., Shrestha, R. P., Stanke, M., Bräutigam, A., Baker, B. J., Banfield, J. F., Garavito, R. M., Carr, K., Wilkerson, C., Rensing, S. A., Gagneul, D., Dickenson, N. E., Oesterhelt, C., Lercher, M. J. and Weber, A. P. (2013). Gene transfer from bacteria and archaea facilitated evolution of an extremophilic eukaryote. *Science* 339(6124): 1207–1210.

Shen, G. and Bryant, D. A. (1995). Characterization of a *Synechococcus* sp. strain PCC 7002 mutant lacking photosystem I. Protein assembly and energy distribution in the absence of the photosystem I reaction center core complex. *Photosynth. Res.* 44(1–2): 41–53.

Shen, G., Canniffe, D. P., Ho, M.-Y., Kurashov, V., van der Est, A., Golbeck, J. H. and Bryant, D. A. (2019). Characterization of chlorophyll *f* synthase heterologously produced in *Synechococcus* sp. PCC 7002. *Photosynth. Res.* 140(1): 77–92.

Shen, J. R. (2015). The structure of photosystem II and the mechanism of water oxidation in photosynthesis. *Annu. Rev. Plant Biol.* 66: 23–48.

Shih, P. M. (2015). Cyanobacterial evolution: Fresh insight into ancient questions. *Curr. Biol.* 25(5): R192–R193.

Sidler, W. (1994). Phycobilisome and phycobiliprotein structures. In: Bryant, D. A. (Ed.) *The Molecular Biology of Cyanobacteria, Advances in Photosynthesis and Respiration* (vol. 1). Springer, pp. 139–216.

Soulier, N., Laremore, T. N. and Bryant, D. A. (2020). Characterization of cyanobacterial allophycocyanins absorbing far-red light. *Photosynth. Res.* 145(3): 189–207.

Spalding, M. H. (2008). Microalgal carbon-dioxide-concentrating mechanisms: *Chlamydomonas* inorganic carbon transporters. *J. Exp. Bot.* 59(7): 1463–1473.

Stanier, R. Y. and Cohen-Bazire, G. (1977). Phototrophic prokaryotes: The cyanobacteria. *Annu. Rev. Microbiol.* 31: 225–274.

Stroebel, D., Choquet, Y., Popot, J. L. and Picot, D. (2003). An atypical haem in the cytochrome b_6f complex. *Nature* 426(6965): 413–418.

Suga, M., Ozawa, S. I., Yoshida-Motomura, K., Akita., F., Miyazaki, N. and Takahashi, Y. (2019). Structure of the green algal photosystem I supercomplex with a decameric light-harvesting complex I. *Nat. Plants* 5(6): 626–636.

Tros, M., Bersanini, L., Shen, G., Ho, M. Y., van Stokkum, I. H. M., Bryant, D. A. and Croce, R. (2020). Harvesting far-red light: Functional integration of chlorophyll *f* into photosystem I complexes of *Synechococcus* sp. PCC 7002. *Biochim. Biophys. Acta Bioenerg.* 1861(8): 148206.

Umena, Y., Kawakami, K., Shen, J. R. and Kamiya, N. (2011). Crystal structure of photosystem II of 1.9 Å. *Nature* 473(7345): 55–60.

Vermaas, W. F. J. (1996). Molecular genetics of the cyanobacterium *Synechocystis* sp. PCC 6803: Principles and possible biotechnology applications. *J. Appl. Phycol.* 8(4–5): 263–273.

Vogel, A. I. M., Lale, R. and Hohmann-Marriott, M. F. (2017). Streamlining recombination-mediated genetic engineering by validating three neutral integration sites in *Synechococcus* sp. PCC 7002. *J. Biol. Eng.* 11: 19.

Vu, T. T., Hill, E. A., Kucek, L. A., Konopka, A. E., Beliaev, A. S. and Reed, J. L. (2013). Computational evaluation of *Synechococcus* sp. PCC 7002 metabolism for chemical production. *Biotechnol. J.* 8(5): 619–630.

Waterbury, J. B. (2006). The cyanobacteria-Isolation, purification and identification. In: Dworkin, M., Falkow, S., Rosenberg, E., Schleifer, K. H. and Stackebrandt, E. (Eds) *The Prokaryotes*. Springer, New York, pp. 1053–1073.

Waterbury, J. B., Watson, S. W., Guillard, R. R. L. and Brand, L. E. (1979). Widespread occurrence of a unicellular, marine, planktonic, cyanobacterium. *Nature* 277(5694): 293–294.

Waterbury, J. B., Watson, S. W., Valois, F. W. and Franks, D. G. (1986). Biological and ecological characterisation of the marine unicellular cyanobacterium *Synechococcus*. In: Platt, T. and Li, W. K. W. (Eds) *Photosynthetic Picoplankton* (vol. 214). Can. Bull. Fish. Aquat. Sci., Ottawa, pp. 71–120.

Whitton, B. A. (2012). *Ecology of Cyanobacteria II: Their Diversity in Space and Time*. Springer, Dordrecht, 760 p.

Whitton, B. A. and Potts, M. (2000). *The Ecology of Cyanobacteria: Their Diversity in Time and Space*. Kluwer Academic Publisher, Dordrecht, London, Boston. 704 p.

Willis, A. and Woodhouse, J. N. (2020). Defining cyanobacterial species: Diversity and description through genomics. *Crit. Rev. Plant Sci.* 39(2): 101–124.

Wolf, B. M. and Blankenship, R. E. (2019). Far-red light acclimation in diverse oxygenic photosynthetic organisms. *Photosynth. Res.* 142(3): 349–359.

Wolk, C. P., Vonshak, A., Kehoe, P. and Elhai, J. (1984). Construction of shuttle vectors capable of conjugative transfer from *Escherichia coli* to nitrogen-fixing filamentous cyanobacteria. *Proc. Natl. Acad. Sci. U. S. A.* 81(5): 1561--1565.

Wolk, C. P., Ernst, A. and Elhai, J. (1994). Heterocyst metabolism and development. In: Bryant, D. A. (Ed.) *The Molecular Biology of Cyanobacteria. Advances in Photosynthesis* (vol. 1). Springer, Dordrecht, pp. 769-823.

Work, V. H., Melnicki, M. R., Hill, E. A., Davies, F. K., Kucek, L. A., Beliaev, A. S. and Posewitz, M. C. (2015). Lauric acid production in a glycogen-less strain of *Synechococcus* sp. PCC 7002. *Front. Bioeng. Biotechnol.* 3: 48.

Xu, Y., Alvey, R. M., Byrne, P. O., Graham, J. E., Shen, G. and Bryant, D. A. (2011). Expression of genes in cyanobacteria: Adaptation of endogenous plasmids as platforms for high-level gene expression in *Synechococcus* sp. PCC 7002. *Methods Mol. Biol.* 684: 273-293.

Yoon, H. S., Hackett, J. D., Ciniglia, C., Pinto, G. and Bhattarya, D. (2004). A Molecular timeline for the origin of photosynthetic eukaryotes. *Mol. Biol. Evol.* 21(5): 809-818.

Zess, E. K., Begemann, M. B. and Pfleger, B. F. (2016). Construction of new synthetic biology tools for the control of gene expression in the cyanobacterium *Synechococcus* sp. strain PCC 7002. *Biotechnol. Bioeng.* 113(2): 424–432.

Zhang, J., Ma, J., Liu, D., Qin, S., Sun, S., Zhao, J. and Sui, S. F. (2017). Structure of phycobilisome from the red alga *Griffithsia pacifica*. *Nature* 551(7678): 57-63.

Zhao, C., Gan, F., Shen, G. and Bryant, D. A. (2015a). RfpA, RfpB, and RfpC are the master control elements of Far-Red Light Photoacclimation (FaRLiP). *Front. Microbiol.* 6: 1303.

Zhao, C., Li, Z., Li, T., Zhang, Y., Bryant, D. A. and Zhao, J. (2015b). High-yield production of extracellular type-I cellulose by the cyanobacterium *Synechococcus* sp. PCC 7002. *Cell Discov.* 1: 15004.

Zhu, Y., Graham, J. E., Ludwig, M., Xiong, W., Alvey, R. M., Shen, G. and Bryant, D. A. (2010). Roles of xanthophyll carotenoids in protection against photoinhibition and oxidative stress in the cyanobacterium *Synechococcus* sp. strain PCC 7002. *Arch. Biochem. Biophys.* 504(1): 86-99.

Chapter 3

Extraction of proteins and other functional components from red seaweed (*Rhodophyta*)

Charlotte Jacobsen, National Food Institute - Technical University of Denmark, Denmark; Alireza Naseri, LACTOSAN A/S, Denmark; and Susan Løvstad Holdt, National Food Institute - Technical University of Denmark, Denmark

1 Introduction

Seaweeds are rich in polysaccharides, minerals, pigments and certain vitamins. Whereas some species are rich in protein and/or polyphenols, seaweeds in general are low in lipids, with maximal levels in certain species amounting up to only 5% (Gosch et al., 2012). However, the lipids contain high levels of health-beneficial, long chain omega-3 polyunsaturated fatty acids (Naseri et al., 2019). The nutritional composition of seaweeds varies with species. This chapter will focus on red seaweeds (*Rhodophyta*).

Red seaweeds contain high amounts of polysaccharides. *Chondrus, Porphyra, Gracilaria and Palmaria* were, for example, reported to contain 36–74% of dry weight (dw) polysaccharides (Holdt and Kraan, 2011). A large proportion of the polysaccharides (mainly cellulose and hemicellulose) is located in the cell wall to provide structure, but mycopolysaccharides and storage polysaccharides are also present (Murata and Nakazoe, 2001; Kumar and Kaladharan, 2007). Polysaccharides in red seaweed include agars, carrageenans, xylans, floridean starch (amylopectin-like glucan), water-soluble

http://dx.doi.org/10.19103/AS.2021.0091.03

sulfated galactan, as well as porphyran as mucopolysaccharides located in the intercellular spaces (Murata and Nakazoe, 2001; Kumar and Kaladharan, 2007). In particular, agars and carrageenans are extracted on commercial scales for numerous applications in food and pharmaceutical products.

In general, red seaweeds have a higher protein content than brown seaweeds but, even among red seaweeds, protein content varies substantially from ca. 4% in, for example, *Eucheuma denticulatum* (Naseri et al., 2019) to 50% in *Pyropia spp.* (formerly known as *Porphyra spp.*) (Holdt and Kraan, 2011). All seaweed proteins contain essential amino acids (EAA) which are available throughout the year (Galland-Irmouli et al., 1999). Despite the high contents of proteins in red seaweeds, there is no commercial production of protein extracted from red, green or brown seaweeds. However, various technologies for extracting seaweed proteins have been reported, and will be reviewed below.

According to Holdt and Kraan (2011), lipid content in red seaweed varies from 0.7% to 3.0% dw in *Chondrus,* from 0.12% to 1.9% dw in *Pyropia,* from 0.4% to 2.6% dw in *Gracilaria* and from 0.2% to 3.8% dw in *Palmaria*. In a recent study, Naseri et al. (2019) investigated lipid content in industrial samples of red seaweeds used for carrageenan production (*Euchema denticulatum, Kappaphycus alvarezii, Chondrus cripsus, Sarcothalia crispate* and *Furcellaria lumbricalis)* and found that it varied between 0.4% and 0.7% of dw. Total content of omega-3 PUFA varied from 4.9% to 23.2% of total fatty acids. Compared with other sources of long chain omega-3 PUFA such as fish (e.g. 9.5–16.9% on average in herring (Jensen et al., 2007)) and microalgae (e.g. 29% on average in *Nannochloroposis salina* (Safafar et al., 2016)), the lipid content of red seaweed is lower and of less commercial interest. It is therefore less economically viable to develop methods for extraction of lipids from seaweed on an industrial scale. For that reason, extraction of lipids will not be further discussed in this review.

Brown seaweeds have a higher content of polyphenols than red seaweeds, but some species of red seaweed also contain polyphenols with antioxidant activity. Jacobsen et al. (2019) provided an overview of reported levels of total phenolic content in *Chondrus crispus* (0.1–0.4% dw.), *Pyropia spp.* (0.3–0.4% dw.) and *Gracilaria spp.* (0.095–1.6% dw.). Antioxidant extracts from brown seaweeds are commercially available and are currently mainly used in cosmetic products.

Like other seaweeds, red seaweeds are also known for their carotenoid content. These are accessory pigments supporting light harvesting by photosynthetic chlorophylls. The main carotenoids in red seaweed are β-carotene (with provitamin A activity), lutein and zeaxanthin (Holdt and Kraan, 2011). *Pyropia* (4500 mg/kg) and *P. palmata* (456 mg/kg) have particularly high contents of β-carotene. Red seaweeds are the only seaweeds that contain phycobiliproteins, which are used as natural colorants for food and cosmetic applications (Holdt and Kraan, 2011). Unlike carotenoids, phycobiliproteins are water soluble. They consist of two types of phycobilins: red phycoerythrobilin and blue phycocyanobilin. Phycoerythrobilin makes up a major proportion of the proteins in red seaweed. For example, *P. palmata* and *Gracilaria tikvahiae* have been reported to contain 1.2% and 0.5% of dw, respectively (Chronakis et al., 2000; Wang et al., 2002).

The aim of this chapter is to provide an overview of methods to extract functional compounds from red seaweed with a particular emphasis on proteins. Methods to extract polysaccharides (carrageenan and agar) and antioxidants (phenolic compounds) will also be briefly discussed. The chapter will only discuss methods that are scalable, and which have been or potentially can be implemented at industrial scale. Methods for extraction of compounds for analytical purposes will only be mentioned when relevant. Due to the increasing focus on sustainability and reduction of waste in the production chain, the chapter will address these issues, focusing on the extraction of multiple compounds from the same biomass (multi-extraction). The chapter also identifies knowledge gaps and areas for further research and development.

2 Extraction of polysaccharides: carrageenan and agar

A major industrial application of red seaweeds is the extraction of agar and carrageenan and therefore these are the only two hydrocolloids (stabilizing agents) discussed in this chapter. This industrial process could be utilized in the future to extract protein from the same seaweed biomass, resulting in a sequential multiple extraction process of both proteins and hydrocolloids. The following issues should be considered for a future multi-extraction process:

- The process should either be carried out before carrageenan extraction under conditions that would not compromise the carrageenan.
- Alternatively, it should be carried out after carrageenan extraction in a way which does not degrade the proteins.

For this reason the extraction of carrageenan is summarized before discussion of extraction of proteins.

2.1 Extraction of carrageenan

Carrageenan is mainly produced from two species, *Kappaphycus alvarezii* and *Eucheuma denticulatum* (followed by *Chondrus crispus* and *Gigartina* species). It is a hydrocolloid widely used in industry, for example, in the manufacture of ice cream, tooth paste and jam. Carrageenans are classified as α-, β-, κ-, λ-, ι-, γ-, μ-, ν- and θ-carrageenans, depending on amount and position of sulfate ester modifications, which also give different gelling properties (Table 1).

Table 1 Different types of carrageenan (Rhein-Knudsen et al., 2015)

Seaweed source	Products	Main chemical structures	Applications
Kappaphycus alvarezii	κ-Carrageenan	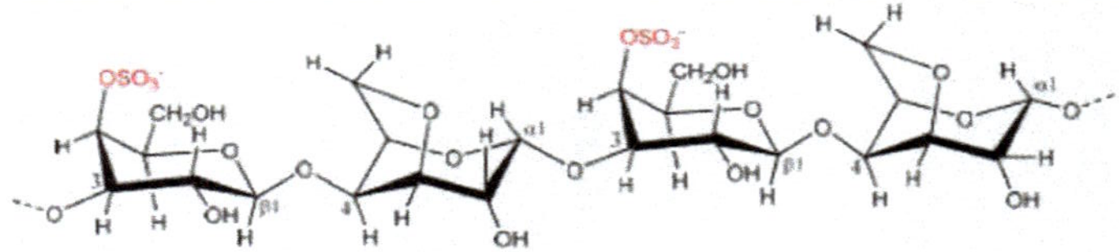	Gelling agent (stiff and brittle gel)
Eucheuma spinosum	ι-Carrageenan	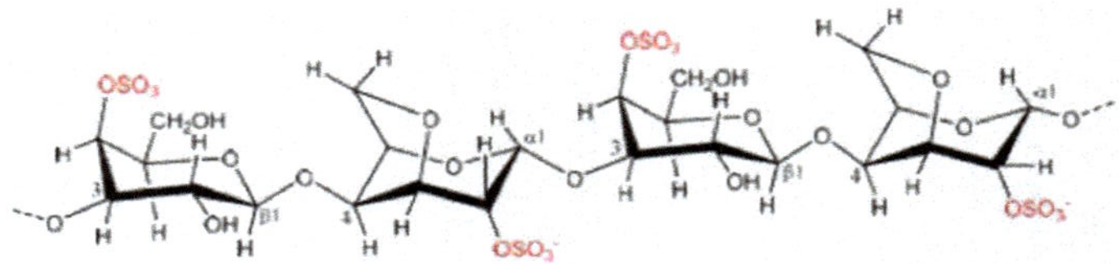	Gelling agent (flexible soft gel)
Gigartina spp. *Chondrus* spp.	λ-Carrageenan	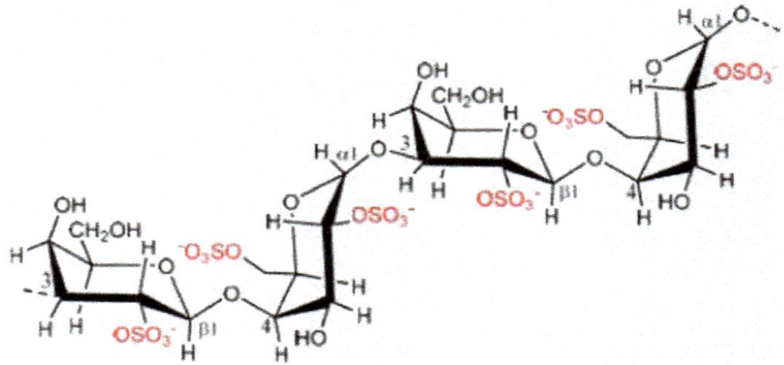	Thickener

Seaweed source	Products	Main chemical structures	Applications
Kappaphycus alvarezii	μ-Carrageenan	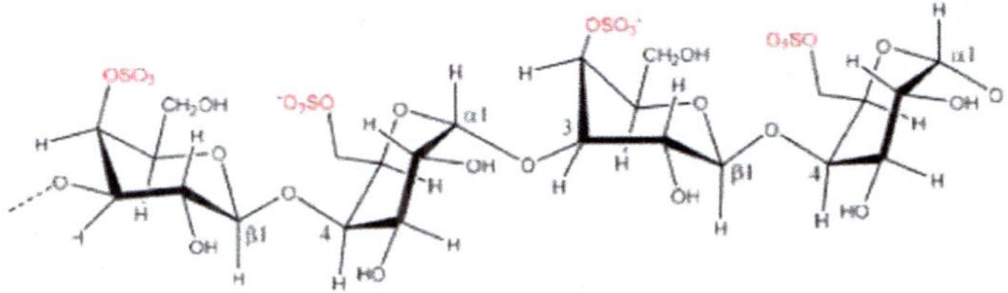	κ-Carrageenan precursor
Eucheuma spinosum	ν-Carrageenan	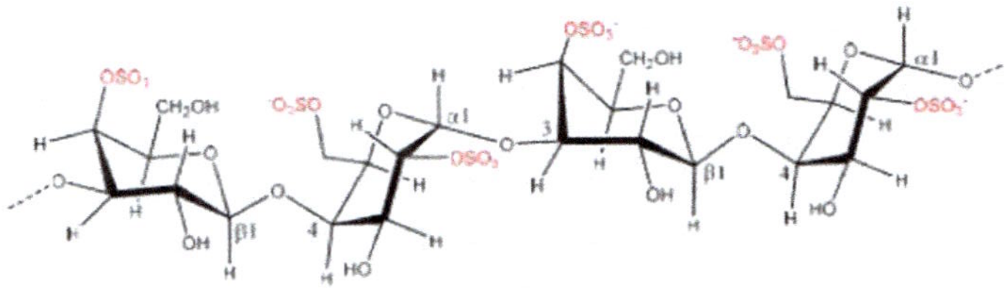	ι-Carrageenan precursor

They all have a base structure of D-galactopyranosyl units bound together by alternating α-1,3 and β-1,4 linkages. In seaweed, carrageenan is located in the cell wall and in intercellular matrix tissue (Yan and Zhang, 2014). The commercially important carrageenans are κ-, ι- and λ- carrageenan (Rhein-Knudsen et al., 2015). In 2014, the total commercial production of carrageenan was 60 000 tons, corresponding to a value of US$ 626 million globally (Rhein-Knudsen et al., 2015). Based on this production number, and assuming a yield of carrageenan extraction at industrial scale of at least 20%, it can be estimated that the total dried seaweed consumption for production was at least 300 000 ton/year (Naseri et al., 2019).

Semi-refined and refined carrageenans are the two main types of commercial carrageenan products. Semi-refined carrageenan is extracted from cleaned seaweeds, which are cooked in an alkali solution. The cooked seaweed is then sun dried, ground and sold as seaweed flour, and mainly used in pet food production (McHugh, 2003; Anisuzzaman et al., 2014).

Production of refined carrageenan starts with alkali extraction of the cleaned seaweed, involving soaking in alkali followed by extraction in boiling water (Fig. 1). Typically KOH is used for κ-carrageenan and NaOH for

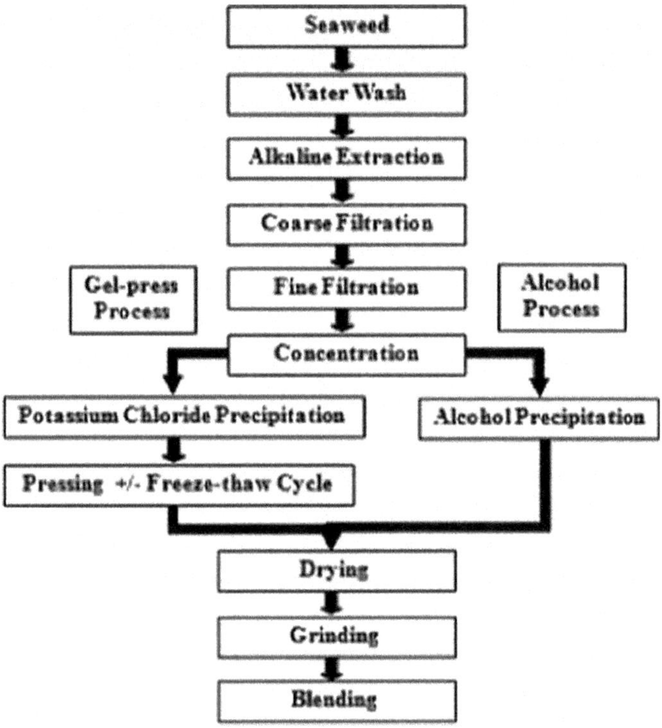

Figure 1 Process for extraction of refined carrageenan (Younes et al., 2018).

ι-carrageenan. The alkali treatment not only extracts carrageenan but also modifies the carrageenan extracted (Istinii et al., 1994). The seaweed residue is then removed by coarse and fine filtration. After this the carrageenan is concentrated and then recovered from the solution by precipitation with isopropyl alcohol. Carrageenan can also be concentrated by potassium chloride precipitation followed by freezing and thawing the gel until water is removed from the carrageenan fibers. Finally, the carrageenan fibers are dried and milled. The resulting carrageenan is practically 'pure' with a white color, and can be used in the food industry for refined products like desserts and puddings (McHugh, 2003; Imeson, 2009).

It is evident from the description above that current industrial practice is not environmentally friendly as it involves the use of chemicals, high temperatures and long reaction times. Moreover, high temperatures and use of chemicals may have negative effects on hydrocolloid functional properties, due to degradation of polysaccharides during extraction (Hernández-Carmona et al., 2013).

Alternative extraction methods, such as enzyme-assisted extraction (Varadarajan et al., 2009; Blanco-Pascual et al., 2014) and microwave-assisted extraction (Vázquez-Delfín et al., 2014) have also been investigated but, to the authors' knowledge, none have been implemented on an industrial scale.

Rhein-Knudsen et al. (2017) have investigated the use of a sustainable substitute for the alkaline solution used for carrageenan extraction. They compared the effect of using cocoa pod husk ash solution or KOH traditionally used for carrageenan extraction on the chemical and rheological properties of κ-carrageenan from *Kappaphycus alvarezii* and the Ghanaian red seaweed *Hypnea musciformis*. Similar extraction yields and successful modification of the hydrocolloids with 3,6-anhydro-galactopyranose and sulfate contents of 38% and 16% were obtained with cocoa pod husk ash and KOH extraction, respectively. Gel strengths of the κ-carrageenans were also similar irrespective of whether KOH or cocoa pod husk solution had been used (Rhein-Knudsen et al., 2017). Recently, Novozymes have developed enzymes particularly for extraction of carrageenan, but whether the technology has been taken up by any industrial producer is not known (Ramani, 2016).

2.2 Extraction of agar

Agar is commercially extracted from *Gracilaria* spp. and *Gelidium* spp. It is used as a gelling agent in the food industry in, for example, confectionary and dairy products and in biotechnological applications such as culture media for microbial growth and for gel-electrophoresis. Agar is chemically similar to carrageenan. However, whereas the 3,6-anhydro-galactopyranosyl is in the D-configuration in carrageenans, agar consists of 3,6-anhydro-L-galactopyranosyl residues. Agar is built up from agarose and agaropectin. Agarose is the neutral unmodified backbone of agar in which around 20% of the dimers hold methyl or sulfate groups. Agaropectin is the modified part of the molecule. Modifications of agaropectin include sulfate, pyruvate and, to a lesser extent, different sugar molecules (Rees, 1969).

The extraction of agar is based on the same principle as that of carrageenan. It starts with a modification and extraction step using an alkali followed by hot water extraction. Agar is recovered by alcohol precipitation, gel pressing, freeze-thawing techniques or syneresis (i.e. expulsion of liquids), depending on seaweed source (Armisen and Galatas, 1987). Native agar (i.e. non-modified agar) can be directly extracted in a similar way in boiling water followed by filtration and alcohol precipitation, gel pressing or, for example, freeze-thawing techniques. In the same way as carrageenan, different alternative, sustainable techniques for carrageenan extraction have been evaluated for agar extraction (Villanueva et al., 2010; Hernández-Carmona et al., 2013).

3 Single extraction of proteins

As mentioned previously in this chapter, there is currently no commercial production of proteins from red seaweeds. However, there is significant future potential for such extraction which could enable better industrial utilization of this resource. The summary of different protein extraction methods below is therefore based on experimental studies reported in the literature. Several studies have focused only on extraction of protein. They did not consider possible co-extraction/multi-extraction of other compounds, for example, polysaccharides including hydrocolloids or polyphenols. These are discussed first.

3.1 Physical extraction with water as extraction solvent

Homogenization, osmotic stress and high shear force are three different extraction technologies, which have been used in different studies to extract protein from seaweeds using ultra-pure water as the extraction solvent. Barbarino and Lourenço (2005) were able to extract 8.9 g protein/100 g dw from *Porphyra acanthophora* using homogenization and aqueous extraction. In another study with *Palmaria palmata*, osmotic stress and high shear force (using an Ultra turrax) were used to extract protein. Almost identical yields were obtained with the two methods. Osmotic stress led to a yield of 39% (6.77 ± 0.22 g/100 g dw), and a high shear force led to a yield of 40% (6.92 ± 0.12 g/100 g dw). There was no significant difference in the amount of total protein (Harnedy and FitzGerald, 2013; Bleakley and Hayes, 2017).

3.2 Chemical and physical extraction

The pH-shift technique, which uses acid or alkaline solution to solubilize protein before recovery at a lower or higher pH, has been used in many studies to maximize protein recovery for fish, seaweed, mussels and soy proteins (Wallace, 2000; Choi and Kim, 2007; Foh et al., 2012; Harrysson et al., 2018; Surasani, 2018). The process is based on the effect of pH on the solubility of the protein in the extracted solution. By decreasing or increasing pH in the protein solution, the side chain charge can change to a strong positive or negative charge that leads to strong repulsive forces between proteins. These strong positive or negative charges are suitable for protein-water interactions and make the protein soluble. The principle of the method is that proteins are solubilized either at the original pH or by adjusting the pH into the alkaline or acid region. Soluble proteins are then recovered in the supernatant after centrifugation. The proteins in the supernatant are then precipitated by adjusting the pH to the electrically neutral point, where the molecule carries no net electrical charge (isoelectric point (pI)). This leads to recovery of the protein in pellet form after the second centrifugation. The main advantages of the pH-shift method are high yields, high reductions in lipid content and improvement in the protein gel-forming ability; lower whiteness and protein denaturation during the processing are the two main disadvantages (Surasani, 2018).

The most common chemicals used in this method are hydrochloric acid (HCl), sodium hydroxide (NaOH), polyethylene glycol (PEG) and potassium carbonate. N-acetyl-L-cysteine (NAC) and 2-mercaptoethanol are also used as a reducing agent to increase the efficiency of extraction of alkaline soluble protein from seaweed. Harnedy and FitzGerald (2013) used NaOH and NAC to extract protein from red seaweed, *P. palmate*; the extract contained 4.16 g protein/100 g dw, which corresponded to a yield of 24% yield. O'Connor et al. (2020) applied three different physical treatments individually to extract protein from four seaweed species. They used high-pressure processing (HPP), laboratory autoclave processing, classical sonication and salting out methods. The four species that were tested were two brown seaweeds, *Fucus vesiculosus* and *Alaria esculenta*, and two red seaweeds, *P. palmata* and *C. crispus*. Yields of protein recovered varied from 3.2% for *F. vesiculosus* pre-treated with HPP to 28.9% protein recovered for *C. crispus* treated with the classical method.

3.3 Enzyme-assisted extraction

As reported above, the yield obtained with extraction processes based on physical or chemical extraction principles was below 30%. This is most likely due to low availability of protein molecules, due to intermolecular interactions (steric hindrance) by anionic cell-wall polysaccharides as well as viscosity issues. Break down of the cell wall polysaccharides by enzymes may therefore be necessary to increase protein extraction efficiency. The use of polysaccharidases enzymes has been reported to provide higher extraction efficiency than conventional methods for different seaweeds (Bleakley and Hayes, 2017).

Fleurence et al. (1995) used κ-carrageenase, β-agarase, xylanase and cellulase to degrade seaweed cell walls and improve protein extraction from *C. crispus*, *G. verrucosa* and *P. palmata* (Fleurence et al., 1995). In general, they found that enzyme hydrolysis contributed to cell wall degradation which increased protein extraction efficiency. They evaluated different seaweed/enzymes combinations (*carrageenase/C. crispus, 3-agarase/Gracilarivae verrucosa, xylanase/P. palmata*). Seaweed/cellulase + specific enzyme (e.g. *C. crispus/carrageenase* + cellulase) systems were also studied. The highest protein yields were found when cellulase was coupled with carrageenase or agarase for an incubation period of 2 h, except for *P. palmata*, for which no improvement in yield was observed when xylanase was coupled with cellulase. The best overall protein yield for *P. palmata* was obtained with xylanase alone with a long incubation time of 14 h. These differences between seaweed species suggest that degradation of cellulose is not imperative for an efficient protein extraction for *Palmaria*.

Extraction of proteins from *P. palmata* was also evaluated by Harnedy and FitzGerald (2013) using the commercial enzymes, Celluclast® 1.5L, Shearzyme® 500L, Ultraflo® or Corolase®. These were used with or without subsequent alkaline extraction using NaOH and N-acetyl-L-cysteine (NAC). The alkaline extraction was also tested without pretreatment with enzymes. Two different physical extraction methods were also evaluated: osmotic shock and high shear force. The protein extraction yields were 24%, 40% and 67% for chemical extraction, physical processing and enzymatic hydrolysis, respectively. The maximum yield of 67% was obtained with a combination of enzymatic hydrolysis and alkaline extraction.

In a recent, study a yield of 80% was obtained when the protease Umamizyme was used for extraction of protein from *P. palmata* (Bjarnadóttir et al., 2018). In this study, the effect of xylanase either alone or in combination with Umamizyme was also evaluated. In terms of dry matter, the highest yield was obtained when Umamizyme was used. The highest protein content in the liquid fraction after extraction was, however, obtained when a combination of xylanase and Umamizyme was used. Some of the proteins in the liquid fraction were in peptide form due to the proteolysis caused by Umamizyme.

The most recent study on protein extraction from *P. palmata* was performed in the authors' lab (Naseri et al., 2020b). The extraction process used consisted of an aqueous/enzymatic treatment followed by alkaline extraction using NAC. The effect of four commercial enzymes, Celluclast® 1.5L, Shearzyme® 500L, Alcalase® 2.4L FG and Viscozyme®L, was evaluated at 0.0%, 0.2% and

0.4% seaweed biomass/solution (*w*/w), at different pH and temperatures. The enzymes were tested individually and in combinations. Proteins recovered in the liquid fraction after centrifugation were precipitated using the pI of the proteins by slowly lowering pH from 13 to 3 using HCl (< 5M). A flow diagram of the process is shown in Fig. 2. The protein extraction efficiency was calculated as the percentage of protein removed from the original seaweed sample by analyzing the protein content of the seaweed before and after extraction.

When enzymes were used individually, Celluclast, Shearzyme and Viscozyme all resulted in extraction efficiencies of ca. 40–45%, which was at the same level as when no enzymes had been added, and extraction only performed with alkaline NAC (Naseri et al., 2020b). These enzymes are all carbohydrases. The results suggest that carbohydrases were not able to break down the cell wall and release intracellular proteins. However, when Alcalase was used, the extraction efficiency almost doubled to ca. 80% suggesting that the cell wall was at least partly broken down, allowing proteins to be released. There was no effect of increasing the concentration of enzymes from 0.2% to 0.4%.

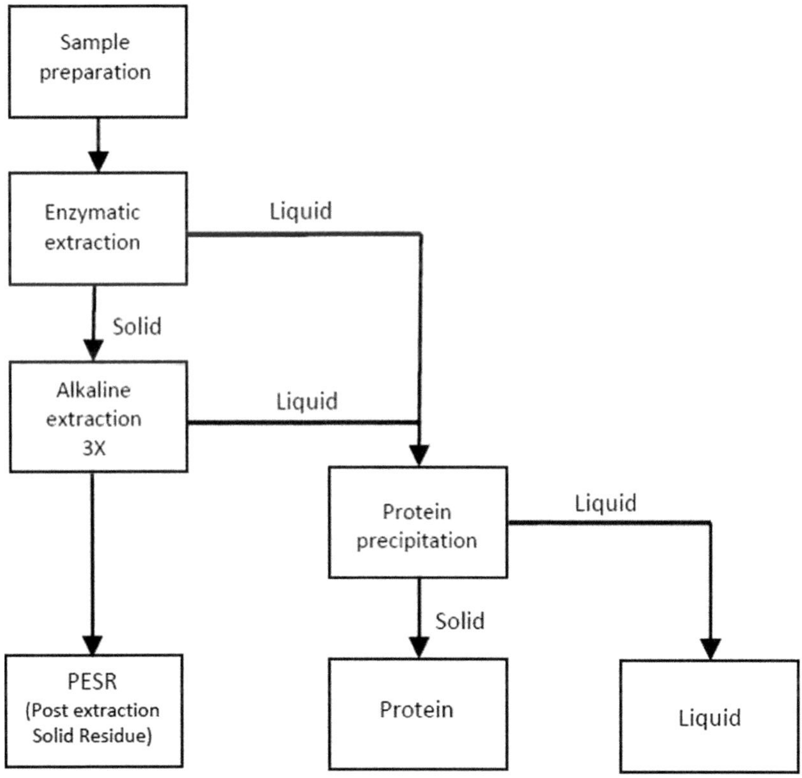

Figure 2 Flow diagram of a sequential process to extract protein from *P. palmata* using enzymes and alkaline extraction assisted by NAC.

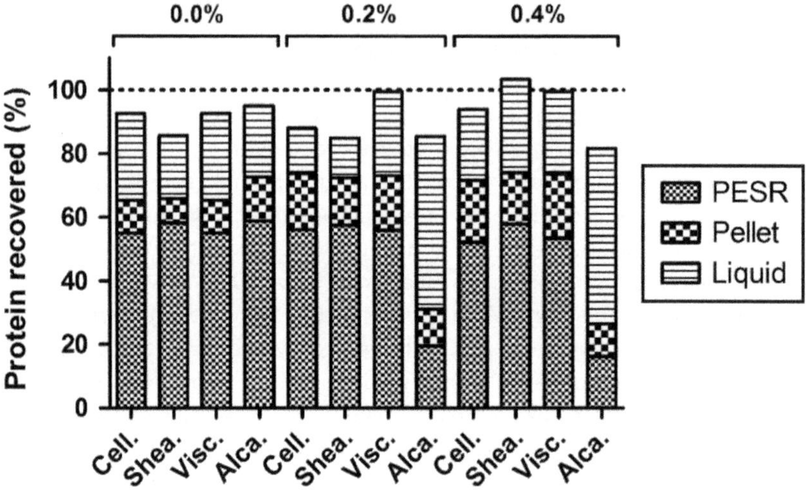

Figure 3 Protein recovered in the post-extraction solid residue (PESR), pellet and liquid fractions obtained using four individual enzymes at three concentrations (0.0%, 0.2%, and 0.4%) ($n = 3$, $p < 0.05$) (Cell: Celluclast®; Shea: Shearzyme®; Visc: Viscozymes®; Alca: Alcalase®). From Naseri et al. (2020b).

Figure 3 shows the distribution of the proteins recovered in the different fractions after extraction. The proportion of protein recovered in the liquid fraction was significantly higher (54.3–55.2%) when Alcalase was used compared with all other enzymes (12.5–29.3%), whereas the proportion recovered in post-extraction seaweed residue was significantly lower. This confirmed that proteins were indeed released from the seaweed to a higher degree when Alcalase was used. This may be because protease activity meant that some of the proteins were hydrolyzed to potentially more bioactive peptides. The proportion of protein recovered in the pellet was also significantly lower when Alcalase was used (10.2–11.6%) compared with all other enzymes (14.9–20.0%). For Alcalase, the highest protein extraction efficiency was obtained at a temperature of 60°C and the lowest efficiency was found at 30°C, which was the lowest temperature evaluated. No effect of pH was seen when different values between 7 and 9 were evaluated.

The enzymes were also evaluated in different combinations at two different pH values (pH 4.5 and 8.0) (Naseri et al., 2020b). As observed from Fig. 4, extraction efficiencies were always highest in combinations including Alcalase. Interestingly, there was no significant effect from altering pH despite the fact the optimum pH for Alcalase is around 8. Celluclast plus Shearzyme and Celluclast plus Viscozyme resulted in significantly higher protein recovered in the pellet fraction (20.4–22.8%) compared with all other treatments (14.4–17.0%). This is most likely related to the wide range of pIs that the different

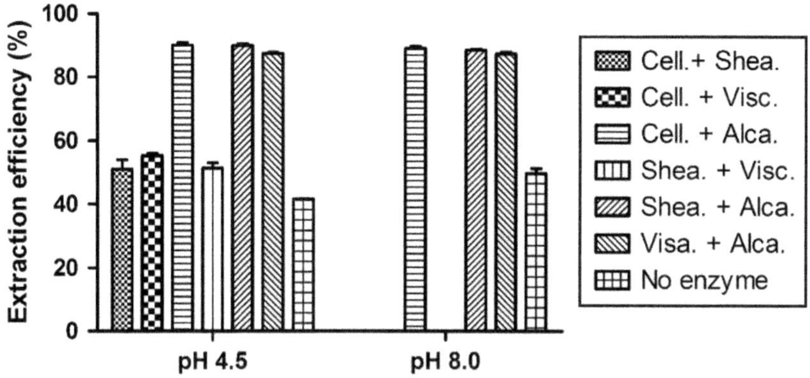

Figure 4 Effect of different enzyme combinations (cocktail) and pH on the protein extraction efficiency (n=3, $p<0.05$) (Cell: Celluclast®; Shea: Shearzyme®; Visc: Viscozymes®; Alca: Alcalase®). From Naseri et al. (2020b).

peptides may have when Alcalase is used, which means that they were not recovered from the liquid, because some of the peptides were resolubilized when pH was lowered.

The amino acid profile was determined in all the fractions after extraction. The maximum content of EAA in the liquid fractions was obtained when Alcalase (500 mg/protein) was used. Interestingly, the total amount of essential amino acids per gram protein was higher in the post-extraction seaweed residue for all enzymatic treatments compared with the raw material (untreated seaweed). The EAA constituted approximately 50% of total amino acids found in the pellet, whereas the EAA ratio varied from 0.2 to 0.44 in the liquid fraction. Both in the liquid and in the pellet, the highest EAA ratio was found when Alcalase alone was used.

Table 2 provides a comparison on the relative proportion of EAA between the study by Naseri et al. (2020b) and previous studies. Generally, the relative proportion of all EAA increased after extraction compared to the raw material. In the study by Bjarnadóttir et al. (2018), the concentration of all EAA in the solid phase increased compared to the raw materials (untreated seaweed) except for methionine, threonine and tryptophan (Bjarnadóttir et al., 2018). However, in the study by Naseri et al. (2020b) one more step to recover protein from the liquid extract was included, and proteins were recovered from three fractions, whereas the other studies only investigated two fractions (solid and liquid). This makes direct comparison between the different fractions difficult. Nevertheless, the high proportion of EAA in the proteins recovered from *P. palmata* suggests that the protein is of high nutritional value. Further studies are, however, needed to evaluate protein digestibility and the potential applications of the protein in food and feed products (Naseri et al., 2020b).

Table 2 Comparison of essential amino acid (EAA) percentage (%) of total amino acids between previous studies and the present study. From Naseri et al. (2020b)

	Study by Naseri et al. (2020b)			Previous studies					
EAA%	Raw material (Untreated)	PESR[a]	Pellet[a]	Galland-Irmouli et al. 1999 (Untreated)	Mæhre et al. 2014 (Untreated)	Bjarnadóttir et al. 2018 (Untreated)	Bjarnadóttir et al. (2018)[b]	Bjarnadóttir et al. (2018)[c]	Mæhre et al. (2016a,b)[d]
Isoleucine	2.8 ± 0.5	4.8 ± 0.6	3.4 ± 0.2	3.7 ± 1.7	4.5 ± 0.1	4.3	5.0	5.3	5.1 ± 0.5
Leucine	6.4 ± 0.0	6.2 ± 0.3	7.6 ± 0.2	7.1 ± 4.0	7.9 ± 0.2	7.2	7.9	8.3	9.1 ± 1.0
Lysine	5.1 ± 0.9	9.4 ± 1.4	4.9 ± 1.0	3.3 ± 2.7	6.2 ± 0.1	5.7	8.0	7.5	5.8 ± 0.4
Methionine	2.1 ± 0.1	1.5 ± 0.0	2.5 ± 0.0	2.7 ± 0.7	2.2 ± 0.1	2.0	1.4	1.6	2.5 ± 0.2
Phenylalanine	3.4 ± 0.2	5.2 ± 0.4	4.3 ± 0.1	5.1 ± 3.2	4.9 ± 0.3	4.7	6.0	5.4	5.4 ± 0.4
Threonine	3.1 ± 0.2	3.2 ± 0.1	3.2 ± 0.1	3.6 ± 3.7	4.9 ± 0.1	4.1	4.0	3.8	5.2 ± 0.4
Valine	3.7 ± 0.5	6.9 ± 1.1	3.8 ± 0.0	6.9 ± 2.7	6.7 ± 0.3	5.4	6.2	6.3	7.3 ± 0.7
Histidine	0.8 ± 0.0	1.9 ± 0.1	1.4 ± 0.0	0.5 ± 0.6	1.3 ± 0.1	1.7	1.9	1.9	1.5 ± 0.1
Tryptophan	n.a.	n.a.	n.a.	n.a.	n.a.	0.9	0.5	0.5	n.a.
Σ EAA%	27.4	39.1	31.1	32.9	38.6	36.0	40.9	40.6	41.8

n.a. not analyzed, EAA: essential amino acid, PESR: post-extraction seaweed residue.
[a] Calculated for *P. palmata* after enzymatic treatment by Celluclast® (0.2%w/w) + Alcalase®.(0.2%w/w) followed by NAC-assisted alkaline extraction.
[b] Calculated for *P. palmata* (solid phase) after enzymatic treatment with Umamizyme.
[c] Calculated for *P. palmata* (solid phase) after enzymatic treatment with Xylanase and Umamizyme.
[d] Calculated for *P. palmata* after enzymatic treatment with 50 U of xylanase and cellulose.

4 Multi-extraction of proteins and carrageenan

As previously mentioned, carrageenan is at present extracted industrially from seaweeds as the sole compound. Until recently there were no studies reporting methods to extract both proteins and carrageenan. However, a few recent studies have looked at effects of pre-treatments before carrageenan extraction. For example, Azevedo et al. (2015) showed that pre-treatment with KOH could increase the gelling properties of *C. crispus* and *Ahnfeltiopsis devoniensis*, probably because proteins and other compounds were pre-extracted by KOH. However, optimization of pre-treatment to obtain high yields of protein was not investigated (Azevedo et al., 2015).

We have recently reported a new multi-extraction process that extracts both protein and carrageenan from one of the major industrially important carrageenan species *E. denticulatum* (spinosum) (Naseri et al., 2020a). The extraction of protein from spinosum upstream of the carrageenan extraction was optimized with regard to temperature, pH, the concentration of enzymes and extraction time. The reason for upstream extraction of protein is to perform extraction before the harsh processing conditions for extraction of carrageenan with high pH and high temperature, which would otherwise harm the protein. The protein extraction process was similar to that shown in the flow diagram for *P. palmata* (Fig. 2). However, in contrast the process for spinosum had to be carried out at room temperature (<30°C) with a pH close to neutral in order to avoid an adverse effect on carrageenan quality, and to avoid gelling during the extraction process.

Celluclast, Shearzyme and Viscozyme were tested individually in concentrations of 0.1%, 0.2% and 0.4% and Alcalase was tested in a concentration of 0.2%. Combinations of Shearzyme and Celluclast were tested in concentrations of 0.1%, 0.2% and 0.4%. Extraction time was also optimized. The individual addition of all three carbohydrases (Viscozymes, Celluclast and Shearzyme) improved protein extraction efficiency compared to when no enzymes were added. The highest protein extraction efficiency was obtained with Viscozymes at a concentration of 0.2% w/w (48.5% protein extracted), whereas the highest extraction efficiency obtained with Celluclast or Shearzyme were ca. 37 and 25%, respectively (both added in a concentration of 0.4%). When proteins were extracted only with alkaline NAC (without addition of enzymes), the protein extraction efficiency was ca. 15%. Viscozymes is a multi-enzyme complex containing arabinase, cellulase, β-glucanase, hemicellulase and xylanase, whereas Celluclast (cellulase) and Shearzyme (endo 1,4-b-xylanase) contained fewer enzymes. This difference most likely explains why Viscozyme was the most efficient enzyme. A combination of Celluclast and Shearzyme performed better than when the enzymes were used individually. It was possible to obtain an extraction efficiency of ca. 39% when these two enzymes were used in a concentration of 0.2%.

When comparing the results to those obtained with the same enzymes for *P. palmata* it is clear that carbohydrases improved protein extraction from spinosum, but not for *P. palmata*. This is due to the fact that that the structure and composition of the cell wall are very different between the two species with *P. palmata* having e.g. floridean starch (Holdt and Kraan, 2011). In line with the findings for *P. palmata,* Alcalase also provided a higher protein extraction efficiency (59.4%) for spinosum compared to all the carbohydrases. An extraction efficiency of 59.4% was obtained at an enzyme concentration of 0.2 % and a pH of 7. The extraction efficiency obtained in spinosum with Alcalase was, however, lower than that obtained in *P. palmata* (ca. 80%). This again highlights the fact that the structure and composition of the cell wall in the two species are quite different. Further studies are needed to understand these differences in detail.

A multi-extraction of carrageenan should not compromise or damage the original product. Naseri et al. (2020a) therefore also investigated the effect of enzyme-assisted protein extraction on carrageenan yield and quality. We found that, when no pre-extraction of protein was performed, the yield of carrageenan was 17.8%, while it increased to 23.8% when only NAC-assisted alkaline extraction (no enzymes) was used (Table 3). Carrageenan yield further increased to ca. 28% when Viscozymes or Alcalase were used. The highest yields were obtained when a combination of Celluclast and Shearzyme was used, whether or not the pH was 6.0 or 7.0. This means that the yield of carrageenan significantly increased in the multi-extraction process compared to the single extraction of carrageenan when the process was carried out in lab scale.

The quality of the extracted carrageenan was also investigated as shown in Table 4. Enzymatic treatment with Celluclast plus Shearzyme did not change gel strength. In contrast, enzymatic treatment with Alcalase or Viscozymes increased gel strength significantly. As these two treatments had the highest protein extraction

Table 3 Effect of protein extraction on the yield (%) of isolated carrageenan of seaweed *Eucheuma denticulatum* treated with different enzymes followed by NAC-assisted alkaline extraction. Yields are given in average ± standard deviation. Different superscripts letters in the same column indicate significant difference at the confidence level of 95% (α = 0.05; n = 3) (Naseri et al., 2020a).

Sample description	Yield (%)
Blank (no enzymatic treatment, no NAC-assisted alkaline extraction)	17.8 ± 0.89 [d]
No enzymatic treatment, only NAC-assisted alkaline extraction	23.8 ± 1.62 [c]
Celluclast 0.2% + Shearzyme 0.2% (pH 6.0) + NAC-assisted alkaline extraction	31.9 ± 4.52 [ab]
Celluclast 0.2% + Shearzyme 0.2% (pH 7.0) + NAC-assisted alkaline extraction	35.5 ± 2.12 [a]
Viscozymes 0.2% + NAC-assisted alkaline extraction	27.6 ± 0.97 [bc]
Alcalase 0.2% + NAC-assisted alkaline extraction	27.8 ± 3.26 [bc]

efficiency, the lowest amount of protein was left in the remaining biomass used for carrageenan extraction. It is likely that the extracted carrageenan had the lowest protein content which seems to have a positive effect on gel strength.

Other quality parameters of carrageenan are also important in industry, and these were also investigated for the evaluation of the multi-extraction process. The maximum gel strength (breaking strength) and breaking distance for the samples were significantly lower for the combination of Celluclast plus Shearzyme, irrespective of the pH, compared to the untreated sample. In contrast, there was no significant difference between the samples treated by only Alcalase or Viscozymes and the untreated sample. Results for turbidity showed that only samples treated with Shearzyme and Celluclast had lower values than the untreated sample whereas all other treatments had no effect. Intrinsic viscosity (IV) and FCC-viscosity are two parameters related to the gel viscosity. There was no difference between the different treatments for these two parameters.

With all these parameters, the study by Naseri et al. (2020a) showed that aqueous extraction with Alcalase at 0.2% *w*/w and pH 7 or Viscozymes at 0.2% *w*/w and pH 7 followed by alkaline NAC extraction were the best treatments to extract protein from spinosum. The treatment with Alcalase or Viscozyme also increases carrageenan yield in the subsequent carrageenan extraction process compared to when proteins were not pre-extracted. Furthermore, treatment with one of these enzymes had no detrimental effect on carrageenan quality. Indeed, quality seemed slightly better for some parameters. The use of the protease Alcalase produces proteins/peptides with low molecular weight which does not happen to the same extent when the carbohydrase Viscozyme is used.

Furcellaria lumbricalis is used commercially in the Baltic Sea for production of the carrageenan-like ingredient furcelleran in a single extraction process. This species has shown a high concentration of protein (28%) and is therefore of interest for possible future multi-extraction of protein and furcellaran (Naseri et al., 2019). A study was also conducted in the author's lab to develop an enzyme-assisted process for the extraction of protein prior to the extraction of furcelleran. A wider range of enzymes than those employed in the two studies discussed above on *P. palmata* and *E. denticulotum* was tested, but none of them could significantly improve protein extraction efficiency (non-published data). In contrast, it was demonstrated that chemical extraction of protein using KOH with good extraction yields was possible. These findings are probably due to a very different composition and structure of the cell walls of *F. lumbricalis* compared to the two other species. This needs further investigation.

5 Extraction of phenolic antioxidants

Several studies have focused on extraction of natural antioxidants from seaweed. Jacobsen et al. (2019) have provided an overview of many of

Table 4 Effect of protein extraction of seaweed *Eurcheuma denticulatum* on the quality of carrageenan gel. Quality is given as averages ± standard deviations. Different superscript letters in the same column indicate significant difference at the confidence level of 95% ($\alpha = 0.05$; $n = 3$). From Naseri et al. (2020a)

Sample treatments	Gel strength 4 mm (g/cm)	Gel strength 8 mm (g/cm)	Gel strength 12 mm (g/cm)	Breaking strength (g/cm)	Breaking distance (mm)	Turbidity (NTU*)	IV	FCC ** viscosity
Blank (No enzymatic treatment, No NAC-assisted alkaline extraction)	8.32 ± 0.13^{bc}	15.6 ± 0.29^{b}	23.6 ± 0.50^{bc}	56.5 ± 3.53^{a}	26.5 ± 0.22^{a}	62.8 ± 25.2^{a}	8.03 ± 0.22^{a}	249 ± 24.7^{a}
NO enzymatic treatment, Only NAC-assisted alkaline extraction, pH=7	8.09 ± 0.01^{c}	14.7 ± 0.14^{c}	21.5 ± 0.57^{c}	36.9 ± 6.71^{b}	22.0 ± 2.33^{c}	43.8 ± 11.2^{abc}	7.83 ± 0.51^{a}	233 ± 33.9^{a}
Celluclast 0.2% + Shearzyme 0.2% + NAC-assisted alkaline extraction, pH=6	8.11 ± 0.42^{c}	14.9 ± 0.61^{c}	21.4 ± 0.01^{c}	34.6 ± 1.62^{b}	22.2 ± 0.92^{c}	33.1 ± 12.0^{bc}	7.51 ± 0.31^{a}	231 ± 49.5^{a}

Celluclast 0.2% + Shearzyme 0.2% + NAC-assisted alkaline extraction, pH=7	8.53 ± 0.01[b]	15.7 ± 0.14[b]	22.6 ± 0.43[c]	35.2 ± 1.50[b]	21.9 ± 0.11[c]	29.6 ± 5.72[c]	7.59 ± 0.24[a]	239 ± 24.0[a]
Viscozyme 0.2% + NAC-assisted alkaline extraction, pH=7	11.6 ± 0.20[a]	19.1 ± 0.32[a]	25.4 ± 2.95[b]	59.0 ± 1.33[a]	24.6 ± 0.60[ab]	48.1 ± 2.14[abc]	7.89 ± 0.02[a]	222 ± 2.1[a]
Alcalase 0.2% + NAC-assisted alkaline extraction, pH=7	11.7 ± 0.31[a]	19.4 ± 0.62[a]	28.3 ± 1.52[a]	59.1 ± 6.41[a]	24.5 ± 0.80[b]	54.2 ± 18.0[ab]	7.48 ± 0.20[a]	218 ± 14.8[a]

*NTU: Nephelometric Turbidity Unit.
**FCC: Food Chemicals Codex.

these studies. Most studies have been carried out on brown seaweeds, while relatively few studies are available on red seaweed. One of these studies was performed by Farvin and Jacobsen (2013). They extracted polyphenols from *Porphyra purpurea, C. cripus, Mastocarpus stellatus, Polysiphonia fucoides* and *Gracilaria vermiculophylla* using either ethanol or water. Yield and content of total phenolics (TPC) are shown in Table 5. *Porphyra purpurea* was found to provide the highest TPC when extracted by water, whereas *P. fucoides* provided the highest TPC when ethanol was used. This ethanol extract had the highest TPC among all extracts (1920 mg/100 g dried seaweed). *Polysiphonia fucoides* extracts also had good DPPH free radical scavenging activity, reducing power and ability to reduce lipid oxidation in a liposome system, whereas the extracts from the other red seaweed species were less efficient. These properties correlated with their high TPC.

Chondrus crispus is the only species important for commercial carrageenan production. This was investigated in the study by Farvin and Jacobsen (2013). This species resulted in a low TPC no matter whether it was extracted by water or ethanol. Interestingly, both water and ethanol extracts from *C. crispus* had better metal chelating properties than those for *P. porphyra*. It is likely that other

Table 5 Total phenolic content of different Danish seaweeds (Farvin and Jacobsen, 2013)

		TPC (mg/100 g dried seaweed)	
Sl. no.	Species	Water	Ethanol
1	*Fucus serratus* (Linnaeus)	411.6 ± 12.2^{ax}	1200.8 ± 9.0^{by}
2	*Fucus vesiculosus* (Linnaeus)	607.7 ± 15.1^{ay}	1045.0 ± 45.8^{bx}
3	*Fucus distichus* (Linnaeus)	293.1 ± 24.9av	862.3 ± 9.3^{bw}
4	*Fucus spiralis* (Linnaeus)	344.4 ± 2.3^{aw}	752.5 ± 13.1^{bv}
5	*Sargassum muticum* (Yendo)	230.8 ± 17.1^{bu}	114.9 ± 30.3^{at}
6	*Saccharina latissima* (Linnaeus)	20.9 ± 0.3^{as}	35.4 ± 1.3^{as}
7	*Laminaria digitata* (Hudson)	11.5 ± 0.3^{as}	32.4 ± 1.0^{as}
8	*Dictyota dichotoma* (Hudson)	104.3 ± 8.7^{ast}	69.5 ± 0.7^{as}
9	*Enteromorpha intestinalis* (Linnaeus)	230.9 ± 20.3^{au}	265.8 ± 8.5^{au}
10	*Ulva lactuca* (Linnaeus)	224.2 ± 3.9^{au}	236.5 ± 7.8^{au}
11	*Palmaria palmata* (Linnaeus)	202.0 ± 11.5^{bu}	76.9 ± 0.9^{as}
12	*Porphyra purpurea* (Roth)	347.9 ± 15.3^{bw}	299.4 ± 5.8^{au}
13	*Chondrus crispus* (Stackhouse)	49.7 ± 2.8^{as}	113.6 ± 8.9^{bt}
14	*Mastocarpus stellatus* (Stackhouse)	83.9 ± 1.5^{as}	102.5 ± 7.4^{at}
15	*Polysiphonia fucoides* (Hudson)	212.7 ± 25.5^{as}	1920.0 ± 61.4^{bz}
16	*Gracilaria vermiculophylla* (Ohmi)	51.4 ± 4.3^{as}	95.2 ± 8.6^{ast}

Values are mean ± SD (n = 3). In TPC, the letters a, b indicates if there is significant difference between water and ethnolic extracts and letters s to z indicates significant difference between species (same letters not significant).

compounds in extracts from *C. crispus*, rather than phenolic compounds, are responsible for metal chelating activity, since neither TPC nor the content of individual phenolic acids was found to correlate with metal chelating activity in any of the seaweed investigated in this study.

Naseri et al. (2019) have investigated the effect of different solvents (ethanol, methanol or distilled water) on extraction of phenolic compounds in *E. denticulatum (spinosum), K. alvarezii, C. cripsus, S. crispate* and *F. lumbricalis*. *Furcellaria lumbricalis* with 19.06 mg GAE /100 g dw had the highest content of total phenolic compounds followed by *C. crispus* and *S. crispata* (5.06 mg GAE/g dw and 4.38 mg GAE/g dw respectively). Especially in the case of *F. lumbricalis*, these values are higher than previously reported values from different red seaweeds. As an example, the following values were reported for *P. palmata* (4.8–5.5 mg GAE/g dw), *C. crispus* (4.0–4.3 mg GAE/g dw), *Meristotheca papulosa* (4.2 mg GAE/g dw) and *Sarcodiotheca gaudichaudii* (4.3 mg GAE/g dw) (Farvin and Jacobsen, 2013; Tibbetts et al., 2016).

Unfortunately, there are no studies available in the literature on multi-extraction of polyphenols from red seaweed. All the studies discussed in this chapter were performed with the sole purpose of extracting antioxidants. Hence, more research is needed to develop methods which integrate the extraction of antioxidants into a multi-extraction process.

6 Conclusion and future trends

Currently, agar and carrageenan are the only ingredients extracted from red seaweeds on a commercial scale. However, red seaweeds contain several other compounds of high value, such as proteins, peptides, antioxidants and pigments. It has been demonstrated recently that it is possible to extract both protein and carrageenan from spinosum in an enzyme-assisted process with no negative effect on carrageenan yields and quality. Such a multi-extraction approach would decrease waste of seaweed resources and produce more high value compounds, but it has not yet been implemented on an industrial scale. The feasibility of a multi-extraction process of carrageenan and protein has been demonstrated for spinosum, but no studies have been reported on multi-extraction of agar and protein from agar-containing red seaweeds such as *Graciliaria spp*.

It is clear from the studies reported in the literature that the same enzyme may not be efficient in all types of red seaweed. For example, it was not possible to obtain satisfactory protein extraction efficiency for *F. lumbricalis* with the same enzymes as used for spinosum. In fact, none of the commercial enzymes evaluated proved efficient for *F. lumbricalis*. There is therefore a need to develop enzyme mixtures targeted specifically for the cell walls of the specific seaweed species of interest. Different enzyme mixtures may be needed for different types of seaweed because of their different cell wall compositions and structures. There is also a need to develop more efficient methods for recovering the released proteins from the extracted protein-rich liquid. To the authors' knowledge, no attempts have been made to extract pigments and antioxidants from red seaweed with a multi-extraction purpose in mind. This deserves more research.

7 Where to look for further information

Further information about seaweeds as a source of proteins can be found here:

- https://foodnationdenmark.com/news/from-seaweed-to-proteins/.
- https://www.dtu.dk/english/news/2019/03/dynamo-theme-4-from-seaweed-to-proteins??id=24c8b5b8-5abc-4615-b93c-abc15e885e7f.
- https://pure.au.dk/portal/en/projects/legumes-and-seaweed-as-alternative-protein-sources-for-sheep(7971a52e-9c97-4e67-9f83-a7fdce7eca77).html.
- https://www.lrfoundation.org.uk/en/impact-stories/safe-seaweed-to-feed-the-planet/.
- https://gfi.org/researchgrants/using-red-seaweed-protein-for-plant-based-meat/.
- https://thefishsite.com/articles/singapore-new-zealand-algae-based-protein-partnership-forged.
- https://www.chalmers.se/en/projects/Pages/Seaweed-as-a-vehicle-for-nutrients-in-a-circular-food-chain--.aspx.
- https://www.macrocascade.eu/.
- https://odr.chalmers.se/bitstream/20.500.12380/256238/1/256238.pdf.

Three PhD theses have also been published on this topic:

- https://munin.uit.no/bitstream/handle/10037/9130/thesis.pdf?sequence=3&isAllowed=y.
- https://core.ac.uk/download/pdf/228339556.pdf.
- https://orbit.dtu.dk/en/publications/valorization-of-red-seaweed-biomasses-towards-future-sustainabili.

8 References

Anisuzzaman, S. M., Bono, A., Krishnaiah, D., Ariffin, B. and Lee, R. (2014). Effect of alkali treatment process parameters on semi refined carrageenan functional groups, *Indian Journal of Chemical Technology* 21(5–6), 386–391.

Armisen, R. and Galatas, F. (1987) Chapter 1. Production, properties and uses of agar. In McHugh, D. J. (Ed.) *Production and Utilization of Products from Commercial Seaweeds*.

Food And Agriculture Organization of the United Nations Rome, FAO Fisheries technical paper 288. Available at: http://www.fao.org/3/x5822e/x5822e00.htm.

Azevedo, G., Torres, M. D., Sousa-Pinto, I. and Hilliou, L. (2015). Effect of pre-extraction alkali treatment on the chemical structure and gelling properties of extracted hybrid carrageenan from Chondrus crispus and Ahnfeltiopsis devoniensis, *Food Hydrocolloids* 50, 150–158. doi: 10.1016/j.foodhyd.2015.03.029.

Barbarino, E. and Lourenço, S. O. (2005). An evaluation of methods for extraction and quantification of protein from marine macro- and microalgae, *Journal of Applied Phycology* 17, 447–460. doi: 10.1007/s10811-005-1641-4.

Bjarnadóttir, M., Aðalbjörnsson, B. V., Nilsson, A., Slizyte, R., Roleda, M. Y., Hreggviðsson, G. Ó., Friðjónsson, Ó. H. and Jónsdóttir, R. (2018). Palmaria palmata as an alternative protein source: enzymatic protein extraction, amino acid composition, and nitrogen-to-protein conversion factor, *Journal of Applied Phycology* 30(3), 2061–2070. doi: 10.1007/s10811-017-1351-8.

Blanco-Pascual, N., Alemán, A., Gómez-Guillén, M. C. and Montero, M. P. (2014). Enzyme-assisted extraction of κ/ι-hybrid carrageenan from Mastocarpus stellatus for obtaining bioactive ingredients and their application for edible active film development, *Food and Function* 5(2), 319–329. doi: 10.1039/c3fo60310e.

Bleakley, S. and Hayes, M. (2017). Algal proteins: extraction, application, and challenges concerning production, *Foods* 6(5), 33. doi: 10.3390/foods6050033.

Choi, J. Y. and Kim, J.-S. (2007). Fish protein recovered using ph shifting method and its physicochemical properties, *Journal of Ocean University of China* 4(3), 224–228. doi: 10.1007/s11802-005-0037-z.

Chronakis, I. S., Galatanu, A. N., Nylander, T. and Lindman, B. (2000). The behaviour of protein preparations from blue-green algae (Spirulina platensis strain Pacifica) at the air/water interface, *Colloids and Surfaces Part A: Physicochemical and Engineering Aspects* 173(1–3), 181–192. doi: 10.1016/S0927-7757(00)00548-3.

Farvin, K. H. and Jacobsen, C. (2013). Phenolic compounds and antioxidant activities of selected species of seaweeds from Danish coast, *Food Chemistry* 138(2–3), 1670–1681. doi: 10.1016/j.foodchem.2012.10.078.

Fleurence, J., Massiani, L., Guyader, O. and Mabeau, S. (1995). Use of enzymatic cell wall degradation for improvement of protein extraction from Chondrus crispus, Gracilaria verrucosa and Palmaria palmata, *Journal of Applied Phycology* 7(4), 393–397. doi: 10.1007/BF00003796.

Foh, M. B. K., Wenshui, X., Amadou, I. and Jiang, Q. (2012). Influence of pH shift on functional properties of protein isolated of tilapia (Oreochromis niloticus) muscles and of soy protein isolate, *Food and Bioprocess Technology* 5(6), 2192–2200. doi: 10.1007/s11947-010-0496-0.

Galland-Irmouli, A. V., Fleurence, J., Lamghari, R., Luçon, M., Rouxel, C., Barbaroux, O., Bronowicki, J. P., Villaume, C. and Guéant, J. L. (1999). Nutritional value of proteins from edible seaweed Palmaria palmata (Dulse), *Journal of Nutritional Biochemistry* 10(6), 353–359. doi: 10.1016/S0955-2863(99)00014-5.

Gosch, B. J., Magnusson, M., Paul, N. A. and de Nys, R. (2012). Total lipid and fatty acid composition of seaweeds for the selection of species for oil-based biofuel and bioproducts, *GCB Bioenergy* 4(6), 919–930. doi: 10.1111/j.1757-1707.2012.01175.x.

Harnedy, P. A. and FitzGerald, R. J. (2013). Extraction of protein from the macroalga Palmaria palmata, *LWT–Food Science and Technology* 51(1), 375–382. doi: 10.1016/j.lwt.2012.09.023.

Harrysson, H., Hayes, M., Eimer, F., Carlsson, N., Toth, G. B. and Undeland, I. (2018). Production of protein extracts from Swedish red, green, and brown seaweeds, Porphyra umbilicalis Kützing, Ulva Lactuca Linnaeus, and Saccharina latissima (Linnaeus) J. V. Lamouroux using three different methods, *Journal of Applied Phycology* 30(6), 3565–3580. doi: 10.1007/s10811-018-1481-7.

Hernández-Carmona, G., Freile-Pelegrín, Y. and Hernández-Garibay, E. (2013). Conventional and alternative technologies for the extraction of algal polysaccharides. In: *Functional Ingredients from Algae for Foods and Nutraceuticals*, 475–516. doi: 10.1533/9780857098689.3.475.

Holdt, S. L. and Kraan, S. (2011). Bioactive compounds in seaweed: Functional food applications and legislation, *Journal of Applied Phycology* 23(3), 543–597. doi: 10.1007/s10811-010-9632-5.

Imeson, A. P. (2009). Carrageenan and furcellaran. In: Phillips, G. O. and Williams, P. A. (Eds) *Handbook of Hydrocolloids* (1st edn.). Woodhead Publishing Limited, pp. 164–185. doi: 10.1533/9781845695873.164.

Istinii, S., Ohno, M. and Kusunose, H. (1994). Methods of analysis for agar, carrageenan and alginate in seaweed, *Bulletin of Marine Science Fisheries Kochi University* 14, 49–55. Available at: https://ir.kochi-u.ac.jp/dspace/bitstream/10126/4331/1/marine14-049.pdf.

Jacobsen, C., Sørensen, A.-D., Holdt, S. L., Akoh, C. C. and Hermund, D. B. (2019). Source, extraction, characterization and applications of novel antioxidants from seaweed, *Annual Review of Food Science and Technology* 10, 541–568.

Jensen, K. N., Jacobsen, C. and Nielsen, H. H. (2007). Fatty acid composition of herring (*Clupea harengus*) – influence of time and place of catch on n-3 PUFA content, *Journal of the Science of Food and Agriculture* 87(4), 710–718.

Kumar, V. and Kaladharan, P. (2007). Amino acids in the seaweeds as an alternate source of protein for animal feed, *Journal of the Marine Biological Association of India* 49(1), 35–40. doi: 10.1007/s10811-005-1641-4.

Mæhre, H. K., Malde, M. K., Eilertsen, K.-E. and Elvevoll, E. O. (2014). Characterization of protein, lipid and mineral contents in common Norwegian seaweeds and evaluation of their potential as food and feed, *J. Sci. Food Agric.* 94, 3281–3290, https://doi.org/10.1002/jsfa.6681.

Mæhre, H. K., Edvinsen, G. K., Eilertsen, K. E. and Elvevoll, E. O. (2016a). Heat treatment increases the protein bioaccessibility in the red seaweed dulse (Palmaria palmata), but not in the brown seaweed winged kelp (Alaria esculenta), *J. Appl. Phycol.* 28, 581–590. https://doi.org/10.1007/s10811-015-0587-4.

Mæhre, H. K., Jensen, I. J. and Eilertsen, K. E. (2016b). Enzymatic pre-treatment increases the protein bioaccessibility and extractability in dulse (palmaria palmata), *Mar. Drugs.* 14. https://doi.org/10.3390/md14110196.

McHugh, D. J. (2003). *A Guide to the Seaweed Industry*. Rome: Food and Agriculture Organization.

Murata, M. and Nakazoe, J. I. (2001). Production and use of marine algae in Japan, *Japan Agricultural Research Quarterly: JARQ* 35(4), 281–290.

Naseri, A., Holdt, S. L. and Jacobsen, C. (2019). Biochemical and nutritional composition of industrial red seaweed used in carrageenan production, *Journal of Aquatic Food Product Technology* 28(9), 967–973. doi: 10.1080/10498850.2019.1664693.

Naseri, A., Jacobsen, C., Sejberg, J. J. P., Pedersen, T. E., Larsen, J., Hansen, K. M. and Holdt, S. L. (2020a). Multi-extraction and quality of protein and carrageenan from commercial spinosum (Eucheuma denticulatum), *Foods* 9(8), 1–14. doi: 10.3390/foods9081072.

Naseri, A., Marinho, G. S., Holdt, S. L., Bartela, J. M. and Jacobsen, C. (2020b). Enzyme-assisted extraction and characterization of protein from red seaweed Palmaria palmata, *Algal Research* 47. doi: 10.1016/j.algal.2020.101849.

O'Connor, J., et al. (2020). Extraction of protein from four different seaweeds using three different physical pre-treatment strategies, *Molecules* 25(8), 2005. doi: 10.3390/molecules25082005.

Ramani, S. (2016). Enzyme technology for handling red seaweeds for carrageenan and more, *22nd International Seaweed Symposium*, June 19–24, Copenhagen. Oral presentation.

Rees, D. A. (1969). Structure, conformation, and mechanism in the formation of polysaccharide gels and networks, *Advances in Carbohydrate Chemistry and Biochemistry* 24(C), 267–332. doi: 10.1016/S0065-2318(08)60352-2.

Rhein-Knudsen, N., Ale, M. T., Ajalloueian, F., Yu, L. and Meyer, A. S. (2017). Rheological properties of agar and carrageenan from Ghanaian red seaweeds, *Food Hydrocolloids* 63, 50–58. doi: 10.1016/j.foodhyd.2016.08.023.

Rhein-Knudsen, N., Ale, M. T. and Meyer, A. S. (2015). Seaweed hydrocolloid production: an update on enzyme assisted extraction and modification technologies, *Marine Drugs* 13(6), 3340–3359. doi: 10.3390/md13063340.

Safafar, H., Hass, M. Z., Møller, P., Løvstad, S. L. and Jacobsen, C. (2016). High-EPA biomass from nannochloropsis salina cultivated in a flat-panel photo-bioreactor on a process water-enriched growth medium, *Marine Drugs* 14(8), 144; doi: 10.3390/md14080144.

Surasani, V. K. R. (2018). Acid and alkaline solubilization (pH shift) process: A better approach for the utilization of fish processing waste and by-products, *Environmental Science and Pollution Research International* 25(19), 18345–18363. doi: 10.1007/s11356-018-2319-1.

Tibbetts, S. M., Milley, J. E. and Lall, S. P. (2016). Nutritional quality of some wild and cultivated seaweeds: nutrient composition, total phenolic content and in vitro digestibility, *Journal of Applied Phycology* 28(6), 3575–3585. doi: 10.1007/s10811-016-0863-y.

Varadarajan, S. A., et al. (2009). Development of high yielding carragenan extraction method from Euc h euma Cotonii using cellulase and Aspergillus niger, Prosiding Seminar Kimia Bersama UKM-ITB VIII, pp. 461–469.

Vázquez-Delfín, E., Robledo, D. and Freile-Pelegrín, Y. (2014). Microwave-assisted extraction of the carrageenan from Hypnea musciformis (Cystocloniaceae, Rhodophyta), *Journal of Applied Phycology* 26(2), 901–907. doi: 10.1007/s10811-013-0090-8.

Villanueva, R. D., Sousa, A. M. M., Gonçalves, M. P., Nilsson, M. and Hilliou, L. (2010). Production and properties of agar from the invasive marine alga, Gracilaria vermiculophylla (Gracilariales, Rhodophyta), *Journal of Applied Phycology* 22(2), 211–220. doi: 10.1007/s10811-009-9444-7.

Wallace, J. S. (2000). Increasing agricultural water use efficiency to meet future food production, *Agriculture, Ecosystems and Environment* 82(1–3), 105–119. doi: 10.1016/S0167-8809(00)00220-6.

Wang, G. C., Sun, H. B., Fan, X. and Tseng, C.-K. (2002). Large-scale isolation and purification of R-phycoerythrin from red alga Palmaria palmata using the expanded bed adsorption method, *Acta Botanica Sinica* 44, 541–546.

Yan, C. and Zhang, W. (2014). Coacervation processes. In: Gaonkar, A. G., Vasisht, N., Khare, A. R. and Sobel, R. (Eds) *Microencapsulation in the Food Industry*. Elsevier Inc., pp. 125–137. doi: 10.1016/B978-0-12-404568-2.00012-1.

Younes, M., Aggett, P., Aguilar, F., Crebelli, R., Filipič, M., Frutos, M. J., Galtier, P., Gott, D., Gundert-Remy, U., Kuhnle, G. G., Lambré, C., Leblanc, J. C., Lillegaard, I. T., Moldeus, P., Mortensen, A., Oskarsson, A., Stankovic, I., Waalkens-Berendsen, I., Woutersen, R. A., Wright, M., Brimer, L., Lindtner, O., Mosesso, P., Christodoulidou, A., Ioannidou, S., Lodi, F. and Dusemund, B. (2018). Re-evaluation of carrageenan (E 407) and processed Eucheuma seaweed (E 407a) as food additives, *EFSA Journal European Food Safety Authority* 16(4), e05238. doi: 10.2903/j.efsa.2018.5238.

Part 2

Cultivation and processing

Chapter 4

Developments in commercial scale farming of microalgae and seaweeds

Sarah E. Loftus and Zackary I. Johnson, Duke University, USA

1 Introduction

Expanding markets for microalgae and macroalgae products have led to increased development of commercial farming operations. While microalgae and macroalgae, or seaweed, have historically been harvested in many parts of the globe, more recent developments seek to improve productivity, decrease production costs, increase scale, and mitigate environmental impacts of cultivation.

Seaweeds have been harvested from coasts around the world for hundreds of years, while seaweed farming became popular in the mid-1900s (Radulovich et al., 2015). About 85% of farmed and harvested seaweed is used as food for people (FAO, 2018). As of 2015, wild harvesting accounted for about 3% of reported seaweed production globally and the remaining was farmed. The top producers of farmed seaweed by country are China, Indonesia, the Republic of Korea, and the Philippines (FAO, 2018). With global seaweed production tripling between 2000 and 2018 (FAO, 2020), commercial-scale operations are expanding to meet growing markets.

Microalgae, seaweed's single-cell counterparts, have also been harvested from the environment for hundreds of years (García et al., 2017), and commercial microalgae farming was established as early as the mid-1900s (e.g. the Taiwan Chlorella Manufacturing Company was established in 1964 (Taiwan Chlorella Manufacturing Company, 2013)). Over the past few decades, the microalgae industry has seen increased interest in nutrient supplements, animal feeds,

http://dx.doi.org/10.19103/AS.2021.0091.07

pigments, cosmetics, and fuel feedstocks. Microalgae farming has been most popular in Asia, though cultivation facilities are located worldwide.

Research on microalgae production for biofuels has followed trends in energy markets, but commercial production of algal fuels has not yet been reached. Research began in the mid-1900s, and in the United States it began in earnest in the 1970s (U.S. DOE, 2016). Currently, many companies that intended to grow microalgae for fuels have either dissolved or switched to producing more profitable food and feed products.

Seaweeds are often farmed in coastal waters on tethered structures, while microalgae are grown in water-filled tanks or tubes on land. Though farmed with different methods, both algae require the same basic necessities: water, nutrients, inorganic carbon, and sunlight (or organic carbon sources for heterotrophically grown microalgae).

Government-supported research programs and privately funded companies have designed cultivation practices and identified areas for research and development for microalgae and seaweed farming. For microalgae, these areas including algal strain selection, crop protection, and pest management such as via the microbiome, inorganic carbon sourcing, water and nutrient recycling, cost-effective harvesting, and post-harvesting treatment and storage before the algae are processed into the final product (U.S. DOE, 2016). For seaweed, ongoing research includes selection of highly productive seaweeds that are not susceptible to disease (FAO, 2018), identifying suitable nearshore and offshore locations for farming (Radulovich et al., 2015), and optimizing the most effective propagation and reproduction techniques. Here we describe current commercial farming strategies for microalgae and seaweed, including achieved developments, and suggest measures needed to expand the algae industry further.

2 Microalgae cultivation

Microalgae are cultivated in either open or closed containers. The algal strain and intended product(s) dictate the specific cultivation strategy. Factors such as climate, water source, nutrient regimen, mode of operation, local predators and pathogens, inorganic (or organic) carbon source, and more must be analyzed for both crop productivity and economic purposes.

2.1 Cultivation systems

Phototrophically cultivated algae are grown in shallow outdoor tanks which are open to the environment, or in closed photobioreactors. Some hybrid systems first grow algae in photobioreactors as inoculum for outdoor ponds (Huntley et al., 2015). Algae grown heterotrophically without sunlight are cultivated in closed bioreactors and generally follow approaches used for the large-scale cultivation of other microbes such as *E. coli* or *S. cerevisiae*.

2.1.1 Open ponds and photobioreactors

Open, outdoor cultivation systems generally have much lower capital and operational costs than closed bioreactors (Singh and Sharma, 2012; Lee and White, 2019). However, since these open systems are not covered, culture conditions can vary more due to weather, and other organisms and non-biologicals (e.g. debris) can enter the culture. Cultures in photobioreactors are therefore easier to control and maintain in the long run and are less likely to be overtaken by other algae, predators, or pathogens (U.S. DOE, 2016). The geographic location of open ponds will determine the important factors of temperature, precipitation, and light levels.

Probably the most common open system design is raceway ponds, or oval-shaped tracks in which algal cultures are circulated using a motorized paddle wheel. Algae are also grown in circular ponds mixed by a revolving arm (Zittelli et al., 2013) or other shapes of ponds that may or may not be mixed. The culture depth is shallow and typically ranges from 10 cm to 60 cm to achieve optimal light exposure, depending on the season and algal strain (Chlorella Industry Co.; U.S. DOE, 2016; Borowitzka and Vonshak, 2017). The flow rate of the culture around the pond can also be adjusted to optimize growth conditions such as light exposure and oxygen concentration. Assessments of the current state of technology have demonstrated annualized productivities of dry weight of >18 $g/m^2/d$, but peak productivities can exceed 30 $g/m^2/d$ for a given month (Huesemann et al., 2020).

Open ponds can vary in size, and because of their lower capital costs, may generally be easier to scale than photobioreactors (U.S. DOE, 2016); for example, raceway production ponds at a farm in Columbus, New Mexico, USA, range in size from 4400 m^2 to 8900 m^2 (White and Ryan, 2015). Raceway ponds can be dug into the earth and fitted with a liner, or built with other materials (Singh and Sharma, 2012). They are constructed on non-arable land so as not to compete with other crops.

Commercial facilities with open cultivation systems may have dozens of ponds, which may be connected to a flow system that also includes smaller inoculation ponds or photobioreactors, harvesting systems, and a holding pond for the used cultivation water. The farm in Columbus, New Mexico, USA, now operated by Qualitas Health, has about 70 raceway ponds in 0.40 km^2 of desert plus 0.02 km^2 for harvesting (White and Ryan, 2015; Qualitas Health, 2017). The company Earthrise has a 0.44 km^2 site in a desert of California, USA, with 37 4850 m^2 ponds (Earthrise Nutritionals LLC, 2019). The Chlorella Industry Company's facility includes 16 circular ponds, whose surface area combined is 0.01 km^2 (Chlorella Industry Co.). Other examples of commercial facilities using open pond systems include the Taiwan Chlorella Manufacturing Company (2013), Cyanotech Corporation in Hawaii, USA (Cyanotech Corporation, 2018),

Far East Bio-Tec Company in Taiwan (Far East Bio-Tec. Co. Ltd, 2020), and EID Parry Limited in South India (Parrys, 2017) (Fig. 1).

Photobioreactors are closed structures that have a higher surface-to-volume ratio than open ponds, and consequently they can sustain higher productivity and density of algal cultures (U.S. DOE, 2016). Many types of photobioreactor designs exist, including bubble column, airlift, flat panel, tubular, helical, bags, and stirred tank reactors (Singh and Sharma, 2012; Zittelli et al., 2013). These

Figure 1 (a) Open raceway ponds of a commercial facility in Kona, Hawaii and (b) close-up of paddlewheels for circulating water and mixing algae.

structures are built from clear material, such as glass, Plexiglas, or plastic, each allowing light to enter (Fig. 2).

Examples of commercial facilities that cultivate phototrophic microalgae in closed systems include Algatech in Israel (Algatech LTD, 2019), Roquette Klotze in Germany (Roquette Klotze GmbH & Co. KG, 2017), Pond Tech in Canada (Pond Tech, 2018), Phycom in The Netherlands (Phycom, 2019), and Algenol in Florida, USA (Algenol). Algatech and Roquette Klotze both use vertical manifold

Figure 2 (Top) Horizontal tubular photobioreactors (Source: author); (bottom) vertical photobioreactors (Source: Huo et al., 2018; reproduced under the terms of a CC BY Licence).

photobioreactors, a series of vertical tubes whose tops and bottoms are each attached by a horizontal tube rather than by U-bends (Zittelli et al., 2013). Pond Tech uses a closed tank system rather than a tubular design (Pond Tech, 2018).

Photobioreactor systems can be located outdoors or in closed facilities such as greenhouses, and are illuminated by sunlight, which also can be supplemented with artificial light. Developments in low-cost light redistribution, such as with lenses and fibers, as well as solar concentration, are ongoing (Zittelli et al., 2013). Cultures in photobioreactors are mixed with a pump or by bubbling gas (Zittelli et al., 2013). Biofilms formed in the reactors need to be cleaned away so that they don't reduce irradiance.

Due to lack of evaporative cooling in some photobioreactor designs, overheating can occur. Either temperature must be well-controlled, an algal strain that is robust to higher temperatures must be grown, or the facility has to be located in a relatively cool climate (U.S. DOE, 2016). To prevent damage to algal cells from shear forces in the photobioreactor, designs must include characteristics such as slower velocities and low-shear pumps (Zittelli et al., 2013), or growing algae that are not fragile.

2.1.2 Fermenters

Microalgae can also be grown heterotrophically in fermenters, which do not require light but do require an additional organic carbon source. Fermenters are closed systems such as photobioreactors, and provide the same controlled conditions and protection against contamination. They can also achieve relatively high biomass concentrations by directly supplying organic carbon instead of relying on algal photosynthesis (U.S. DOE, 2016). While heterotrophic cultivation may not be realistic for producing algal fuels due to the requirement for organic carbon, it is feasible for producing higher value products such as food. For example, Duplaco in The Netherlands grows algae with sugars inside 650 L fermenters, after a scale-up process to create the inoculum (Duplaco, 2018). Corbion Biotech based in the USA grows algae in fermentation tanks with sugars from corn, sugar cane, and sugar beets (Corbion Biotech, 2020). Triton Algae Innovations also grows algae in fermenters (Triton Algae Innovations). Veramaris utilizes corn stover, wheat, and beet sugars to heterotrophically cultivate algae for omega-3 fatty acids for aquaculture feed (Veramaris).

2.2 Cultivation conditions and operation

2.2.1 Mode of operation

Cultivation systems can be operated in either batch or continuous modes. In batch modes, the entire culture is harvested and then the pond or bioreactor is filled with inoculum and growth medium. The batch mode can help prevent the algae from being overtaken, infected, or consumed by other organisms, given the relatively short cultivation period (Huntley et al., 2015). In continuous modes, a fraction of the culture is continuously harvested and growth medium is added to replace the lost volume. The semi-continuous mode is similar to the continuous mode, except that the partial harvests take place at defined intervals. Partial harvesting allows the unharvested culture to act as inoculum, saving resources in culture scale-up (Borowitzka and Vonshak, 2017). For example, the Columbus, New Mexico algae farm was operated semi-continuously, with up to 40% of the pond volume collected for each harvest (White and Ryan, 2015).

2.2.2 Light and temperature

Light and temperature are two of the primary factors affecting biomass productivity of phototrophically cultivated microalgae. The majority of commercial farms are operated outdoors and use sunlight (Borowitzka and Vonshak, 2017), since using artificial light precludes all but very high-value bioproducts (Blanken et al., 2013). Choosing a suitable geographic location for a farm is therefore critical, since this determines temperature and insolation (cumulative light energy per area), which change daily and seasonally. Regions at low latitudes and low elevation are generally considered optimal for light, temperature, and length of the growing season (Moody et al., 2014). Geographic locations with suitable temperature and light profiles for growing microalgae should also align with accessible resources, have low risk of natural hazards, and not compete with other food crops (U.S. DOE, 2016).

Several commercial open-pond facilities are located in desert regions, such as Qualitas Health in New Mexico, USA; Algatech in the Arava desert in Israel; and Earthrise in southern California, USA. Other farms are located in more tropical locations such as Cyanotech in Hawaii, USA; EID Parry in South India; and the Taiwan Chlorella Manufacturing Company and the Far East Bio-Tec Company, both in Taiwan.

Within cultivation systems, the culture density, culture depth (or width in some photobioreactors), and turbulence affect light attenuation (Borowitzka, 2016). Algal cells experience repeating patterns of brightness and darkness as they move from the surface to the bottom or interior of the culture while circulating (Borowitzka and Vonshak, 2017). Light absorption by algal cells can be optimized by changing these factors, but generally depth is kept shallow (e.g., less than 30 cm). Light variability may also be important as the so-called 'flashing light' effect has been shown to substantially increase productivity in raceway ponds (Zhang et al., 2015).

Microalgal cultivation facilities must receive enough light to maximize algal productivity without causing photoinhibition. Different algal strains have different irradiance optima and tolerances, and at some levels of irradiance, the photosynthetic apparatus becomes less efficient and excessive oxidative stress from photosynthetic byproducts can damage them (U.S. DOE, 2016; Morris et al., 2011). While in some cases photoinhibition can induce accumulation of desired algal compounds such as lipids (U.S. DOE, 2016), detrimental photoinhibition should be avoided. When continuous cultures are harvested or when batch cultures are inoculated, care must be taken to maintain a high enough culture density so that algae are not exposed to an inhibiting irradiance (Borowitzka and Vonshak, 2017). A cell's irradiance tolerance can also vary based on other growth conditions such as temperature. Heating a culture during cooler times of the day or year, when cells are more susceptible to photoinhibition, may be necessary (Borowitzka, 2016).

Microalgae also have an optimal temperature range, and algal strains suited to the given climate and/or season must be selected. The temperature of open ponds is regulated via evaporative cooling, unlike in photobioreactors which can require controlled temperature systems (U.S. DOE, 2016).

2.2.3 pH

Algal photosynthesis draws down carbon dioxide, increasing the pH of the culture medium. A higher pH results in less bioavailable inorganic carbon as carbon dioxide. pH can be monitored using sensors and controlled using automated systems that add carbon dioxide to lower the pH and supply inorganic carbon. Some microalgae can use both carbon dioxide and bicarbonate as inorganic carbon sources, and grow best in the pH range of 7–8 (Borowitzka, 2016). Other strains of algae that prefer alkaline conditions are grown under high pH conditions (pH>9) and may offer advantages in terms of inorganic carbon supply and pest control (Canon-Rubio et al., 2016).

2.2.4 Salinity

Water salinity must be monitored because algae have salinity tolerances. Salinity is affected by the water source, additives, evaporation, the extent of water recycling, and, in open systems, precipitation. Algae with a wide salinity tolerance are therefore preferred (Borowitzka, 2016). Since evaporation will concentrate salts in the water, it is necessary in some cases to periodically divert a portion of the water for disposal or to an evaporation pond. This water can be replaced with the original water supply or with freshwater, though in many cases adding freshwater may not be a sustainable solution (U.S. DOE, 2016). The farm in Columbus, New Mexico, which reuses post-harvesting water and has an evaporation rate of about 2 m per year, diverts less than 1% of the post-harvesting water to an evaporation pond and replaces it with the original well water supply (White and Ryan, 2015).

2.2.5 Crop protection

All open cultivation systems are potentially exposed to predators (Ganuza et al., 2016; Yuan et al., 2019), pathogens (Carney and Lane, 2014), or competing 'weeds' (Lee and White, 2019) such as bacteria, viruses, rotifers, amoebas, ciliates, fungi, and other microalgae (White and Ryan, 2015) (Fig. 3). The presence of certain pests is influenced by geographic location, season, the algal strain, and culture conditions. Pests are more likely to appear after rain, high winds, or abnormal temperatures (Lee and White, 2019). As with any agricultural crop, protocols are necessary to protect and monitor the crop to limit losses.

Methods to prevent crop infestations include growing an algal strain that can tolerate extreme conditions such as high pH and salinity, which are generally intolerable to other microorganisms, or operating ponds in batch mode to prevent proliferation of pests, pathogens, and weed algae. If pests are problematic, crop protection generally proceeds to identifying the pest, quantify its impact on production (or other key variables), identifying the source (or vector), and exploring potential removal / remediation approaches. A proactive pest management strategy such as pest avoidance is generally preferred over direct interventions (e.g. additive approaches such as reactive pesticides) that may limit downstream applications (e.g. food, nutraceuticals, etc.). Integrated pest management includes monitoring for and an understanding of the pests' life cycle, optimizing pond operation, and suppressing pests via physical actions such as cleaning the equipment or temporarily stopping mixing (Lee and White, 2019).

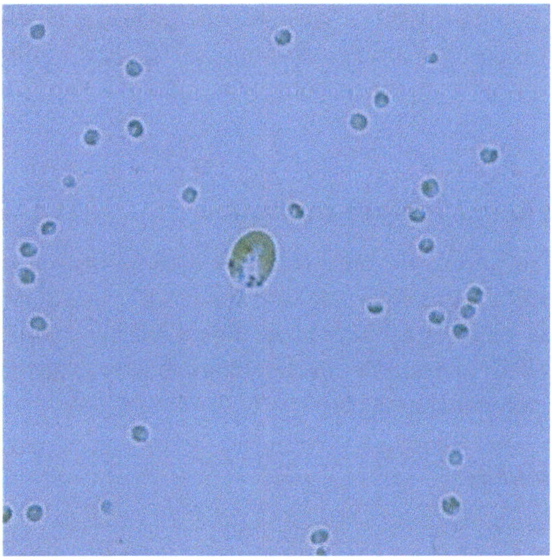

Figure 3 Microscope image of microalgae with pest (Source: author).

A first step in monitoring algal health includes checking the microalgae daily for any signs of physical changes such as color, foaming, or smell. Optical signatures such as spectroradiometry (Reichardt et al., 2014) or fluorometry as well as microscopy can also be used to identify crop changes that might be indicative of a pest invasion (Lee and White, 2019).

When algae crops are changing due to an unknown pest, a sample of the culture can be studied in the facility's lab to learn how the crop and pest react to various conditions and treatments (Lee and White, 2019). The pest can be studied further to determine its life cycle, optimal temperature range, and more, to predict how it may propagate in algal cultures over time. In the first years of open-pond cultivation, several new pests are likely to be found each year (Lee and White, 2019).

Pest-specific monitoring methods must then be found or designed. Monitoring methods of known pests depend on the ease of pest detection. While certain pests such as rotifers may be detected visually (Lee and White, 2019), other pests such as bacteria require molecular methods such as qPCR (Carney and Lane, 2014).

When a pest invasion occurs, the farm operators must balance the costs of crop treatment and crop loss, and only treat the crop if the pest is at a sufficient concentration to cause damages worth more than the treatment (Lee and White, 2019). Adding chemical agents is a common treatment (Carney and Lane, 2014; Lee and White, 2019). Proactively adding relatively small doses of biocide can reduce the total volume needed to prevent infestation, and would be less toxic to the algae crop (Lee and White, 2019).

Ongoing developments in commercial algal crop protection include improving the detection limit of pest assays, taking advantage of natural biological control agents (Carney and Lane, 2014), and implementing crop protection strategies when designing growth medium composition (Lee and White, 2019).

2.3 Resources for microalgae cultivation

Water, nutrients, and inorganic carbon are the major operational resource inputs for photoautotrophic microalgae cultivation, and their costs and environmental impacts must be considered alongside their effects on algal productivity. These resources will affect conditions described above such as salinity, pH, and potential for pest contamination.

2.3.1 Water sources

Water source is determined by geographic location, and freshwater consumption is a major factor affecting the environmental sustainability of a commercial facility. Using non-freshwater sources (e.g. saline groundwater, seawater) and reusing cultivation water after harvesting are therefore important goals in commercial algae farming (U.S. DOE, 2016). Water sources must be suitable for the algae crop and for consumers of the intended final product. The source may also be treated such as by filtration and UV light (Huntley et al., 2015) or with ozone or other forms of treatment (Lu et al., 2020).

In deserts, some commercial farms use saline well water that is not suitable for terrestrial crops (White and Ryan, 2015), and others such as Earthrise farms use river water (Earthrise Nutritionals LLC, 2019). On the coast, Cyanotech grows *Arthrospira* in freshwater mixed with seawater pumped from 2000 feet underground (Cyanotech Corporation, 2018).

Several other nutraceutical companies use groundwater for cultivation, such as Chlorella Industry Company (Chlorella Industry Co.), Roquette Klotz (Roquette Klotze GmbH & Co. KG, 2017), and E.I.D. Parry Limited (Parrys, 2017).

Most commercial microalgae farms recycle cultivation water (White and Ryan, 2015; Borowitzka and Vonshak, 2017; Cyanotech Corporation, 2018; Earthrise Nutritionals LLC, 2019), since this eliminates the need for wastewater treatment, reduces production costs, and recycles residual nutrients (Thomassen et al., 2016). Recycling cultivation water after harvesting can also lead to accumulation of certain micronutrients, pests (Yuan et al., 2019), organic matter, and dissolved solids if not carefully monitored and/or treated (White and Ryan, 2015; Borowitzka, 2016). Not all microalgae are suitable for growing in their reused cultivation water, due to potential inhibitory exudates (Loftus and Johnson, 2019; Loftus et al., 2020). Therefore, preliminary tests need to be conducted along with the intended harvesting method (Loftus and Johnson, 2017). Whether water is recycled or not, the salinity of open pond systems operated in semi-continuous modes will increase over time due to evaporation, therefore additional source water is needed to offset this loss (U.S. DOE, 2016).

2.3.2 Nutrients

Major inorganic nutrient requirements of microalgae include nitrogen and phosphorus, as well as trace metals and vitamins. Some commercial farms take advantage of trace metals already present in ground and ocean water sources (Borowitzka, 2016), and agricultural fertilizers have been used for major nutrients (Lee and White, 2019). Liquid forms of nutrients are preferable to powders to allow for slower mixing and to protect workers' health (Lee and White, 2019).

Growth medium composition can be tailored to the composition of the algae crop and its nutrient demands, to prevent accumulation of ions in semi-continuous cultures and in cultures recycling the cultivation water (Hadj-Romdhane et al., 2012). Residual nutrients are also recycled when cultivation water is reused, reducing nutrient additions but requiring monitoring of water chemistry (Lee and White, 2019).

2.3.3 Inorganic carbon

Photoautotrophic algae cultivation requires an inorganic carbon source, which must be supplemented for adequate supply to high-density cultures. Sensors allow automatic carbon dioxide gas distribution (Huntley et al., 2015; Pond Tech, 2018).

Bicarbonate can also be added as an inorganic carbon source, but this can increase alkalinity and ion concentrations, and can lead to precipitation of other compounds. However, bicarbonate can be supplied to the commonly grown *Arthropspira* cyanobacterium, which thrives in high pH and highly alkaline conditions (Borowitzka, 2016). For example, Cyanotech adds food-grade baking soda to *Arthrospira* cultivation medium (Cyanotech Corporation, 2018), though Earthrise supplies carbon dioxide gas to *Arthrospira* (Earthrise Nutritionals LLC, 2019).

2.3.4 Labor

Not all microalgae cultivation processes can be automated, and skilled staff are required for successful operations (Borowitzka and Vonshak, 2017). Commercial farms have research staff to provide recommendations on how the farm should be operated to optimize algae production (Lee and White, 2019). Operational staff conduct sample collection, pest monitoring, nutrient addition, harvesting, and cleaning of the cultivation system. Worker safety must be protected amongst concentrated nutrients and chemical agents (Lee and White, 2019), and during physical labor in hot conditions. A study that modeled a ~111 ha open pond facility estimated 41 full-time-equivalent employees including laborers, scientists, contractors, maintenance workers, and administrators (Beal et al., 2015).

3 Seaweed cultivation

Seaweed cultivation is a larger and more established industry compared to cultivation of microalgae, and most cultivation systems have less requirements because they leverage natural resources (U.S. DOE, 2016). As multicellular organisms, seaweeds are also relatively complex compared to microalgae, and different species have different life stages and optimal growth conditions. The specific seaweed species and the intended final products inform how and where the seaweed is grown.

3.1 Cultivation systems

Like microalgae, seaweeds absorb water, nutrients, and inorganic carbon from surrounding seawater. With the exception of *Sargassum*, they are secured in place with an anchor-like structure called the holdfast (Radulovich et al., 2015). Seaweeds are typically farmed on structures like ropes and nets in the ocean, but like microalgae they can be cultivated in seawater-filled tanks on land, though these facilities are typically smaller in scale (Radulovich et al., 2015) (Fig. 4).

3.1.1 Propagation and reproduction

Seaweeds can have intricate life cycles that involve both sexual and asexual reproduction, including haploid and diploid life stages, which may be initiated by seasonal and/or environmental signals (Redmond et al., 2014). Growing seaweeds begins with either asexual reproduction from spores or sexual reproduction from seaweed gametes. Vegetative propagation can also occur, depending on the species, in which a piece of the seaweed regrows (Radulovich et al., 2015; Hurtado et al., 2019).

Spores, fertilized eggs, or seaweed pieces are attached to structures like twine, ropes, and nets, the process of which is called seeding (Hurd et al., 2014). Seeding allows more control of the propagules, but is also more time-intensive and costly than leaving a fraction of the established seaweed crop in the ocean to regrow after harvesting (Radulovich et al., 2015). Maintaining propagules also requires more work in nursery or laboratory facilities (Radulovich et al., 2015),

Figure 4 Harvesting rope-grown kelp off the coast of Maine, USA (Source: © Atlantic Sea Farms, Maine, USA (https://atlanticseafarms.com/)).

though growers may buy spores or seeded twine and nets from a hatchery (Hurd et al., 2014).

Cloning via vegetative propagation and asexual spores limits genetic diversity in a seaweed crop, which may pose higher risks of susceptibility to pathogens and pests (Redmond et al., 2014). Research and development in genetic improvement via breeding is ongoing (Redmond et al., 2014; Hwang et al., 2019). To protect the genetic diversity of plants in the wild, regulations may also require that seaweed seed be collected from a certain number of native plants within a certain radius of a farm. For example, the U.S. state of Alaska requires collection of 50–60 kelp plants in the wild within a 50 km radius of a kelp farm to create seed stock for the farm (Stopha, 2020).

3.1.2 Coastal and offshore farms

Seaweeds are mostly farmed in nearshore or coastal waters. Seaweeds are attached to ropes or nets that are tethered to the seafloor by pegs or poles (Radulovich et al., 2015). In fixed systems, the ropes and nets stay in place so the seaweed experiences a range of depths depending on the tide. In floating or semi-floating systems, enough sag is left in the ropes so that the seaweed is always floating at the surface (Hurd et al., 2014). A strategy that is under development involves attaching seaweed directly to the seafloor where shallow depths allow sufficient sunlight (Radulovich et al., 2015).

Farms vary in size depending on location and configuration, with larger and more established farms typically located in eastern Asia. Smaller yet expanding coastal farms up to about 0.1 km^2 are located in areas such as Europe (Campbell et al., 2019) and the United States (Kim et al., 2019). Small, family-operated farms are also located throughout areas such as the Philippines and Fiji where barriers to entry are low (Foscarini and Prakash, 1990) (Radulovich et al., 2015).

Offshore farming is also under development (Kim et al., 2019). In deep offshore waters, floating ropes and nets can be kept in place by weights resting on the seafloor and by buoys floating on the ocean surface (Radulovich et al., 2015).

3.1.3 Contained tanks

Some farms based on land grow seaweed in tanks or ponds filled with seawater (Fig. 5). The seaweed can grow freely unattached in tumble culture (Radulovich et al., 2015), or attached to a substrate such as rope. Conditions can be better maintained in tanks compared to ocean farming. Tank cultivation is only performed for small-scale farms and/or for high-value products because of relatively high production costs (Hurd et al., 2014; U.S. DOE, 2016).

Figure 5 Land-based seaweed tanks (Source: © Acadian Seaplants Limited, Canada (https://www.acadianseaplants.com/)). [Photo of land-based seaweed tanks; DOI: 10.1016/B978-012374473-9.00757-8].

One example of a small-scale farm that uses its seaweed product directly as fresh food is Monterey Bay Seaweeds in California, USA. The seaweeds are grown in open tanks with recirculating seawater (Monterey Bay Seaweeds).

3.2 Cultivation conditions

Selecting the proper farm location is a critical factor that will affect the environmental conditions and coexisting organisms to which seaweeds are exposed (Bruhn et al., 2016). Matching a suitable seaweed species for the given location is essential (Radulovich et al., 2015).

3.2.1 Light

As with microalgae, light availability is determined by seaweed depth and other factors such as season and time of day (Hwang et al., 2007). Water turbidity, such as from microalgae and suspended solids, can limit light and negatively affect biomass yields (Bruhn et al., 2016; van der Molen et al., 2018).

Seaweeds can have different optimal irradiances, depending on their ecological niches and life stage (Hwang et al., 2007). For example, one study found that *Gracilaria cornea* from Florida, USA, had optimal photosynthetic performance at 100–800 μmol photon m^{-2} s^{-1}, similar to other tropical seaweeds, and exhibited photoinhibition above 900 μmol photon m^{-2} s^{-1} (Dawes et al., 1998). In an experiment testing optimal cultivation depth of *Sargassum fulvellum* on the coast of South Korea, the species showed optimal growth at 1 m, corresponding to average irradiances of 845 and 332 μmol photons m^{-2} s^{-1} in the summer and winter, respectively. However, the optimal depth of younger *S. fulvellum* during the nursery stage (defined as a length of 0.5-2 cm) was 1.5 m, corresponding to an average irradiance of 488 μmol photons m^{-2} s^{-1} in the spring (Hwang et al., 2007).

3.2.2 Climate and temperature

The top seaweed-producing countries are located in both temperate and tropical geographic locations (Radulovich et al., 2015). Spore formation and reproduction can require certain seasons and/or temperatures depending on the species. Seeded nets and ropes are also placed in the ocean under certain temperature conditions (Hurd et al., 2014). For example, the optimal temperature for *Pyropia* species decreases as the seaweed gets older; therefore cultivation and harvesting must be timed accordingly (Hurd et al., 2014).

For tropical and subtropical seaweeds, the optimal temperature can range from 15°C to 35°C (Dawes et al., 1998). Models fitted to laboratory data suggest that for *Kappaphycus alvarezii*, gross photosynthetic rate is optimized at 31.1°C and the relative growth rate is optimized at 30°C. *K. alvarezii* cannot survive at temperatures above 36°C (Terada et al., 2016). Optimal temperatures for kelp are lower, ranging from 5°C to 10°C, so kelp farms in the northern U.S. states are typically seeded in fall and harvested in spring (Rinker et al., 2021). If the seawater is already near a species' maximum temperature threshold, then rising sea surface temperatures may negatively affect the crop (Chung et al., 2017).

3.2.3 Salinity

Different seaweed species also may have optimal salinity ranges. For example, *Kappaphycus* and *Eucheuma* require salinities above 30 ppt (Hurd et al., 2014), and tropical and subtropical seaweeds have optimal ranges of roughly 20–40 ppt (Dawes et al., 1998). Since freshwater can kill seaweed, farms should not be located directly near a river mouth (Foscarini and Prakash, 1990).

3.2.4 Water movement

Water must be moving around seaweed to supply a sufficient flux of nutrients, yet must not be too fast to harm the plants or pull up the seeded structures (Foscarini and Prakash, 1990). An optimal water velocity range is 20–40 m min^{-1}, and should not exceed 60 m min^{-1} (Radulovich et al., 2015). Seaweed farms should be arranged so that the current flows through the farm instead of perpendicular to the ropes and nets (Foscarini and Prakash, 1990).

Increased storm frequency and intensity in the future may lead to greater nutrient availability from runoff and mixing in coastal seaweed farms, but wave forces and runoff can also lead to destruction of farms, pollutant exposure, and increased turbidity (Chung et al., 2017).

3.2.5 Nutrients

Nutrients can be provided from natural processes such as ocean upwelling and other water sources (Radulovich et al., 2015). Farms can also take advantage of anthropogenic nutrient sources such as nearby marine finfish farms (Bruhn et al., 2016; Kim et al., 2019).

Seaweeds grown in nutrient-rich water have higher biomass quality (e.g. higher protein content) (Bruhn et al., 2016). For example, *Pyropia* biomass quality is optimal above a nitrogen concentration of 15 μM (Hurd et al., 2014).

In nutrient-polluted areas, seaweeds could be grown specifically for nutrient removal purposes. This seaweed may not be suitable for producing human food or animal feed, but could have value in other markets such as fertilizers or water quality trading programs. A case study modeling seaweed farming in the Gulf of Mexico found that farming for nutrient assimilation could be cost-effective (Racine et al., 2021).

In offshore areas where nutrients are less plentiful near the surface, systems are being developed that expose seaweeds to higher nutrient concentrations by either moving them to deeper waters at night, or creating artificial upwelling systems. One study used a buoy elevator to suspend giant kelp at 9 m depth during the day and at 80 m depth at night off the coast of California, USA, which improved biomass production and protein content (Navarrete et al., 2021). Another study developed an air-lifted upwelling system to cultivate brown kelp in Aoshan Bay in China. Kelp grown in the upwelling section had greater leaf lengths, widths, and weight than kelp grown in a non-upwelling area (Fan et al., 2020).

3.2.6 Crop protection

When growing in the ocean, seaweeds are exposed to a range of organisms that can infect them, consume them, or otherwise reduce their growth. These can include pathogenic bacteria, fungi, microalgae, plants and invertebrates that grow on seaweed, and consumers such as snails, crabs, fish, and sea turtles (Hurd et al., 2014; Radulovich et al., 2015; Ingle et al., 2018).

Unlike in contained microalgae systems, no pesticides, antibiotics, or other chemical agents are generally used in seaweed farming (Radulovich et al., 2015), though problems with pests can occur. Workers can physically clean the seaweed and manually remove pests such as sea urchins (Hurd et al., 2014). Another crop protection strategy is to locate seaweeds so that they are submerged by seawater only part of the time, based on tides (Radulovich et al., 2015). Weed algae growing on the seaweed do not do well in the dry environment (Hurd et al., 2014). Nets can also be frozen to get rid of pests (Hurd et al., 2014).

Pests and pathogens can also become more prevalent under certain environmental conditions. For example, one study found that biofouling of *Saccharina latissima* was positively correlated with temperature (Bruhn et al., 2016).

3.3 Cultivation resources

Seaweed cultivation requires relatively few resources, including human labor, a seawater source, and an initial investment in materials such as ropes and nets. Seaweed farming does not require freshwater and does not require land that could otherwise be used for terrestrial crops (Radulovich et al., 2015). It also does not require fertilizers or pesticides, which are often banned because of environmental concerns (Hurd et al., 2014; Radulovich et al., 2015; Campbell et al., 2019). However, nursery facilities that cultivate seaweed propagules may need to add nutrients to the seawater, (Redmond et al., 2014).

3.3.1 Labor

Labor is one of the top resources required for seeding, cultivating, and harvesting seaweeds. Seaweed farming provides income to many coastal communities, improving standards of living (Radulovich et al., 2015). In areas such as Europe where labor costs are relatively high, reducing manual labor requirements is an ongoing effort (Campbell et al., 2019). Automating what is done through manual labor may also increase productivity (Hayashi et al., 2017) and reduce exposure to hazards such as dangerous marine animals or sea conditions (Radulovich et al., 2015).

4 Conclusion and future trends

While commercial seaweed farms are established worldwide, and more microalgae farms are gaining commercial status, developments to improve the scale, productivity, and environmental sustainability of algae farms is a burgeoning effort.

A broad strategy to improve productivity includes continuously selecting the most suitable algae for the given environmental conditions and intended products. For seaweeds, this would involve increasing genetic diversity of the crop through sexual reproduction and breeding, as opposed to solely using vegetative propagation and asexual spore formation (Robinson et al., 2013; Hwang et al., 2019). Since seaweed cultivation in ocean farms means exposing the seaweed to local ecosystems, introducing non-native species is often avoided (Campbell et al., 2019).

For microalgae, selection entails outdoor trials that expose algae to various conditions and pests to determine their performance in harsh environments

(Borowitzka, 2013). Given the short time scale of microalgae reproduction, random genetic mutations can lead to selection of the most suitable phenotype within a large-scale culture over time, though other considerations such as product specifications must also be taken into account (Borowitzka, 2016).

Many studies are also researching genetic engineering approaches to create microalgal strains with desirable product traits that are also suitable for outdoor cultivation (Borowitzka and Vonshak, 2017). Some found that genetically engineered strains posed no risk to the native surrounding microalgal communities (Szyjka et al., 2017). However, cultivating genetically modified algae would require testing and approval (U.S. DOE, 2016), and microalgae nutraceutical companies such as Earthrise and Cyanotech specifically advertise their products as 'non-GMO', likely for customer approval.

In addition to selecting algae crops that are more resistant to pests and pathogens, improving pest management strategies would also help improve biomass yields. Integrated pest management is a major area of development for both microalgae and seaweeds, as described by Lee and White (2019), and Ingle et al. (2018), respectively.

Increasing the scale of cultivation while keeping production costs manageable is also an area of development. For example, this includes choosing effective and efficient harvesting methods, recycling water and nutrients for microalgae cultivation (U.S. DOE, 2016), and automating labor-intensive practices for seaweed farming (Campbell et al., 2019). Long-established lessons from the agriculture industry can be applied further to inform algae cultivation practices (Radulovich et al., 2015; Lee and White, 2019).

Development of algae cultivation strategies can be quickened by open access to research data and information (U.S. DOE, 2016), such as sharing information on crop protection strategies (Lee and White, 2019). For seaweeds, there is also a need for collaboratively developing best practices and policies to manage environmental impacts across farms throughout the world (Rebours et al., 2014).

Future developments that can expand seaweed farm distribution include offshore farming (U.S. DOE, 2016; Kim et al., 2019) and investigating additional locations globally that are suitable for growing seaweed (Radulovich et al., 2015). Finally, the effects of climate change on factors such as temperatures and storm frequency will affect both microalgae and seaweed farming (Chung et al., 2017), and these effects should be studied to plan for and mitigate any negative effects on algae productivity.

5 Where to look for further information

The biennial publication from the FAO on fisheries and aquaculture (most recently FAO (2020) The State of World Fisheries and Aquaculture 2020,

Sustainability in Action) provides current trends on macroalgae production (microalgae are currently not broken out) and additional insight on industry challenges and advancements. Annual reports from the National Renewable Energy Laboratory (NREL) provide similar updates of the technology and challenges associated with growing microalgae at testbed sites (most recently Davis and Kline 2021). There are a number of organizations that advance the development of both macro- and microalgae, including:

- The Algae Biomass Organization, which promotes the development of viable commercial markets for renewable and sustainable commodities derived from algae. https://algaebiomass.org/.
- The Phycological Society of America (PSA) to promote research and teaching in all fields of phycology https://www.psaalgae.org/.
- The Federation of European Phycological Societies (FEPS) to promote the study, conservation and application of algae, to support educational activities on matters involving algae https://www.feps-algae.org/.
- The Seaweed Hub collaboration provides a framework to share information, address challenges, identify needs and opportunities https://seaweedhub.org/.

Additional centers of excellence in algae research and commercialization include:

- AlgaePARC in the Netherlands http://www.algaeparc.com/.
- The US Department of Energy including NREL https://www.nrel.gov/.
- The Arizona Center for Algae Technology and Innovation https://azcati.asu.edu/.
- The University of Marine's Center for Cooperative Aquaculture Research https://umaine.edu/cooperative-aquaculture/.
- The Scottish Association for Marine Science seaweed farms and nursery facilities https://www.sams.ac.uk/.

6 Acknowledgements

U.S. Department of Energy grant numbers DE-EE0007091, DE-EE0008518 and DE-EE0009278 to ZIJ and U.S. Department of Agriculture (through Cornell University) grant number 2019-69012-29905.

7 References

Algatech LTD (2019). Our technology. Available at: https://www.algatech.com/our-technology/.

Algenol Algae Growing Systems. Available at: https://www.algenol.com/growing-systems/.

Beal, C. M., Gerber, L. N., Sills, D. L., Huntley, M. E., Machesky, S. C., Walsh, M. J., Tester, J. W., Archibald, I., Granados, J. and Greene, C. H. (2015). Algal biofuel production for fuels and feed in a 100-ha facility: a comprehensive techno-economic analysis and life cycle assessment. *Algal Research* 10: 266–279.

Blanken, W., Cuaresma, M., Wijffels, R. H. and Janssen, M. (2013). Cultivation of microalgae on artificial light comes at a cost. *Algal Research* 2(4): 333–340.

Borowitzka, M. A. (2013). Species and strain selection. In: *Algae for Biofuels and Energy* Borowitzka, M. A. and Moheimani, N. R. (Eds). Dordrecht: Springer Netherlands, pp. 77–89.

Borowitzka, M. A. (2016). Algal physiology and large-scale outdoor cultures of microalgae. In: *The Physiology of Microalgae* Borowitzka, M. A., Beardall, J. and Raven, J. A. (Eds). Cham: Springer International Publishing, pp. 601–652.

Borowitzka, M. A. and Vonshak, A. (2017). Scaling up microalgal cultures to commercial scale. *European Journal of Phycology* 52(4): 407–418.

Bruhn, A., Tørring, D. B., Thomsen, M., Canal-Vergés, P., Nielsen, M. M., Rasmussen, M. B., Eybye, K., Larsen, M., Balsby, T. and Petersen, J. (2016). Impact of environmental conditions on biomass yield, quality, and bio-mitigation capacity of Saccharina latissima. *Aquaculture Environment Interactions* 8: 619–636.

Campbell, I., Macleod, A., Sahlmann, C., Neves, L., Funderud, J., Øverland, M., Hughes, A. D. and Stanley, M. (2019). The environmental risks associated with the development of seaweed farming in Europe–prioritizing key knowledge gaps. *Frontiers in Marine Science* 6: 1–22. DOI: 10.3389/fmars.2019.00107.

Canon-Rubio, K. A., Sharp, C. E., Bergerson, J., Strous, M. and De la Hoz Siegler, H. (2016). Use of highly alkaline conditions to improve cost-effectiveness of algal biotechnology. *Applied Microbiology and Biotechnology* 100(4): 1611–1622.

Carney, L. T. and Lane, T. W. (2014). Parasites in algae mass culture. *Frontiers in Microbiology* 5: 278.

Chlorella Industry Co., L. Kyushu Chikugo Plant. Available at: https://www.chlorella.co.jp/en/research/factory/.

Chung, I. K., Sondak, C. F. A. and Beardall, J. (2017). The future of seaweed aquaculture in a rapidly changing world. *European Journal of Phycology* 52(4): 495–505.

Corbion Biotech, I. (2020). Innovation. Available at: http://algavia.com/how-we-innovate/.

Cyanotech Corporation (2018). Cyanotech. Available at: https://www.cyanotech.com/.

Davis, R. and Bruno, K. 2021. *Algal Biomass Production via Open Pond Algae Farm Cultivation: 2020 State of Technology and Future Research*. Golden, CO: National Renewable Energy Laboratory. NREL/TP-5100-79931. Available at: https://www.nrel.gov/docs/fy21osti/79931.pdf.

Dawes, C. J., Orduña-Rojas, J. and Robledo, D. (1998). Response of the tropical red seaweed Gracilaria cornea to temperature, salinity and irradiance. *Journal of Applied Phycology* 10(5): 419–425.

Duplaco (2018). Microalgae production process. Available at: https://duplaco.com/duplaco-algae/?lang=en.

Earthrise Nutritionals LLC (2019). Earthrise. Available at: https://www.earthrise.com/ecofriendly-farm.

Fan, W., Zhang, Z., Yao, Z., Xiao, C., Zhang, Y., Zhang, Y., Liu, J., Di, Y., Chen, Y. and Pan, Y. (2020). A sea trial of enhancing carbon removal from Chinese coastal waters by stimulating seaweed cultivation through artificial upwelling. *Applied Ocean Research* 101: 102260.

FAO (2018). *The Global Status of Seaweed Production, Trade and Utilization*. Rome: Fish and Agriculture Organization, p. 120.

FAO (2020). *The State of World Fisheries and Aquaculture 2020. Sustainability in Action*. Rome: Fish and Agriculture Organization.

Far East Bio-Tec. Co. Ltd (2020). Facility. Available at: https://www.febico.com/en/page/Facility/facility.html.

Foscarini, R. and Prakash, J. (1990). *Handbook on Eucheuma Seaweed Cultivation in Fiji*. Suva, Fiji: Ministry of Primary Industries, Fisheries Division and South Pacific Aquaculture Development Project Food and Agriculture Organization of the United Nations.

Ganuza, E., Sellers, C. E., Bennett, B. W., Lyons, E. M. and Carney, L. T. (2016). A novel treatment protects Chlorella at commercial scale from the predatory bacterium Vampirovibrio chlorellavorus. *Frontiers in Microbiology* 7: 848.

García, J. L., de Vicente, M. and Galán, B. (2017). Microalgae, old sustainable food and fashion nutraceuticals. *Microbial Biotechnology* 10(5): 1017–1024.

Hadj-Romdhane, F., Jaouen, P., Pruvost, J., Grizeau, D., Van Vooren, G. and Bourseau, P. (2012). Development and validation of a minimal growth medium for recycling Chlorella vulgaris culture. *Bioresource Technology* 123: 366–374.

Hayashi, L., Reis, R. P., dos Santos, A. A., Castelar, B., Robledo, D., de Vega, G. B., Msuya, F. E., Eswaran, K., Yasir, S. Md., Ali, M. K. M. and Hurtado, A. Q. (2017). The cultivation of Kappaphycus and Eucheuma in tropical and sub-tropical waters. In: *Tropical Seaweed Farming Trends, Problems and Opportunities: Focus on Kappaphycus and Eucheuma of Commerce* Hurtado, A. Q., Critchley, A. T. and Neish, I. C. (Eds). Cham: Springer International Publishing, pp. 55–90.

Huesemann, M. H., Edmunson, S. J., Gao, S., Dale, T., Negi, S., Laurens, L., Pienkos, P., Knoshaug, E., Davis, R., Klein, B., Lane, T., Timlin, J., Reichardt, T. A., Poorey, K., Barry, A. and McGowen, J. (2020). *DISCOVR: Development of Integrated Screening, Cultivar Optimization, and Verification Research*. DOI: 10.2172/1676405.

Huntley, M. E., Johnson, Z. I., Brown, S. L., Sills, D. L., Gerber, L., Archibald, I., Machesky, S. C., Granados, J., Beal, C. and Greene, C. H. (2015). Demonstrated large-scale production of marine microalgae for fuels and feed. *Algal Research* 10: 249–265.

Huo, S., et al. (2018). Bioaccumulation of *Chlorella Zofingiensis* G1 cultures grown outdoors in photobioreactors. *Frontiers in Energy Research*. DOI: 10.3389/fenrg.2018.00049.

Hurd, C. L., Lobban, C. S., Bischof, K. and Harrison, P. J. (Eds) (2014). Seaweed mariculture. In: *Seaweed Ecology and Physiology*. Cambridge: Cambridge University Press, pp. 413–439. DOI: 10.1017/CBO9781139192637.011.

Hurtado, A. Q., Neish, I. C. and Critchley, A. T. (2019). Phyconomy: the extensive cultivation of seaweeds, their sustainability and economic value, with particular reference to important lessons to be learned and transferred from the practice of eucheumatoid farming. *Phycologia* 58(5): 472–483.

Hwang, E. K., Baek, J. M. and Park, C. S. (2007). Assessment of optimal depth and photon irradiance for cultivation of the brown alga, Sargassum fulvellum (Turner) C. Agardh. *Journal of Applied Phycology* 19(6): 787–793.

Hwang, E. K., Yotsukura, N., Pang, S. J., Su, L. and Shan, T. F. (2019). Seaweed breeding programs and progress in eastern Asian countries. *Phycologia* 58(5): 484–495.

Ingle, K. N., Polikovsky, M., Chemodanov, A. and Golberg, A. (2018). Marine integrated pest management (MIPM) approach for sustainable seagriculture. *Algal Research* 29: 223-232.

Kim, J., Stekoll, M. and Yarish, C. (2019). Opportunities, challenges and future directions of open-water seaweed aquaculture in the United States. *Phycologia* 58(5): 446-461.

Lee, P. A. and White, R. L. (2019). Agronomic practices for photoautotrophic production of algae biomass. In: *Grand Challenges in Algae Biotechnology* Hallmann, A. and Rampelotto, P. H. (Eds). Cham: Springer International Publishing, pp. 111-156.

Loftus, S. E., Hunt, D. E. and Johnson, Z. I. (2020). Reused cultivation water from a self-inhibiting alga does not inhibit other algae but alters their microbiomes. *Algal Research* 51: 102067.

Loftus, S. E. and Johnson, Z. I. (2017). Cross-study analysis of factors affecting algae cultivation in recycled medium for biofuel production. *Algal Research* 24: 154-166.

Loftus, S. E. and Johnson, Z. I. (2019). Reused cultivation water accumulates dissolved organic carbon and uniquely influences different marine microalgae. *Frontiers in Bioengineering and Biotechnology* 7: 101.

Lu, Z., Loftus, S., Sha, J., Wang, W., Park, M. S., Zhang, X., Johnson, Z. I. and Hu, Q. (2020). Water reuse for sustainable microalgae cultivation: current knowledge and future directions. *Resources, Conservation and Recycling* 161: 104975.

Monterey Bay Seaweeds People and Process. Available at: http://www.montereybayseaweeds.com/people-process.

Moody, J. W., McGinty, C. M. and Quinn, J. C. (2014). *Global Evaluation of Biofuel Potential from Microalgae*. Proceedings of the National Academy of Sciences.

Morris, J. J., Johnson, Z. I., Szul, M. J., Keller, M. and Zinser, E. R. (2011). Dependence of the cyanobacterium *Prochlorococcus* on hydrogen peroxide scavenging microbes for growth at the Ocean's surface. *PLoS ONE* 6(2): e16805.

Navarrete, I. A., Kim, D. Y., Wilcox, C., Reed, D. C., Ginsburg, D. W., Dutton, J. M., Heidelberg, J., Raut, Y. and Wilcox, B. H. (2021). Effects of depth-cycling on nutrient uptake and biomass production in the giant kelp Macrocystis pyrifera. *Renewable and Sustainable Energy Reviews* 141: 110747.

Parrys (2017). Locations. Available at: http://www.eidparry.com/locations/.

Phycom (2019). Phycom. Available at: https://phycom.eu/.

Pond Tech (2018). Technology. Available at: https://www.pondtech.com/technology/.

Qualitas Health (2017). Our Farms. Available at: https://www.qualitas-health.com/farms.

Racine, P., Marley, A., Froehlich, H. E., Gaines, S. D., Ladner, I., MacAdam-Somer, I. and Bradley, D. (2021). A case for seaweed aquaculture inclusion in U.S. nutrient pollution management. *Marine Policy* 129: 104506.

Radulovich, R., Neori, A., Valderrama, D., Reddy, C. R. K., Cronin, H. and Forster, J. (2015). Farming of seaweeds. In: *Seaweed Sustainability: Food and Non-Food Applications* Tiwari, B. K. and Troy, D. (Eds). San Diego: Elsevier Science & Technology, pp. 27-59.

Rebours, C., Marinho-Soriano, E., Zertuche-González, J. A., Hayashi, L., Vásquez, J. A., Kradolfer, P., Soriano, G., Ugarte, R., Abreu, M. H., Bay-Larsen, I., Hovelsrud, G., Rødven, R. and Robledo, D. (2014). Seaweeds: an opportunity for wealth and sustainable livelihood for coastal communities. *Journal of Applied Phycology* 26(5): 1939-1951.

Redmond, S., Green, L., Yarish, C., Kim, J. and Neefus, C. (2014). *New England Seaweed Culture Handbook-Nursery Systems*. Connecticut Sea Grant CTSG-14-01. 92 pp.

Available at: https://opencommons.uconn.edu/cgi/viewcontent.cgi?article=1000&context=seagrant_weedcult.

Reichardt, T. A., Collins, A. M., McBride, R. C., Behnke, C. A. and Timlin, J. A. (2014). Spectroradiometric monitoring for open outdoor culturing of algae and cyanobacteria. *Applied Optics* 53(24): F31–F45.

Rinker, M., Airhart, K., Anderson, D., Garavelli, L., Caruso, O., Grear, M., et al. (2021). Kelp energy products and marine renewable energy for coastal Alaska communities (report No. PNNL-31092). Pacific Northwest National Laboratory (PNNL).

Robinson, N., Winberg, P. and Kirkendale, L. (2013). Genetic improvement of macroalgae: status to date and needs for the future. *Journal of Applied Phycology* 25(3): 703–716.

Roquette Klotze GmbH & Co. KG (2017). Algomed. Available at: https://www.algomed.de/en/cultivation/.

Singh, R. N. and Sharma, S. (2012). Development of suitable photobioreactor for algae production–a review. *Renewable and Sustainable Energy Reviews* 16(4): 2347–2353.

Stopha, M. (2020). Alaska kelp farming: the blue revolution [WWW document]. Available at: https://www.adfg.alaska.gov/index.cfm?adfg=wildlifenews.view:article&articles_id=949.

Szyjka, S. J., Mandal, S., Schoepp, N. G., Tyler, B. M., Yohn, C. B., Poon, Y. S., Villareal, S., Burkart, M. D., Shurin, J. B. and Mayfield, S. P. (2017). Evaluation of phenotype stability and ecological risk of a genetically engineered alga in open pond production. *Algal Research* 24: 378–386.

Taiwan Chlorella Manufacturing Company (2013). Taiwan chlorella. Available at: http://www.taiwanchlorella.com/.

Terada, R., Vo, T. D., Nishihara, G. N., Shioya, K., Shimada, S. and Kawaguchi, S. (2016). The effect of irradiance and temperature on the photosynthesis and growth of a cultivated red alga Kappaphycus alvarezii (Solieriaceae) from Vietnam, based on in situ and in vitro measurements. *Journal of Applied Phycology* 28(1): 457–467.

Thomassen, G., Egiguren Vila, U., Van Dael, M., Lemmens, B. and Van Passel, S. (2016). A techno-economic assessment of an algal-based biorefinery. *Clean Technologies and Environmental Policy* 18(6): 1849–1862.

Triton Algae Innovations Process Overview. Available at: https://www.tritonai.com/process.

U.S. DOE (2016). National algal biofuels technology review. In: Barry, A., Wolfe, A., English, C., Ruddick, C. and Lambert, D. (Eds). U.S. Department of Energy, Office of Energy Efficiency and Renewable Energy, Bioenergy Technologies Office. Available at: https://www.energy.gov/sites/default/files/2016/06/f33/national_algal_biofuels_technology_review.pdf.

van der Molen, J., Ruardij, P., Mooney, K., Kerrison, P., O'Connor, N. E., Gorman, E., Timmermans, K., Wright, S., Kelly, M., Hughes, A. D. and Capuzzo, E. (2018). Modelling potential production of macroalgae farms in UK and Dutch coastal waters. *Biogeosciences* 15(4): 1123–1147.

Veramaris What We Do. Available at: https://www.veramaris.com/what-we-do-detail.html#fermentative.

White, R. L. and Ryan, R. A. (2015). Long-term cultivation of algae in open-raceway ponds: lessons from the field. *Industrial Biotechnology* 11(4): 213–220.

Yuan, D., Yao, M., Wang, L., Li, Y., Gong, Y. and Hu, Q. (2019). Effect of recycling the culture medium on biodiversity and population dynamics of bio-contaminants in Spirulina platensis mass culture systems. *Algal Research* 44: 101718.

Zhang, Q., Xue, S., Yan, C., Wu, X., Wen, S. and Cong, W. (2015). Installation of flow deflectors and wing baffles to reduce dead zone and enhance flashing light effect in an open raceway pond. *Bioresource Technology* 198: 150–156.

Zittelli, G. C., Biondi, N., Rodolfi, L. and Tredici, M. R. (2013). Photobioreactors for mass production of microalgae. In: *Handbook of Microalgal Culture* Richmond, A. and Hu, Q. (Eds). Chichester: John Wiley & Sons, Ltd., pp. 225–266. DOI: 10.1002/9781118567166.ch13.

Chapter 5

Developments in algal processing

Schonna R. Manning and Reuben D. Gol, University of Texas at Austin, USA

1 Introduction

Algae are primarily photosynthetic organisms that are genetically, morphologically, and biochemically diverse with representatives occurring on more than a dozen major branches of the tree of life. Cyanobacteria, also known as blue-green algae, are photosynthetic prokaryotes that arose roughly around 3.5 billion years ago to begin oxygenic life, and they later gave rise to several lineages of eukaryotic algae, including microscopic and macroscopic forms. The major eukaryotic algal groups include, but are not limited to, green (Chlorophyta), red (Rhodophyta), brown (Ochrophyta), and golden algae (Haptophyta and Bacillariophyta). While microalgae, including cyanobacteria, occur in fresh water, brackish, and marine habitats, worldwide, macroalgae (i.e. seaweeds and kelps) are primarily marine.

The large-scale cultivation of algae, algaculture, is becoming more widespread for the generation of renewable algal biomass for food, feedstocks, and high-value specialty products. Presently, there are around 160 000 described species of micro- and macroalgae (algaebase.org), yet less than 1% of these have been explored for their nutritional potential. Historically, the majority of research has focused on macroalgae, which have been grown for more than a thousand years for human consumption and livestock feed as well as specialty chemicals. Most seaweeds are consumed directly and they are common ingredients in soups, salads, and sushi, for example, kombu (*Saccharina japonica*), wakame (*Undaria pinnatifida*), hijiki (*Sargassum*

http://dx.doi.org/10.19103/AS.2021.0091.09

fusiforme), dulse (*Palmaria palmata*), and nori (*Porphyra* spp.) (White and Wilson, 2015).

In 2012, global seaweed producers reported nearly 21 million metric tons of biomass generated from chlorophytes (~22 000 tons), ochrophytes (~8 500 000 tons), and rhodophytes (~12 400 000 tons) across 291 species and 43 countries (see White and Wilson, 2015, for a comprehensive list). Biomass yields increased to 25 million metric tons in 2014, which later grew to more than 30 million tons in 2016, and yields have increased each year. The global seaweed market for 2020 was valued around US$16.7 billion, and it is predicted to be worth US$30.2 billion by the year 2025 (Research and Markets, 2020b). Table 1 outlines genera from each major group of macroalgae, including their products and applications. Despite the tremendous diversity of macroalgae, approx. 98% of seaweed production comes from just six genera, including the brown kelps, *Laminaria* and *Undaria*, and a variety of red seaweeds, for example, *Eucheuma*, *Gracilaria*, *Kappaphycus*, and *Porphyra* (Pereira and Yarish, 2008).

Table 1 Genera of macroalgae and microalgae from the major algal phyla, including products and applications (adapted from White and Wilson, 2015; Barkia et al., 2019; others)

Phylum and genera of algae	Uses
<u>Seaweeds</u>	
Chlorophyta: *Caulerpa*, *Chaetomorpha*, *Codium*, *Enteromorpha*, *Ulva*	agricultural feed, food, medicine, paper, specialty chemicals
Ochrophyta: *Ascophyllum*, *Durvillaea*, *Ecklonia*, *Fucus*, *Himanthalia*, *Laminaria*, Macrocystis, *Saccharina*, *Sargassum*, *Undaria*	agricultural feed, alginates, carrageenan, food, medicine
Rhodophyta: *Chondrus*, *Eucheuma*, *Gelidium*, *Gigartina*, *Gracilaria*, *Hypnea*, *Kappaphycus*, *Palmaria*, *Porphyra*	agar, agricultural feed, carrageenan, food, medicine, paper
<u>Microalgae</u>	
Bacillariophyta: *Chaetoceros*, *Phaeodactylum*	agricultural feed, biomaterials, omega fatty acids
Cyanobacteria: *Arthrospira* (formerly *Spirulina*)	agricultural feed, bioactive natural products, nutritional supplements, phycocyanin
Chlorophyta: *Botryococcus*, *Chlorella*, *Chlorococcum*, *Dunaliella*, *Haematococcus*, *Tetraselmis*	agricultural feed, bioactive natural products, carotenoids, hydrocarbons, nutritional supplements, omega fatty acids
Haptophyta: *Isochrysis*	agricultural feed, bioactive natural products, omega fatty acids
Ochrophyta: *Nannochloropsis*	agricultural feed, bioactive natural products, carotenoids, nutraceutical, omega fatty acids
Rhodophyta: *Porphyridium*, *Rhodella*	bioactive natural products, phycocyanin, phycoerythrin

Microalgae have been mass cultivated for many decades as feed for aquaculture, and various species of microalgae have attracted attention as synthetic biofactories for the production of foods, nutraceutical supplements, and bioactive compounds, including essential fatty acids and antioxidants (Benedetti et al., 2018; Khan et al., 2018). By comparison, the global microalgal market generated only ~20 000 metric tons of biomass in 2015, and annual yields have not changed substantially (Tredici et al., 2016). While there are more than 100 000 described species of microalgae, only a few genera from each major group have been explored for their nutritional potential (Table 1).

Global microalgal biomass production was valued around US$3.4 billion in 2020 and this is projected to grow to US$4.6 billion by 2027 (Research and Markets, 2020a). Biomass from *Arthrospira* (formerly known as *Spirulina*) and *Chlorella* comprise roughly two-thirds of the current market for food and nutraceutical applications, and there are smaller markets for the production of algal proteins and natural pigments. With the predicted shortage in nutritional protein for a growing global population, certain genera have been proposed as alternatives to help fulfill this 'protein gap' – strains of *Chlorella*, *Euglena*, *Scenedesmus*, and *Acutodesmus* are known to produce up to 56% protein w/w, while strains of *Arthrospira* can produce more than 60% protein w/w (see Becker, 2007, for review). The global algal protein market was US$691 million in 2018 and it is expected to grow to more than US$740 million by 2025 in support of alternative plant-based protein isolates (Grand View Research, 2019). The pigment market is also expected to grow as more algae are being exploited for their diverse natural pigments to supplant artificial food colorants (Novoveská et al., 2019); the carotenoid market is expected to expand from US$1.5 billion in 2019 to more than US$2 billion by 2026 (Markets and Markets, 2020a).

While algae grow faster than terrestrial crops, the mass cultivation of macroalgae and microalgae has emergent problems, for example, disease/contamination, low productivity, and culture crashes (Benemann et al., 1987; Stephens et al., 2012; Wang et al., 2013; da Silva and Reis, 2015). Thus, current research has invested in methods for 'crop protection' to improve biomass yields in outdoor microalgal cultivation systems. While open ponds are more common and economical for commodity products, closed photobioreactor (PBR) systems are gaining popularity for the cultivation of algae for specialty products.

There are notable differences in the cultivation of macroalgae and microalgae that can impact harvesting and downstream processing decisions. Aside from wild-harvested seaweeds, the cultivation of macroalgae is sustained through the establishment of near-shore seaweed gardens and systems of long-lines with buoys and anchors in the open ocean (Bak et al., 2018). Seaweed gardens are maintained by tying juvenile seaweeds to ropes in nurseries that can be accessed at low tide. In the open ocean, kelp sporophytes are seeded on spools of string that can be spiral-cast with long lines and harvested once mature (Kim et al., 2017). Other systems use land-based ponds for the cultivation of specialty seaweeds for higher-value products (Pereira et al., 2013). Similarly, the mass cultivation of microalgae is maintained in large open raceway ponds or enclosed PBRs (Tredici et al., 2016). Unlike macroalgae, harvesting microalgal cultures requires large volumes of water to be removed to concentrate the biomass to a paste or slurry.

Downstream processing, from harvesting to product extraction, can account for more than 75% of the costs of algal production (Kapoore et al., 2018). Despite the obvious differences in scale, similar downstream processes can be used for the treatment of bulk biomass. And numerous industrial processes currently exist for the treatment of terrestrial feedstocks that can be readily applied to algal biomass. Current methods for processing algae are detailed in Fig. 1.

Given the diversity of the algae and their products, there is no universal set of processing methods that has been adopted across the industry for harvesting, dewatering, drying, cellular disruption, and nutrient recovery (Mata et al., 2010). Moreover, certain products are sensitive to light, heat, pressure, and/or

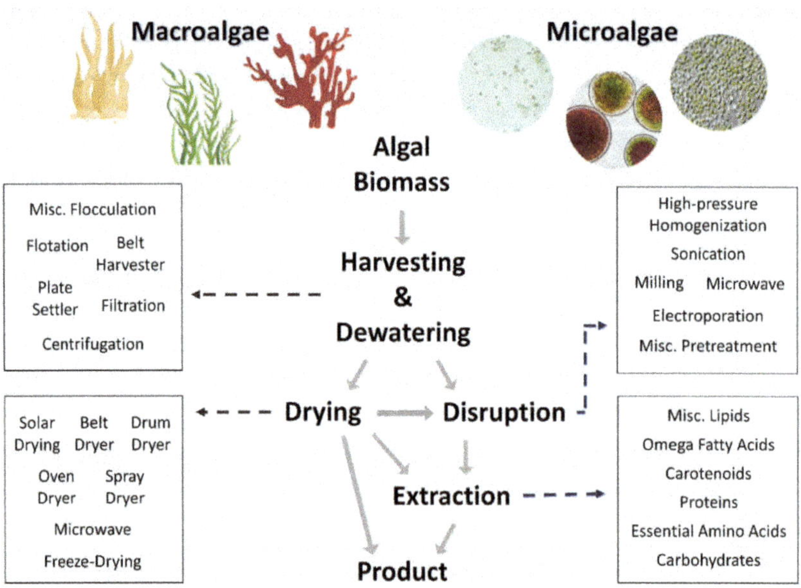

Figure 1 Macroalgae and microalgae are grown for the production of biomass and bioproducts. Algal biomass is harvested and dewatered using various mechanisms and dried further if needed. Whole biomass may be utilized in products or the biomass may be disrupted for the recovery of different nutrient classes. Algal biocomponents can be damaged by certain treatments, for example, excessive heating, thus it is important to select processing methods that minimize effects to the biomass and extracted products.

oxygen, that is, carotenoids, proteins, fatty acids, etc. Thus, many factors must be considered (and often tested) when determining which methods are feasible.

This chapter outlines various developments in algal processing technologies with a focus on processing methods that preserve biomass integrity and minimize the degradation of valuable biocomponents. Relevant case studies are presented with a discussion of capabilities, scalability, and limitations, providing a foundation for the selection of compatible processing technologies. There is a growing body of literature on these topics, and the authors regret they are unable to include all of these references due to page limitations.

2 Harvesting and dewatering

Harvesting is the initial step after cultivation to collect the algal biomass for further processing. The harvesting of macroalgae follows a relatively simple process of cutting or removing the biomass from the growth lines and collecting the leaf-like structures (i.e. blades). Once large enough, seaweeds are harvested from off-shore gardens by cutting above the meristematic region, allowing for regeneration and continued growth. Kelps harvested from long-line systems can be line-dried or cut from the rope for drying and processing. Unlike seaweed gardens that regenerate over time, the seeding of long lines is repeated each growing season (see Kim et al., 2019, for review).

By comparison, harvesting microalgae follows a much more complicated set of processes and the methods selected are largely dependent on the unique characteristics of the algal cells. The small size of the cells (ranging from 0.5 μm to 200 μm), the negligible difference in density between the cells and water, the release of extracellular organic matter (EOM), and the negative surface charge of the cells all contribute to the challenge of removing residual water, while also maintaining biomass quality and favorable cost and energy balances. EOM is typically composed of proteins, polysaccharides, and miscellaneous polymeric substances that are released from the algal cells into the growth medium, which increases in concentration as the cells age (Baroni et al., 2020). These substances can cause fouling of filtration membranes and affect cell-surface charges, which can negatively impact dewatering efforts. These challenges have resulted in dewatering technologies that typically have high capital costs and account for a large portion of the energy consumption in the entire process life-cycle, while still creating a bottleneck and limiting the scalability of commercial microalgae production.

Microalgal cultures are typically grown at <0.1% total suspended solids (TSS) to eliminate shading and maximize light penetration into cells in the growth medium, resulting in large volumes of culture for harvesting and dewatering processes, with relatively low biomass outputs (Gerardo et al., 2015). A microalgal suspension is initially harvested and thickened to slurry of 2–7% TSS (Uduman et al., 2010). The secondary dewatering stage produces a cake or sludge, which is around 15–25% TSS; a third, optional stage, can be used to dry the biomass to >90% TSS. Best practices in industry may utilize a combination of various harvesting and dewatering methods to achieve the desired results.

Microalgae possess certain morphological and chemical features that allow cells and filaments to stay in suspension. These are the main drivers influencing separation of the algal cells from the aqueous medium and water itself (Gerardo et al., 2015). The size, shape, and density of the cells influence the settling rate while the surface charge and electrical interactions between algal cells, as well as interactions between the cells and their growth medium, influence the stability of the suspension. The intensity of the surface charge varies by species, ionic concentration and pH of the growth medium, and other environmental factors. These parameters can be altered through flocculation, electro-processes, or other methods, improving the dewatering characteristics of conventional harvesting methods.

Experts in algaculture suggest an ideal commercial harvesting and dewatering system would be compatible with a wide range of microalgal strains, maximize the recycling of the growth medium, and process large volumes of culture while minimizing processing time, impacts on biomass quality, and capital and operating costs (Barros et al., 2015; Vandamme 2017; Marrone et al., 2018). This system would maximize algal concentration in the output stream and easily integrate with cultivation facilities to minimize pumping and power requirements.

While there is no one-size-fits-all system, there are several commercial processes currently established for the harvesting and dewatering of microalgae that provide a variety of options based on mechanical, chemical, biological, and electrical methods. Thus the remainder of this section will present methods currently employed for harvesting and concentrating microalgal cultures, including gravity sedimentation, centrifugation, filtration, flotation, and the use of coagulants to improve harvesting and dewatering efforts (Fig. 2). The advantages, disadvantages, and costs of each method are presented in Table 2.

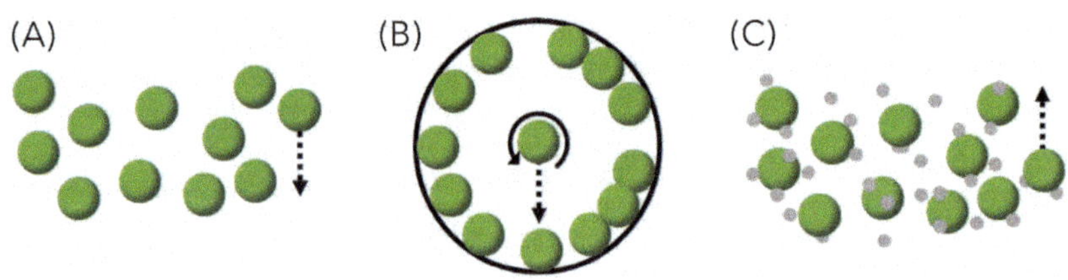

Figure 2 Harvesting and dewatering mechanisms used to concentrate microalgal cells include, but are not limited to, gravity sedimentation (A), centrifugation (B), and dissolved-air flotation (C). The arrow indicates the directional forces and movement of the algal cells.

Table 2 Comparison of harvesting and dewatering technologies used for the concentration of microalgal biomass. Each method has advantages and disadvantages as well as associated capital and operating costs. With the exception of centrifugation, post-harvest biomass may need further dewatering before further processing, that is, drying, disruption, and extraction

Process	Advantages	Disadvantages	Costs
Gravity sedimentation	Low energy inputs; equipment can be modified for improved performance	Cells take a long time to settle; prolonged periods can result in biochemical changes; low biomass recovery efficiency; low harvesting/dewatering efficiency	$
Centrifugation	Highly efficient and reliable for harvesting, dewatering, and concentrating biomass	Equipment is expensive and requires routine maintenance; high energy inputs	$$$$$
Filtration	Gentle harvesting method; removes up to 60% of media/water	Membranes are subject to biofouling; requires frequent maintenance; reduced efficiency with small cells; low dewatering efficiency	$$$$
Flotation	Efficiently separates very small cells from media/water	Needs a compatible collection mechanism (e.g. belt harvester); poor method for larger cells; low dewatering efficiency	$$$
Flocculation	Metal salts are inexpensive; effective for a broad range of cell sizes; can be combined with other harvesting methods	Salts accumulate with the biomass; some coagulants are cost-prohibitive or toxic	$$$

2.1 Gravity sedimentation

The simplest method for harvesting microalgae is gravity sedimentation. This occurs when the microalgal cells are denser than water, allowing them to settle out of suspension over time. The settling of algal cells can be characterized by the settling velocity with respect to a variety of characteristics of the algal cell and density of the suspension fluid, which is derived from Stokes' Law shown in Equation 1 (after Svarovsky, 2000):

$$u_p = \frac{d^2(\rho_p - \rho_f)g}{18\mu_f} \tag{1}$$

where u_p is the settling velocity of the algal cell (m/s), d is the diameter of the algal cell (m), ρ_p is the density of the algal cell (kg/m^3), ρ_f is the density of the fluid (kg/m^3), g is the gravitational acceleration (m/s^2), and μ_f is the viscosity of the fluid ((N*s)/m^2).

After growth, the algal culture can be transferred to and held in an undisturbed holding pond or vessel. Through Equation 1, the algal cells with a higher density than the growth medium settle to the bottom of the pond or vessel where they are harvested by accumulating a concentrated layer of algal biomass. Gravity sedimentation of microalgae is viewed as a simple, energy-efficient, and inexpensive method, with low capital and operating costs, but it is time-consuming, requires large amounts of land area, and yields a low concentration of algae (Rawat et al., 2011). Furthermore, this process may take days, which can lead to the loss of target product(s) in the biomass. This method is most effective when processing cells >10 μm in diameter.

The efficiency of gravity sedimentation has been improved with modified settling tanks. Plate settlers or lamella-type settling tanks equipped with flat, inclined plates promote increased contact between the algal cells as they are sinking (Fouad et al., 2016). In general, modifications of settling tanks aim to promote the contact and aggregation of smaller algal cells to increase the diameter of the particles, and thus their settling rates, improving all aspects of the harvesting process.

2.2 Centrifugation

Centrifuges can be used to harvest a wide variety of microalgal species and do not require any additional steps or added chemicals. Centrifugation follows the same underlying principle as gravity sedimentation, using Equation 1, except the rate of sedimentation is accelerated by a centrifugal force instead of gravitational force. Acceleration speeds in commercial-scale centrifuges typically reach 10000 *g*-force and, in some cases, upward of 14000 *g*-force (Heasman et al., 2000). Due to the high centrifugal forces involved, cell damage is common, resulting in slightly diminished biomass quality for certain final products (Milledge and Heaven, 2011).

Centrifuges require extremely high energy inputs, as well as high capital and operating costs for the equipment. Nevertheless, centrifugation produces the best dewatering results of all known harvesting methods (Dassey and Theegala, 2013; Heasman et al., 2000). Centrifugation is often used following various harvesting methods to further dewater the biomass and generate slurry, cake, or sludge, which can save costs and energy inputs due to reduced volumetric inputs. The strong dewatering parameters make centrifuges a favorable commercial option, especially for low-volume, high-value products. The high capital costs, operating costs, and energy requirements limit the scalability of this method, pushing the algal industry to pursue cheaper and more energy-efficient methods (Dassey and Theegala, 2013; Coons et al., 2014).

Stacked-disk, continuous-flow centrifuges are commonly used to harvest and dewater microalgae for high-value products (Grima et al., 2003). A stacked-disk centrifuge typically contains a shallow cylindrical bowl with numerous closely spaced metal cones that rotate at high speeds (Milledge and Heaven, 2011). The solids and liquids within an algal culture are separated into the narrow flow channels between the stacked cones, or disks, due to the varying densities of the materials; the higher density solids travel to the outer cones for collection and the lower density water concentrates in the center of the centrifuge.

Evodos® has created its own stacked-disk centrifuge that provides superior concentration performance at reduced energy inputs when compared to an industrial stacked-disk centrifuge. The Evodos-style centrifuge differs from the standard bowl-style centrifuge by how the disks are stacked and oriented; plates are stacked horizontally and aligned vertically along a central rotor compared to the vertical stack of the bowl-type centrifuge (Evodos, 2018; Yu, 2012). As the rotor reaches higher speeds, the higher density solids travel outward through the channels between the plates and cake along the outer shell of the system; the lower density water concentrates in the center around the rotor. The cake buildup on the outer shell requires a discharge stage to collect the biomass, creating a semi-continuous system, but the Evodos-style centrifuge can achieve upward of 31.5% TSS consuming only 0.95 kWh/m^3 of water removed compared to the bowl-style centrifuge that can achieve upward of 22% TSS while consuming 8 kWh/m^3 of water removed (O'Connell et al., 2013).

2.3 Filtration

Filtration can be used as a harvesting and/or dewatering method that involves the flow of an algal culture across (or through) a filter or membrane medium wherein the driving force can be gravity, vacuum, or pressure. The common methods of filtration include direct-flow, or dead-end, filtration and crossflow, or tangential-flow, filtration (Fig. 3). While filtration can be an effective means for harvesting and dewatering microalgae, the main performance issues are clogging of the membrane pores (or membrane fouling) and frequent maintenance. Membrane fouling is a phenomenon that is induced by pore clogging, cake formation, and adsorption of organic foulants (Babel and Takizawa, 2010; Chiou et al., 2010; Zou et al., 2011; Zhang et al., 2013a,b).

Historically, direct-flow filtration has been the most commonly used method. Direct-flow filtration can be divided into two categories: surface and deep-bed filtration. Surface filtration is a common method where the solids are deposited on the membrane and form an initial thin layer, building a cake that grows thicker and decreases the flux over time (Chen and Hsiau, 2009). Deep-bed

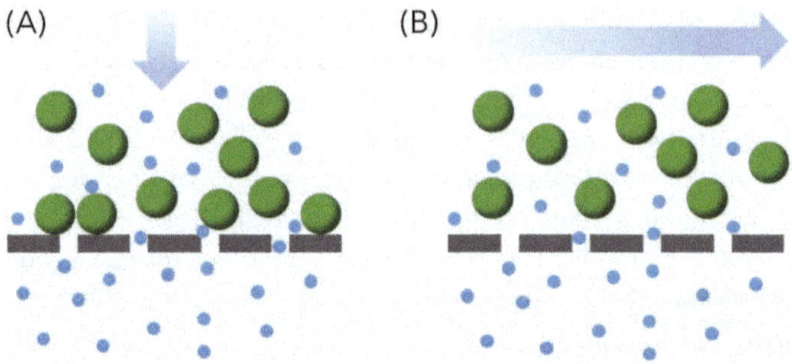

Figure 3 Membrane filtration types include dead-end, or direct-flow, filtration (A) and crossflow, or tangential-flow, filtration (B). Source: Adapted from: Kirschner et al. (2018).

filtration is where the solids are deposited within the filter-bed matrix of a depth filter. The algal cells flow into the filter medium and are retained within the filter bed. This variation of direct-flow filtration requires frequent backwashing due to severe clogging and fouling (Kuberkar and Davis, 2000; Chinu et al., 2010; Plantier et al., 2013).

As opposed to direct-flow filtration, crossflow filtration is a relatively new method that has shown potential as a continuous harvesting/dewatering process. Crossflow filtration has shown reduced fouling and clogging of the membrane and achieves higher throughputs (Bhave et al., 2012; Wilson, 2012; Gerardo et al., 2015). With the fluid velocity tangential to the membrane, the inherent shear forces assist in removing the algal cells deposited on the membrane. Furthermore, the efficiency of filtration can be improved by altering the composition of membrane polymers to reduce biofouling and downtime needed for maintenance (Gol, 2020; Huang et al., 2020; Zhao et al., 2020).

2.4 Flotation

Flotation is a method that can be used to harvest, dewater, and concentrate microalgae where cells are encouraged to float to the surface and are collected by skimmers or belt harvesters (Teixeira and Rosa, 2006; OriginClear, 2015). The process of flotation is induced by small air bubbles that are released into cultures, which adhere to algal cells, effectively reducing their particle density to less than that of the growth medium and allowing them to float to the surface (Singh et al., 2011). Flotation can be easily scaled to commercial settings and maintains high throughputs, low costs, and low space requirements, making flotation one of the more promising harvesting techniques (Ndikubwimana et al., 2016). This method is especially useful when dealing with smaller cell sizes <10 μm in diameter.

The most common method is dissolved-air flotation (DAF), which relies on the higher solubility of air in water at higher pressures (Show et al., 2015).

With this process, a liquid stream saturated with pressurized air is added to the DAF unit and gently mixed with the algal culture. As the pressure stabilizes to atmospheric levels, the dissolved air is released from the liquid, forming small micro-bubbles that adhere to the microalgae in suspension, raising them to the surface for collection, for example, belt harvester. Due to the smaller algal cell sizes processed using DAF, the addition of a chemical or biological coagulant (see next section) is common to improve yields (Kwon et al., 2014; Yuan et al., 2015; de Souza Leite et al., 2019).

2.5 Flocculation

Coagulation and flocculation are two phenomena that rely on the ability to manipulate the cell-surface charge (zeta potential) of each individual algal cell. Typically, the zeta potential of algal cells fluctuates between −2 mV and −75 mV, changing with cell age and culture conditions (Taki et al., 2008; Gerardo et al., 2015; Kim and Kwak, 2017). This negative charge is due to the deprotonation of carboxylic acid and amino groups on the cell surface at pH >5 (Hadjoudja et al., 2010; Brady et al., 2014). The intensity of the surface charge can also be influenced by the strain of algae, pH, and ionic concentration of the growth medium, including other environmental factors.

The neutralization of the algal cell-surface charge occurs following the introduction of a coagulant to the algal slurry, which is agitated to promote cell-to-cell interactions and adhesion to create larger cell aggregates, a process known as flocculation (Fig. 4). As algal cells encounter each other, the smallest cells flocculate together to create larger masses, effectively increasing the particle size/weight and improving harvesting characteristics.

Three different devices are used for flocculation: electrostatic patching, bridging, and sweep flocculation (see Vandamme et al., 2017, for review). Electrostatic patching employs a cationic polymer (coagulant) that binds to a negatively charged algal cell, effectively reversing the cell-surface charge to positive and allowing for further adhesion to other negatively charged cells. Bridging occurs when a polymer or colloid binds to the surface of two different algal cells, forming a bridge. Sweep flocculation is a phenomenon typically seen with inorganic coagulants (as opposed to organic coagulants), wherein metal salts precipitate throughout the slurry, creating a mass of minerals and algal cells.

Inorganic coagulants are typically multivalent cationic metal salts, such as aluminum sulfate, ferric chloride, and ferric sulfate (Papazi et al., 2010; Şirin et al., 2012). These metal salts dissociate in the growth medium and form polyhydroxy complexes when pH levels are optimal, ionizing the growth medium and neutralizing the surface charges of the microalgal cells (Grima et al., 2003). The effectiveness of metal salts is highly dependent on the electronegativity and

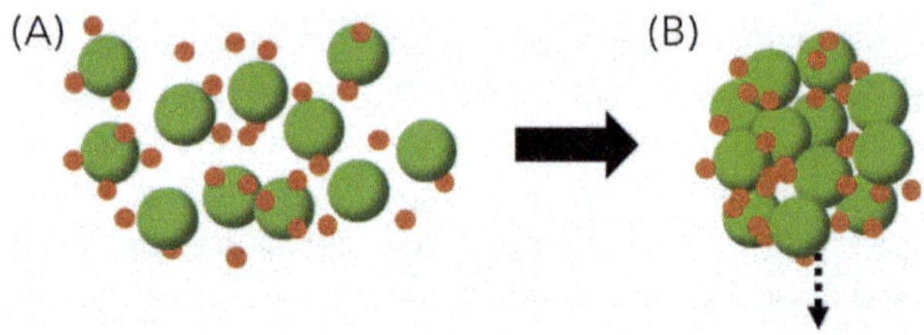

Figure 4 Chemical and biological coagulants (orange) can be introduced to cultures to neutralize the cell-surface charges of the algal cells (green) and increase cell-to-cell interactions (A). Ultimately, the cells aggregate to form larger, heavier particles that settle out of suspension, a process called flocculation (B).

the solubility of the chosen coagulant (Maeda et al., 2002; Papazi et al., 2010). Higher solubilities allow for quicker dissociation into the growth medium and higher electronegativities allow for faster coagulation of the algal cells (Maeda et al., 2002).

The pH of the growth medium is another factor in the mechanics of flocculation. At low pH, the metal hydroxides cause coagulation by charge neutralization (Gao et al., 2010). At high pH, the metal hydroxides precipitate on the algae, physically linking the cells, and sweep flocculation takes place (Duan and Gregory, 2003). Aluminum salts have shown the best characteristics, with respect to cost, dosage, pH, and biomass quality, but cell lysis occurs, releasing EOM and causing other unwanted issues (Papazi et al., 2010). Lime, that is, calcium hydroxide, shows strong flocculation characteristics without cell lysis but only works with cultures containing >10 mg/L of magnesium and produces contaminated biomass due to excessive calcium buildup (Schlesinger et al., 2012; Phasey et al., 2017). For these reasons, the microalgal industry has shifted toward the use of organic coagulants.

Organic coagulants for microalgae are typically derived from a specific subset of polymers called polyelectrolytes. Due to the inherent negative charge of the microalgae, cationic polyelectrolytes are the polymers of interest for microalgal coagulation. Other cationic polymers that are currently under study include chitosan, cellulose, and man-made fibers (Zhuang et al., 2014; Barros et al., 2015; Roselet et al., 2015). Cationic polymers neutralize the zeta potential of the microalgal cells by adsorbing onto the cell walls (Vandamme et al., 2015). The flocculation mechanisms involved are typically bridging and electrostatic patching. In marine microalgal cultures, the polymers can shrink, and coagulation effects are greatly reduced, requiring 5–10× more coagulant than freshwater cultures (Schlesinger et al., 2012). For effective flocculation using cationic polyelectrolytes, the salinity of the growth medium should be less than 5 ppt. Organic coagulants work well in freshwater conditions at pH 7.0, with decreasing performance at higher salinities and suboptimal pH (Divakaran and Pillai, 2002; Show et al., 2015).

Biopolymers, such as chitosan made from the exoskeletons of crustaceans, have gained more attention due to concerns over potential toxicity and risks from polyelectrolytes (Costa et al., 2014). While chitosan has slightly less flocculation capacity when compared to cationic polyelectrolytes, it has shown similar results with the benefit of no toxicity (Renault et al., 2009; Rashid et al., 2013). However, biopolymers typically require higher dosages than cationic polyelectrolytes and often become too expensive for commercial growth systems (Roselet et al., 2015).

The optimal dosage for all coagulants depends on the concentration of the algae and tends to follow a logarithmic function as the concentration of the algae increases. Metal salts typically require upward of 20 mg/L (or more), while biopolymers are closer to 10 mg/L, and polyelectrolytes are between 1 mg/L and 10 mg/L (Roselet et al., 2015). A low dosage of polymer can create a weak bridging between cells; similarly, too much polymer causes electrostatic hindering on the algal cells, reducing the bridging effects. Moreover, the release of EOM from algal cells can interfere with the process, requiring higher dosages of coagulant (Garzon-Sanabria et al., 2012).

The dosage of coagulant also depends on ionic strength, pH, and nutrient composition of the growth medium. Increases in the molecular weight of the polymer coagulant can reduce the optimal dosage needed but higher costs may be involved (Garzon-Sanabria et al., 2012; Granados et al., 2012). Due to the variation in the factors listed above, it can be difficult to maintain optimal coagulant dosages in commercial settings. The high costs involved with coagulant material, the high sensitivity to pH and other environmental factors, and the risk of contamination have pushed the algae industry to devise alternative methods.

Autoflocculation is another type of flocculation that does not involve the addition of coagulants and is induced as a response to environmental stresses. The main trigger for autoflocculation in an algal culture is typically achieved when cultures exceed pH 9.0 (Horiuchi et al., 2003). After growth, the algal cultures are transferred to a holding vessel and the CO_2 supply is cut off, encouraging the algae to remove CO_2 from the growth medium and increasing the pH as the concentration of oxygen increases. Thus, autoflocculation can provide a low-cost, low-energy, non-toxic method with the capability to recycle the growth medium. However, at higher pH levels, precipitates form, for example, calcium phosphates, calcium carbonates, and magnesium hydroxides, which begin to aggregate the algal cells through bridging and sweep flocculation (Brady et al., 2014). While the newly formed salts are non-toxic, they can increase the ash content of the harvested biomass, requiring further processing such as mild acidification and rinsing of the biomass after harvesting (Vandamme et al., 2015). Furthermore, the process is also slow and unreliable, with the best results found for marine cultures, since the ions in marine waters react more readily to high pH levels (Ummalyma et al., 2016; Yang et al., 2016).

Bioflocculation is another approach to sustainable coagulation and flocculation using microorganisms, including bacteria, fungi, and microalgae that release biopolymers into algal cultures. These biopolymers act as coagulants and induce electrostatic patching and bridging to promote flocculation. The optimal dosage for bioflocculants relies on many variables, such as species, ion concentration of the growth medium, EOM, and temperature (Oh et al., 2001; Ummalyma et al., 2016). However, a co-culture of the desired algal species with a foreign bacteria, fungi, or flocculating microalgae has a high risk of contamination or other negative interactions, such as nutrient competition (Gerardo et al., 2015). While this method is sustainable and derived from renewable resources, the variability or unreliability in the process is often too risky for most commercial applications.

Note that several coagulants introduce challenges in the biosafety of both the biomass and the growth media during harvesting and dewatering stages. Metal salts may provide efficient means for coagulating and flocculating biomass, but they are not eco-friendly, contaminate the biomass, and limit the ability to recycle the media. Lime also contaminates the biomass through excessive calcium buildup. Alternatively, bioflocculation introduces its own issues, when considering biosafety, as the biomass can be contaminated with bacteria, fungi, and/or foreign substances that may not be suitable for certain end products.

3 Drying algal biomass

Following harvesting and dewatering, macroalgal and microalgal biomass can be dried further to less than 10% water (w/w) using thermal treatments, that is, solar dryers, convection ovens, belt dryers, rotary drum dryers, spray dryers, and commercial microwaves, as well as non-thermal methods (Fig. 5). The moisture content of the final biomass and integrity of biochemical components depend heavily on the process(es) selected. The least expensive methods include passive drying and direct solar drying. However, controlled drying methods are recommended to minimize exposure, oxidation, and loss of targeted products (see Show et al., 2015, for review). The advantages, disadvantages, and relative costs of common drying methods are presented in Table 3.

3.1 *Solar drying*

Solar drying is a widely used industrial process for the preparation of dried foods that utilizes sunlight to evaporate water from biomass. This technique exposes biomass to solar irradiance in an open-air format, that is, outdoors under direct sunlight. Alternatively, the drying process may take place in a

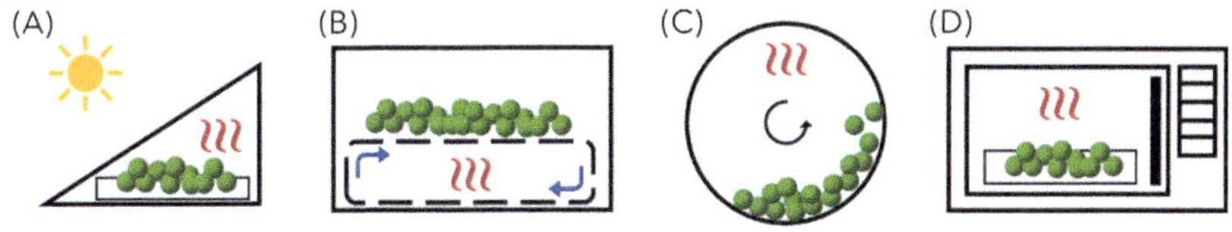

Figure 5 Algal biomass can be dried to less than 10% moisture w/w using solar dryers (A), belt dryers (B), drum dryers (C), and microwave (D), among other thermal and non-thermal methods.

semi-enclosed structure, that is, a solar dryer chamber, that is designed to insulate sunlight, causing rapid heating while constantly exchanging air (see Fig. 5A). There are two approaches used to desiccate biomass in solar drying chambers: direct and indirect. With direct solar drying, biomass is arranged around the drying chamber and the biomass is heated, directly. Whereas, with indirect solar drying, biomass is placed on light-colored trays arranged on a black material that absorbs the sunlight, heating the air in the chamber, rather than the biomass itself.

Table 3 Comparison of methods used to dry algal biomass, including advantages, disadvantages, and associated costs to purchase and operate equipment. Drying efficiency varies across method, taxon, and initial moisture content. Excessive heating and high temperatures are common disadvantages for all thermal methods that can burn biomass and damage bioactive compounds. Moreover, drying equipment can be expensive to purchase, operate, and may require frequent maintenance

Process	Advantages	Disadvantages	Costs
Solar	Low energy inputs; inexpensive construction	Poor control of temperature can lead to overheating or prolonged drying; time-consuming	$
Oven	Controlled temperature/moisture content	Biomass needs to be mixed often to permit even drying	$$
Belt	Constant movement dries biomass evenly; programmed chambers/cells	Needs additional equipment (e.g. trays) for drying microalgal biomass	$$$
Drum	Constant movement of biomass helps to dry evenly	May be difficult to collect dry biomass off drum walls	$$$
Spray	Dries biomass very rapidly; generates fine powders; high productivity in small footprint	High energy inputs; specialized equipment is complex and expensive to maintain	$$$$
Microwave	Uses commercial equipment to dry biomass	High energy inputs; may impact nutritional quality	$$$
Freeze-drying	Effectively removes water and minimizes cellular activity; preserves biochemical content	High energy inputs; equipment is expensive and requires routine maintenance	$$$$

Commercially important seaweeds are commonly solar dried for packaging and distribution. Poblete et al. (2018) dried *Gracilaria chilensis* in a closed solar dryer and achieved optimal chamber conditions when the seaweeds were distributed horizontally; there was a baffle present to exchange fresh air, and a fan was employed for the continuous movement of air. Up to 67.6% moisture was removed with open-air drying at 21.7°C. However, solar drying reduced moisture by 86.1% at 41.3°C in a shorter period of time without prolonged exposure and excessive temperatures. In another study, solar-dried *Durvilleae antartica* biomass maintained the highest essential amino acid and omega fatty acid levels when compared to other thermal drying methods (Uribe et al., 2020). While there are many commercially available solar dryers, the design is relatively simple and can be easily constructed. Prakash et al. (1997) developed an inexpensive solar chamber and dried *Arthrospira* and *Scenedesmus* biomass. Starting with 90% moisture content, their solar dryer produced ~140 g of dry microalgal biomass/m^2 with less than 10% moisture in less than 5 h. Such dryers can be easily scaled for increased biomass processing.

3.2 Oven drying

Algal biomass can be dried in commercial ovens at a set temperature until the target moisture content is achieved. Uribe et al. (2020) found that convection oven-dried kelp *Durvillaea antarctica* had the highest phytochemical and vitamin content, and the biomass maintained its potent antioxidant activity, when compared to other thermal drying methods. In another study, cultures of the microalga *Chlorella vulgaris* were applied in thin films and the composition and color of the biomass were compared for treatments dried at temperatures from 40°C to 140°C for various periods of time (Hosseinizand et al., 2018). While some biomass pigments, carbohydrates, and lipids were adversely impacted at both low- and high-temperature treatments, ultimately, it was determined that the best range for drying *Chlorella vulgaris* was between 40°C and 80°C to minimize the loss of products.

3.3 Belt drying

Belt dryers have adjustable temperature chambers, also known as cells, for the desiccation of algal biomass. Commercial belt dryer systems can be operated in batch or continuous mode, wherein wet algae are fed onto a perforated conveyor belt that carries the biomass through several drying and cooling chambers (see Fig. 5B). Each chamber is equipped with a ventilation fan and heat exchanger to flow through and over the biomass until dried to the specified moisture level. Trică et al. (2020) used thin-layer convection drying at 50–80°C to simulate belt drying of *Cystoseira barbata*, a type of brown macroalgae.

While belt drying is common for macroalgae, this process may also be used for microalgae. Cultures of *Arthrospira* are often dried to a thick paste using a belt dryer, aided by heat and vacuum to achieve >15% solids (Borowitzka, 1998).

3.4 Drum drying

Drum dryers, also known as rotary dryers, are designed to dehydrate algal biomass in a special kiln where heat is introduced into a rotating drum (see Fig. 5C). Drum dryers can be operated in batch or semi-continuous mode, removing and adding biomass as needed. This process has been used for decades to dry red, green, and brown seaweeds, and this method can also be expanded to strains of microalgae. *Scenedesmus* cultures were dehydrated by a drum dryer to a thin film of ~25% solids, which also lysed the cells and destroyed microbial contaminants (Soeder and Pabst, 1975; Mohn and Soeder, 1978; Soeder, 1980). This process has also been demonstrated as a method for drying filtered and frozen pastes of *Arthrospira* (Silva et al., 2019).

3.5 Spray drying

Spray drying is a process used to rapidly dry wet algae to powders by exposing the biomass to hot gas for short periods of time. Lin (1985) reported that spray-dried microalgae generated different particle sizes and morphologies when comparing unicellular and filamentous genera; *Chlorella*, a spherical unicell, formed individual particles composed of a few thousand cells that contained a void space in the center, while filaments of *Arthrospira* adhered together and intertwined, forming long spheroid structures.

As with any thermal drying process, the biomass is exposed to elevated temperatures during the spray drying process, and certain biocomponents may be compromised. The diatom, *Phaeodactylum tricornutum*, was spray-dried to examine the stability of lipids and carotenoids (Ryckebosch et al., 2011). In this study, spray-dried biomass was less susceptible to lipolysis than freeze-dried microalgae. However, spray-dried cultures were more susceptible to oxidation, which was attributed to the degradation of antioxidant compounds during the process. Similarly, Orset et al. (1999) evaluated the effects of spray drying the green microalga *Dunaliella salina*. While spray-dried powders of *D. salina* were unstable in the presence of light and oxygen, the addition of tertiary butylhydroquinone (TBHQ) helped to minimize the degradation of β-carotene and intercellular components.

3.6 Microwave drying

Commercial microwaves can also be used to dry algal biomass using existing equipment (see Fig. 5D). López-Hortas et al. (2018) used a microwave at

50 W, 300 W, and 800 W to dry the brown kelp, *Laminaria ochroleuca*. While microwave drying was very effective, the power and duration of the treatment were dependent on the water content, which can vary, and over drying could lead to burning of the biomass. Microwave can also be used to dry microalgal cultures. Al Rey's group (2016) used a microwave to dry *Chlorella vulgaris* biomass. They tested three different microwave intensity levels (300 W, 600 W, and 900 W) to completely dry 10 g , 20 g, and 30 g of algal biomass and found that 20 W/g gave a lower specific energy requirement without compromising the lipid content.

3.7 Freeze-drying

An alternative to thermal drying, lyophilization, also known as freeze-drying, is a commercial process for preserving food products that removes water from frozen biomass by sublimation under vacuum and heat. The brown seaweed, *Himanthalia elongata*, was freeze-dried and the biomass was later sonicated for the isolation of functional proteins (Garcia-Vaquero et al., 2017). Freeze-drying was also preferred to maintain pigments in *Durvilleae antartica* biomass when compared to thermal drying methods (Uribe et al., 2020).

While freeze-drying can remove nearly all water from biomass, this process can still generate viable cells post-lyophilization in some microalgal cultures (Holm-Hansen, 1964). Of note, 13 strains of cyanobacteria were able to recover after lyophilization when rehydrated (Corbett and Parker, 1976). However, species of diatoms (e.g. *Chaetoceros* sp. and *Phaeodactylum tricornutum*) were not able to survive the freeze-drying process likely due to irreversible cell damage (Cordero and Voltolina, 1997), which can lead to oxidation of intracellular components.

Once dried, biomass may be processed further to generate uniform powders and disrupt cells for products or in preparation for nutrient extraction. Regardless of the drying method(s) selected, it is advised that overheating biomass be minimized to avoid damaging nutritional content. Dried materials should be stored in the dark, sealed, and refrigerated to maintain the integrity of the biomass and chemical composition.

4 Biomass disruption

Macroalgal biomass can be challenging to disrupt due to their chemically stable and structurally complex cell walls, and microalgal biomass is notoriously difficult to rupture due to their small cell sizes and chemical characteristics (Chisti and Moo-Young, 1986; Domozych, 2019). Overall, algae have complex polysaccharide cell wall matrices (e.g. cellulose) that can interfere with lysis and extraction as well as the digestibility of the biomass and the absorption of their components (Barkia et al., 2019). They also possess numerous interior membranes and compartments (e.g. chloroplast) that must be penetrated for complete disruption.

There are several processes that are available for the disruption of fresh, dried, frozen, and freeze-dried biomass. Numerous methods use aggressive treatments to disintegrate the cell walls using chemicals (i.e. strong acids and strong bases) and/or subjecting biomass to high temperatures and pressures (e.g. autoclave). However, this section will focus on mild disruption conditions to preserve the integrity of proteins, lipids, and carbohydrates as well as bioactivity. Techniques include mechanical methods such as solid shear and liquid shear, as well as non-mechanical methods and pretreatments that aid in the disruption of algal biomass (Fig. 6) (for review, see Günerken et al., 2015; Kapoore et al., 2018; Maneein et al., 2018; Wang et al., 2019). A comparison of methods used to disrupt algal biomass is shown in Table 4.

4.1 Mechanical: solid shear

Solid shear processes reduce dried biomass to small uniform particles in the micron and nanometer size range for product applications and/or the extraction of nutrients. High-pressure homogenization methods include industrial-scale processes such as knife milling, ball milling, and bead milling, among other grinding mechanisms. This industrial process is usually employed to generate fine powders for terrestrial biomass (e.g. grains and flours). And, these processes have also been applied to algal biomass.

Knife mills use sharp blades to cut biomass into fine particles, which can be accomplished using an industrial blender. This method is especially useful for larger feedstocks, that is, seaweeds and kelps. Species of dried seaweeds, including greens, reds, and browns, can be reduced to particle sizes less than 500 microns by knife mill for subsequent extraction. While knife milling works well for macroscopic algae, it is not an efficient method for the disruption of microalgal biomass.

Ball mills grind biomass into small particles by the repetitive impact of large heavy balls inside a rotating drum (see Fig. 6A). Iqhrammullah et al. (2020) used a ball mill to grind solar-dried *Chaetomorpha linum*, a species of green seaweed, for the extraction of polyols. And Montingelli et al. (2016) found that ball milling of oven-dried *Laminaria* was more effective when compared to other disruption methods. Ball mills have also been used for the disruption of freeze-dried microalgal biomass, including the chlorophytes *Chlorella sorokiniana*, *Haematococcus pluvialis*, and *Chromochloris zofingiensis* (Taucher et al., 2016). There were varying degrees of disruption efficacy, which depended on the strain and cell robustness. Thus, the milling process must be optimized for each alga (e.g. ball size, number of cycles).

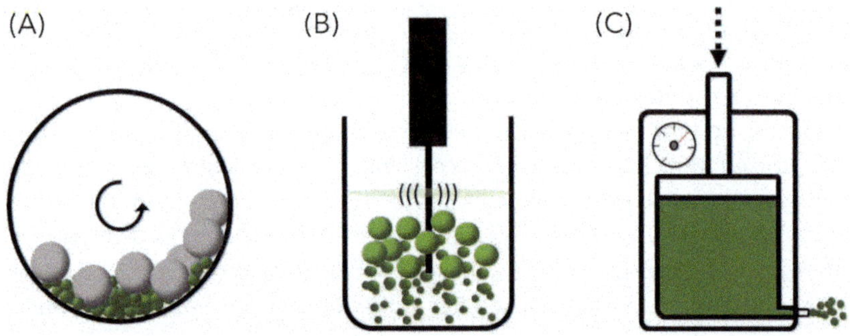

Figure 6 Algal biomass can be disrupted by solid shear and liquid shear mechanisms. Milling is a common homogenization process to refine dried algae by solid shear forces, such as ball milling (A), wherein the algae are crushed to small particles by heavy balls in a rotating drum. Resuspended algae, for example, fresh, dried, or freeze-dried, are disrupted and dispersed into smaller particles using ultrasonication (B). High-pressure homogenization can be used for cultures/concentrates to generate liquid shear using a cell press (C) to compress and rapidly decompress algal cells, causing disruption.

Table 4 Comparison of methods used to disrupt algal biomass, including solid shear, liquid shear, as well as non-mechanical processes. Advantages, disadvantages, and associated costs are presented for each method. Disruption efficiency varies widely across methods and species

Process	Advantages	Disadvantages	Costs
Milling	Produces fine, uniform particles and powders	Some milling methods are not compatible with microalgal biomass	$$
High-pressure homogenization	Generates nanometer-sized particles	Equipment is expensive; high energy inputs	$$$$
Microwave	Disrupts cells by heating intracellular water	High energy inputs; excessive heat can damage nutrient content	$$$
Electroporation	Transient exposure to electric fields creates pores and lyses cells	High energy inputs; equipment can be expensive; does not work well with saline media	$$$
Freeze-thaw	Uses intracellular water to generate ice crystals to disrupt cells	High energy inputs; low efficiency with certain algal cells; time-consuming	$$
Osmotic shock	Salts are inexpensive; low energy inputs	Excess salts may remain with biomass; time-consuming	$
Solvent-assisted	High disruption efficiency; low energy inputs	Biomass may be contaminated with potentially toxic solvent(s)	$$
Enzyme-assisted	Highly selective and processive; high-efficiency cell disruption	Some enzymes can be expensive; enzymes must be disposed after use	$$$$

Alternatively, bead mills use micron-sized inert beads to disrupt cells and grind dry or wet biomass materials to smaller and smaller particles. Postma et al. (2017) evaluated the disruption of microalgae, including strains of *Chlorella vulgaris*, *Neochloris oleoabundans*, and *Tetraselmis suecica*, wherein the biomass was processed with different bead sizes, ranging from 0.3 mm to 1.0 mm. They found that smaller beads (0.3–0.4 mm) had the best results for the recovery of proteins, lipids, and carbohydrates. While this process is typically used with microalgal biomass, this method may also be used to disrupt macroalgal biomass for improved nutrient recovery (Nobles and Manning, 2019).

4.2 Mechanical: liquid shear

Liquid shear processes utilize wet biomass, that is, cell cultures and concentrates, which overcomes expensive and prolonged dewatering and drying steps. These methods include high-pressure homogenization, sonication, and explosive decompression to disrupt and disintegrate cell walls. The efficacy of disruption is dependent on the strain of algae as well as the intensity and duration of the treatment(s).

Disruption of biomass by sonication utilizes a probe or bath-based equipment that emits wave energy that agitates cells in liquid cultures (see Fig. 6B). These sonic waves induce cavitation events inside the cells caused by an increase in pressure and temperature, which ultimately leads to the disintegration of cells into particles and aids with component extraction (Shi et al., 2007). Frequencies >20 kHz are typically used, leading to the process also being known as ultrasonication.

Sonication has been applied for the disruption of many kinds of microalgae and the recovery of various products. Wang et al. (2014a) used sonication to disrupt cultures of *Scenedesmus dimorphus* and *Nannochloropsis oculata* using low- and high-frequency conditions; the 3.2 MHz (100 W) setting was more effective at disrupting *S. dimorphus*, whereas the 20 kHz (40 W) setting was more effective at disrupting *N. oculata*. However, the use of both high- and low-frequency treatments was even more effective. The difference in sonication frequency and recovery efficacy is likely related to the inherent differences in the size and cell wall composition of these two species. Thus, the conditions must be tuned for the alga.

Kurokawa et al. (2016) also used sonication to lyse *Nannochloropsis* as well as cultures of *Chaetoceros* (a genus of diatom, Bacillariophyta). In their study, biomass suspensions treated with frequencies from 2.2 MHz to 4.3 MHz resulted in the best lysis. However, it should be noted that the lysis efficiency was dependent on species of algae, the sonication frequency, and the density of the suspension. While sonication methods are highly effective at cellular disruption, it should be noted that this process can generate heating of the suspension with prolonged treatments and higher frequencies, which can diminish the quality of certain products.

Liquid high-pressure homogenization is a mechanical disruption method that subjects cell cultures, slurries, and resuspensions to high-pressure conditions by physical compression or the introduction of pressurized gas. This process can utilize commercial milk homogenizers, microfluidizers, and cell presses to lyse algal cells, which operate between 2000 psi and 50 000 psi. For example, French-style cell presses subject algal cultures to high pressures up to 40 000 psi, driven by a hydraulic piston (Fig. 5C). Homogenization can also be achieved by the introduction of nitrogen gas into culture slurries, which permits the diffusion of gas across cell membranes using pressures greater than 5000 psi (e.g. Parr pressure chamber). Ultimately, cultures are released from a small-bore opening to generate high shear forces and rapid decompression, rupturing cells to generate nano-sized particles.

Homogenates can also be recirculated to achieve more complete lysis and reduce the size of particles in suspension. However, there is no single pressure or number of cycles that works for all algae, and the pressure needed for complete disruption depends on the strain characteristics and viscosity of the culture (Samarasinghe et al., 2012; Safi et al., 2014). Spiden et al. (2013) was able to achieve 50% lysis per cycle for fresh cultures of *Nannochloropsis* sp. at ~30 000 psi, *Chlorella* sp. at ~15 000 psi, and *Tetraselmis suceica* at ~2500 psi. While, Halim et al. (2012) found that it took five cycles at greater than 12 000 psi to achieve 74% lysis in cultures of *Chlorococcum*. Cha et al. (2012) used up to 20 000 psi to rupture *Chlorella ellipsoidea* cells, which reduced the particle size nearly sevenfold, from approx. 2.5 μm to around 360 nm. Tibbetts et al. (2017) resuspended freeze-dried *Chlorella vulgaris* in demineralized water for homogenization at 25 000 psi at 75 mL/min, and they determined that three cycles were adequate since no further improvements were found with additional cycles.

Safi et al. (2014) used high-pressure liquid homogenization of dried biomass to evaluate the release of soluble proteins in a variety of commercially relevant microalgae. They found that homogenization at 2700 bar for two cycles released up to 25% more protein than traditional alkaline treatments. The efficiency of lysis and release of protein was dependent on the algal strain and cell robustness. *Haematococcus pluvialis* released the smallest amount of protein (~21.2%) when compared to *Pophyridium cruentum* (51.2%), which can be explained by the recalcitrant cell walls of *H. pluvialis*. While homogenization can improve lysis and nutrient recovery, prolonged exposure to high-pressure conditions can lead to metabolite degradation (Safi et al., 2013).

While homogenization has been shown to be effective at disrupting a range of algal cell types, conditions must be determined empirically for each strain of algae, the form of the biomass (e.g, fresh, frozen, dried, and freeze-dried), as well as the density of the suspension. It is also important to note that increased pressures and recirculation of homogenates can generate excess heat and reduce bioactivity, thus it is recommended to circulate culture homogenates through a chilled bath to minimize heating and the degradation of targeted biocomponents.

4.3 Mechanical: other

Microwaves can also be used to disrupt wet algal biomass for enhanced extraction (Pasquet et al., 2011; Biller et al., 2013; Rakesh et al., 2015; Rokicka et al., 2020). By comparison to other techniques, cell disruption by microwave is considered a simple method that is less time-consuming (Lee et al., 2010). Microwave treatments applied at 2.45 GHz caused the heating and evaporation of water and polar compounds in wet biomass, creating expansion and disruption of cells (Kadam et al., 2013). Cheng et al. (2013) demonstrated that increased microwave treatments led to an increase in the cell wall thickness and pore diameter, increasing cellular disruption for temperatures ranging from 80°C to 120°C. And while McMillan et al. (2013) achieved the best disruption with laser (96.5% lysis), microwave treatments (94.9% lysis) were found to be the best choice for both the highest disruption efficiency and lowest energy consumption.

Electroporation is another disruption method that exposes liquid algal cultures to electric fields of varying intensity for a duration of time that leads to an increase in membrane permeability (Zbinden et al., 2013). Electroporation methods are designed to be reversible or non-reversible, wherein the cell membrane is able or unable to repair itself following the process. One such method is the pulsed electric field (PEF), which interacts with the surface charges on the cells of the algae to distort and disrupt the cells. The PEF utilizes short, high-voltage electrical pulses in a chamber designed to lyse cell cultures for improved product recovery (for review, see Joannes et al., 2015; Kempkes, 2016). The intensity and duration of the pulse can be tuned to improve the lysis efficiency among various strains. PEF technology has also been shown to reduce common pests found in commercial cultivation ponds using the same principles to inactivate bacterial contaminants without compromising the algae or bioproducts. Because the principles of cell disruption using PEF are based on charged particles in solution, this process works best with freshwater cultures to avoid interfering charges in brackish or saline media.

4.4 Non-mechanical

Increased biomass disruption can be achieved by using non-mechanical methods, that is, freezing and thawing biomass, osmotic shock, solvents, and/or enzymes, to aid in the disintegration of the cell walls, enhancing lysis, and improving the exposure of cellular contents for extraction (see Günerken et al., 2015; Kapoore et al., 2018; Phong et al., 2018, for review). Freezing wet biomass followed by thawing (i.e. freeze-thaw) allows for the production of intracellular and extracellular ice crystals that lyse microalgal cells and enhance disruption (Sfriso et al., 2018). Rakesh et al. (2015) found that osmotic shock from the addition of 5–15% NaCl was effective at disrupting cells for various strains of *Botryococcus* as well as a strain of *Chlorella sorokiniana*.

Enzyme-based cell lysis is considered an environmentally friendly method for the disruption of algal biomass. Enzymes target specific cell wall compounds for enhanced disruption; for example, cellulases and lipases are employed to hydrolyze cellulose and glycerolipids, respectively (Phong et al., 2018). Moreover, enzyme treatments are preferred over the use of organic solvents that can contaminate the residual biomass for other downstream applications. Soto-Sierra et al. (2018) used a protease and an autolysin for the disruption of *Chlamydomonas reinhardtii* for the release of proteins and lipids. While enzyme treatments can be highly effective, they tend to be cost-prohibitive at scale. Furthermore, certain enzymes may alter co-products and their bioactivity. Thus, selected enzyme treatments should consider the targeted product(s).

Disruption methods may also be coupled with extraction methods for the isolation of algal nutrients. For example, Alavijeh et al. (2020) combined bead milling with hydrolytic enzymatic treatments to improve the cellular disruption for a strain of *Chlorella vulgaris*. They found enzymatic hydrolysis of bead milled biomass was maximized using lipase at pH 7.4, 37°C for 24 h, and the recovery of biocomponents was 44% less for biomass that was not milled. Thus demonstrating that the disruption method and the degree of disruption are important factors for nutrient recovery.

5 Nutrient recovery

Whole biomass may be used for nutraceutical and feedstock applications, although it is often desirable to isolate higher-value proteins, lipids, and carbohydrates for use in defined formulations. To maximize the recovery of components, the biomass should have minimal exposure to harsh treatments (e.g. excessive heating, strong acids, or bases) that could potentially denature proteins and inactivate bioactive compounds. Furthermore, extraction schemes should aim to recover multiple components to enhance the value and utility of algal biomass. For example, recovering lipids first would generate a protein- and carbohydrate-rich residual biomass that could be further processed for the extraction of additional nutrient classes (see example in Fig. 7).

The isolation of biocomponents from algal biomass may follow classical extraction and recovery techniques used with traditional feedstocks. And studies have shown that nutrient recovery can be enhanced by combining compatible biomass disruption and extraction methodologies, that is, sonication-assisted extraction (SAE), microwave-assisted extraction (MAE), and supercritical fluid extraction (SFE). Thus, this section will highlight traditional methods as well as disruption-assisted extraction techniques that are currently employed for the recovery of lipids, carbohydrates, and proteins from algal biomass. The advantages, limitations, and costs of owning and operating these technologies are presented in Table 5.

5.1 Lipids

Algal oils, including neutral and polar lipids, are derived from cell membranes and lipid storage bodies; the composition of glycerolipids, fatty acids, and carotenoid pigments are highly variable across species of micro- and macroalgae (Kumari et al., 2013; Mimouni et al., 2018; Li-Beisson et al., 2019). Typically, macroalgae contain very little oil (1–5% w/w), and the biomass has a higher proportion of carbohydrates and proteins (El Maghraby et al., 2015). By comparison, certain microalgae are well known for their high lipid content (20–70% w/w), which are mass cultivated for biofuels-based research and as sources of nutritional oils, that is, carotenoids and omega fatty acids (Sun et al., 2018; Tiwari and Kiran, 2018; Sati et al., 2019).

Traditionally, lipids are recovered from biomass using organic solvents, and the solvent is removed under heat and vacuum, leaving the extracted oil. While ethanol is a popular food-grade, nontoxic solvent used for the extraction of algal lipids, ethanol and other short-chain alcohols are more selective to polar lipids, including phospholipids, galactolipids, and xanthophylls. Neutral lipids, for example, triglycerides, carotenoids, and hydrocarbons, generally require more nonpolar solvents like hexanes or chloroform for recovery. However, coalescence principles aid in the broader extraction of total lipids. While there are amphipathic solvents like acetone and 2-ethoxy ethanol that are capable of recovering both polar and neutral lipids from wet and dried algal biomass (Jones et al., 2012), there is a movement away from solvent-based recovery to

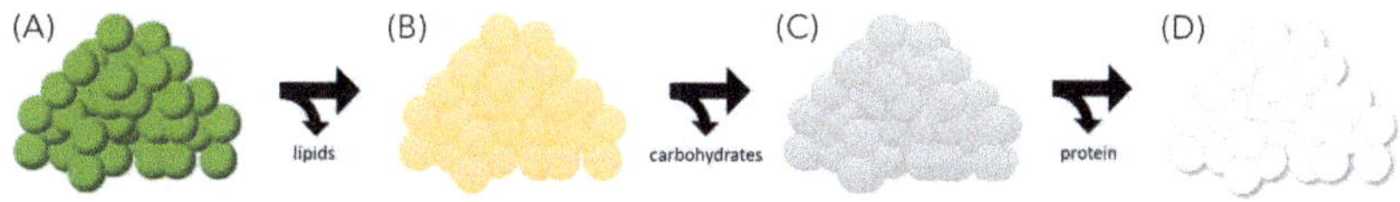

Figure 7 Algal biomass contains multiple nutrient classes that can be extracted (A). In this simplified example, the extraction of lipids yields residual biomass that is rich in proteins and carbohydrates, including insoluble fiber (B). This 'algal flour' can be further processed by resuspending in water for the extraction and isolation of carbohydrates using alcohol-based precipitation, resulting in biomass that contains primarily proteins and fiber (C). Proteins can be recovered by alkaline hydrolysis and isolated by salt precipitation, leaving biomass that is mostly insoluble fiber, residual salts, and minerals (D).

Table 5 Comparison of traditional and enhanced extraction methods employed for the recovery of nutrients from algal biomass (e.g. lipids, carbohydrates, and proteins), including advantages, disadvantages, and related costs. These methods have a common disadvantage since crude extracts may require further isolation steps, for example, chromatography, to achieve product purity

Process	Advantages	Disadvantages	Costs
Solvent-based extraction	Recovers a wide range of polar and nonpolar compounds (i.e. lipids)	Biomass may be contaminated with potentially toxic solvent(s); time-consuming	$
Solvent-based precipitation	High-efficiency recovery of carbohydrates from cell extracts	Biomass may be contaminated with solvent(s); time-consuming	$$
Salt-based precipitation	High-efficiency recovery of bulk proteins from cell extracts	Dialysis needed to remove salts; time-consuming	$$
Sonication-assisted extraction	Enhances cell disruption; high-efficiency extraction in aqueous and organic solvents	High energy inputs; health hazard due to prolonged ultrasonic frequencies	$$$
Microwave-assisted extraction	Enhances cell disruption and release of nutrients	High energy inputs; may impact nutrient quality, for example, denatures protein	$$
Supercritical fluid extraction	Pressure/temperature can be tuned for the extraction of targeted products; solvent-less process	Specialized equipment is expensive; requires routine maintenance	$$$$$

minimize the exposure of the lipids and biomass to potentially toxic reagents, limiting their applications. Thus, food-grade processes aim toward solvent-less extraction, for example, coalescence of algal oils with surrogate oils or SFE (Meireles, 2013).

More recently, MAE and SAE have been used to enhance the disruption and recovery of lipids from numerous strains of algae. In their study, Rakesh et al. (2015) combined osmotic shock and microwave treatments to improve the recovery of neutral lipids from strains of *Botryococcus* sp. and *Chlorella sorokiniana*. Rokicka et al. (2020) also compared the recovery of lipids from *Botryococcus braunii* and *Chlorella vulgaris* and determined that MAE was much more efficient than SAE. MAE was also found to be the best method for the recovery of chlorophyll-*a* and fucoxanthin in the diatom *Cylindotheca closterium* (Pasquet et al., 2011). Of note, microwave treatments were shown to increase the saturated fatty acids extracted from the diatom *Chaetoceros calcitrans*, while SAE increased the recovery of polyunsaturated fatty acids but lowered the content of monounsaturated fatty acids (Nogueira et al., 2018). Thus, MAE and SAE can be utilized for the recovery of lipids, although methods should be optimized for the species/strain of alga and target product(s).

Another method, SFE, employs CO_2 at high temperatures and pressures for the extraction of lipids from dry or wet algal biomass. At the critical point (31°C, 74 bar) and above, liquid CO_2 becomes a supercritical fluid that is capable of permeating and extracting lipophilic components from algal biomass. Supercritical CO_2 behaves like a nonpolar solvent (e.g. hexane) at moderate temperature and pressure, whereas, at higher temperatures and pressures, it acts more like a polar solvent (e.g. ethanol). Thus, by tuning the temperature and pressure conditions, supercritical CO_2 can be used to extract a wide range of lipid classes and valuable co-products.

SFE has been shown to work well with a variety of algae, although the capital and operational costs can be expensive. Thus, most applications using SFE target higher-value nutraceutical products, such as omega fatty acids, for example, docosahexaenoic acid (DHA) and eicosapentaenoic acid (EPA), and carotenoids, for example, astaxanthin (Solana et al., 2014; Hosseini et al., 2017). For review, see Meireles (2013) for an extensive list of algae and lipids that have been extracted using SFE, including operating conditions and co-solvents, for example, ethanol and methanol.

5.2 Carbohydrates

Algae synthesize a wide array of structural and storage carbohydrates; the composition of cell walls and storage polysaccharides are highly diverse and specific to the major algal lineages (see Domozych, 2019, for review). These include many gelatinous compounds associated with algae and their biofilms, including extracellular polymeric substances (EPS) and numerous sulfated polysaccharides (SPS), for example, ulvans, rhamnans, fucoidans, and alginates (Pomin and Mourão, 2008; Wang et al., 2014b; Torres et al., 2019).

The carbohydrate content varies widely among genera of macro- and microalgae. Carbohydrates can account for greater than 75% of the dry weight as shown with strains of *Ascophyllum*, *Palmaria*, and *Porphyra* (Holdt and Kraan, 2011). Many seaweeds have been mass cultivated for decades for the production of polysaccharide-based hydrocolloids that are widely used in industry for example, agar, agarose, and carrageenan (Qin, 2018). Microalgae typically contain more modest quantities of carbohydrates. However, certain microalgae have been cultivated for their innate or enhanced synthesis of polysaccharides. And there are now several genera of microalgae being exploited for their bioactive polysaccharides, including sulfated polysaccharides from *Porphyridium* and calcium spirulan from *Arthrospira* (Raposo et al., 2013).

Soluble carbohydrates can be recovered from algal biomass using traditional extraction methods. Traditionally, polysaccharides are isolated by resuspending disrupted biomass in water to extract soluble carbohydrates into the aqueous suspension, which is followed by the separation of the biomass, bulk precipitation of sugars using a polar solvent, for example, methanol or isopropyl alcohol, and membrane fractionation (Torres et al., 2019). In the absence of acid hydrolysis, carbohydrates remaining with the biomass are associated with glycosides and insoluble fiber.

MAE and SAE technologies have been applied to improve the recovery of polysaccharides from a variety of seaweeds. Several studies have shown that MAE enhanced the recovery of fucoidans from milled kelp biomass in strains of *Ascophyllum nodosum* and *Ecklonia radiata* (Lorbeer et al., 2015; Garcia-Vaquero et al., 2020). Lorbeer et al. (2015) determined that the duration of treatment was critical for the extraction of high-molecular weight fucoidan, wherein longer MAE treatments yielded lower molecular weight fucoidan and co-extracted contaminants, for example, laminarin. Similarly, Shi et al. (2007) found that the duration of SAE treatments, as well as the temperature and ethanol concentration, impacted yields of polysaccharides from defatted biomass of *Chlorella pyrenidosa*. Thus, several parameters should be tested to determine the best operating conditions for the recovery of carbohydrates using MAE and SAE methodologies.

5.3 Proteins

Algae produce a suite of structural and catalytic proteins – this fraction can range from 4% to 25% of the dry weight in seaweeds, although proteins can account for greater than 70% of microalgal biomass dry weight (see Becker et al., 2007, for review). Like plant-derived proteins, algal proteins have high nutritional value, including essential amino acids that can be extracted and purified from algal biomass employing similar processes used for industrial feedstocks.

Traditional methods extract proteins into aqueous buffers, and fractions are recovered by precipitation with ammonium sulfate, followed by dialysis to desalt. This classical approach can be used to extract total proteins or the isolation of specific protein fractions. Sfriso et al. (2018) were able to achieve 95–98% extraction of phycoerythrin, a phycobiliprotein, in a variety of red seaweeds, including species of *Agardhiella*, *Gracilariopsis*, *Gracilaria*, *Polysiphonia*, and *Pyropia*; phycoerythrin extraction and recovery were achieved by homogenization of fresh biomass, followed by a freeze-thaw cycle and resuspension in 1 mM EDTA (pH 9). In another study, the aqueous extraction (pH 7) of proteins from freeze-dried biomass of the brown seaweed *Himanthalia elongata* yielded five discrete proteins as verified by SDS-PAGE gel electrophoresis (Garcia-Vaquero et al., 2017).

Protein extraction can be enhanced with SAE, although MAE can lead to protein denaturation and loss of bioactivity. Kazir et al. (2019) developed a procedure for the isolation of food-grade algal protein concentrates derived from seaweeds. Comparing five procedures, they determined that dried, homogenized biomass resuspended in 10% NaOH (w/v) coupled with multiple rounds of SAE provided the best yields for species of *Ulva* and *Gracilaria*, seaweeds that possess highly digestible proteins with antioxidant activities. These treatments yielded 70% and 86% proteins, respectively.

In some cases, nutrients extracted from microalgae were improved with PEF (Kempkes, 2016). Biomass treated with PEF increased the recovery of pigments and proteins from strains of *Arthrospira* and *Chlorella* (Toepfl and Heinz, 2008). Aouir et al. (2015) found that the recovery of proteins using PEF was similar to results using sonication for the recovery of phycocyanin, another phycobiliprotein. However, extracts from PEF-based treatments were found to be purer and had fewer co-extracted compounds. This process can be used to aid in the release of biocomponents, although Postma et al. (2016) found that protein recovery using PEF alone was incomplete.

Lastly, it is important to note that multiple disruption and extraction processes may be needed to maximize nutrient recovery. While there are no standardized procedures for processing algal biomass, these studies provide the foundation for making informed decisions, considering the unique physical and chemical properties of the algae as well as the integrity of the targeted end products.

6 Product considerations

Algae can be used as whole biomass and/or as functional ingredients extracted for the production of food, feedstocks, and nutraceutical products. Seaweeds are broadly available in the food market, and more microalgal products are introduced every year, yet there are few microalgae that are widely approved for direct consumption. However, biomass and extracts (e.g. omega fatty acids, astaxanthin, and α-carotene) have been readily incorporated into tablet and capsule forms for nutraceutical supplements, and there are numerous algae-related products available in local markets. Industrial markets are presently seeking liquid, flake, sheet, and powder forms of algal biomass for a variety of product and feed applications (Markets and Markets, 2020b).

Algal blendstocks are common for feedstock formulations, wherein a small percentage of algal biomass (1–15%) is mixed with traditional feed to supplement the diets of domestic livestock. It was found that cultures of *Chlamydomonas reinhardtii* could be used directly as feed without cell disruption; however, sonication was needed to improve the bioavailability of nutrients from *Chlorella vulgaris* biomass in digestive studies; cultures that were treated by sonication doubled the bioavailability of carotenoids, for example, β-carotene and lutein, to greater than 10% and 15%, respectively (Gille et al., 2016). Cha et al. (2012) also documented that untreated *Chlorella ellipsoidea* biomass had much less bioavailable carotenoids, and other nutrients, than biomass that was disrupted by microfluidization. Research has shown that adding small amounts of algae to feed has demonstrated probiotic and other benefits, including improved gut health, while also providing added beneficial nutrients in the by-products, that is, meat, eggs, and milk (see Barkia et al., 2019; Saadaoui et al., 2020, for review). As a result, this has created a new market of algae-enhanced commodity products, for example, astaxanthin eggs and DHA and EPA eggs and milk.

It is important to note that while algae do produce numerous antioxidant compounds, the biomass is (or was recently) living and perishable, which requires proper storage and protection against oxidation to prevent losses. In some cases, antioxidants such as TBHQ and tocopherol (vitamin E) are added as preservatives. Moreover, biomass and extracted products should be stored in opaque containers to reduce photooxidation, purged with inert gas (e.g. with nitrogen) to displace oxygen, tightly sealed and refrigerated/frozen whenever possible. Combined, these measures should help to extend the shelf life of algal biomass and products prior to formulation.

Finally, aside from the biosafety concerns mentioned for each processing method, it should be emphasized that not all algae are viable as nutritional sources and many species are known to produce potent toxins, including neurotoxins (e.g. anatoxins and saxitoxins from cyanobacteria). Therefore, it is recommended that each alga be thoroughly researched, selected, and carefully screened using molecular markers and metabolic profiling to determine the potential for toxin presence/production. Moreover, lab and scaled cultures should be closely monitored, for example, minimally by microscopy, to ensure that harmful algal species do not invade cultures creating additional biosafety challenges.

7 Where to look for further information

The field of algaculture is highly interdisciplinary, requiring knowledge in biology, chemistry, and engineering to assess culture characteristics, evaluate biomass composition, and develop compatible processing solutions. As this chapter demonstrates, different strains of algae possess unique physical and chemical features that may impact their suitability for certain processes, and some strains may require specialized equipment to improve cell disruption and product recovery. Thus, it is recommended that researchers compile primary literature across scientific disciplines to get a more complete assessment of emergent challenges for each alga.

Numerous journals focus on the biology of algae, their biochemistry, and the utility of algal biomass for food, feed, fuels, and specialty products, including, but not limited to, the *Journal of Applied Phycology*, *Algal Research*, *Phycologia*, *Energies*, *Biomolecules*, *Marine Drugs*, and *Biotechnology and Bioengineering*. Currently, there are dozens of reviews in the literature covering topics on algal processing, from harvesting to product extraction, many of which are mentioned in this chapter. Moreover, each review contains 100s of articles that can be mined for additional information.

There are several groups dedicated to the community of scientists working with algae in the areas of cultivation, analysis, and process development. Organizations of note include the Algae Biomass Organization (algaebiomasssummit.org), European Algae Biomass Association (eaba -association.org/en), and the congress on Algal Biomass, Biofuels, and Bioproducts (elsevier.com/events/conferences). These groups hold annual scientific meetings wherein they invite speakers to present significant advances and provide a venue for attendees to interact with other scientists in the field.

U.S. national laboratories publish information on biomass research as well as collaborations with academic and industrial partners:

- National Renewable Energy Laboratory (www.nrel.gov).
- Los Alamos National Laboratory (www.lanl.gov).
- Sandia National Laboratories (www.sandia.gov).
- Pacific Northwest National Laboratory (www.pnnl.gov).
- Iowa National Laboratory (www.inl.gov).

8 References

Alavijeh, R. S., Karimi, K., Wijffels, R. H., van den Berg, C. and Eppink, M. 2020. Combined bead milling and enzymatic hydrolysis for efficient fractionation of lipids, proteins, and carbohydrates of *Chlorella vulgaris* microalgae. *Bioresource Technology* 309, 123321.

Al Rey, C. V., Mayol, A. P., Ubando, A. T., Biona, J. B. M. M., Arboleda, N. B., David, M. Y., Tumlos, R. B., Lee, H., Lin, O. H., Espiritu, R. A. and, Culaba, A. B. and Kasai, H., 2016. Microwave drying characteristics of microalgae (*Chlorella vulgaris*) for biofuel production. *Clean Technol. eEnviron. pPolicy* 18(8), 2441-2451.

Aouir, A., Amiali, M., Kirilova-Gachovska, T., Benchabane, A. and Bitam, A. 2015. The effect of pulsed electric field (PEF) and ultrasound (US) technologies on the extraction of phycobiliproteins from *Arthrospira platensis*. In *2015 IEEE Canada International Humanitarian Technology Conference (IHTC2015)* (pp. 1-4). IEEE.

Babel, S. and Takizawa, S. 2010. Microfiltration membrane fouling and cake behavior during algal filtration. *Desalination* 261(1-2), 46-51.

Bak, U. G., Mols-Mortensen, A. and Gregersen, O. 2018. Production method and cost of commercial-scale offshore cultivation of kelp in the Faroe Islands using multiple partial harvesting. *Algal Research* 33, 36-47.

Barkia, I., Saari, N. and Manning, S. R. 2019. Microalgae for high-value products towards human health and nutrition. *Marine Drugs* 17(5), 304. https://doi.org/10.3390/md17050304.

Baroni, É, Cao, B., Webley, P. A., Scales, P. J. and Martin, G. J. O. 2020. Nitrogen availability and the nature of extracellular organic matter of microalgae. *Industrial and Engineering Chemistry Research* 59(15), 6795–6805.

Barros, A. I., Gonçalves, A. L., Simões, M. and Pires, J. C. M. 2015. Harvesting techniques applied to microalgae: a review. *Renewable and Sustainable Energy Reviews* 41, 1489–1500.

Becker, E. W. 2007. Microalgae as a source of protein. *Biotechnology Advances* 25(2), 207–210.

Benedetti, M., Vecchi, V., Barera, S. and Dall'Osto, L. 2018. Biomass from microalgae: the potential of domestication towards sustainable biofactories. *Microbial Cell Factories* 17(1), 173.

Benemann, J. R., Tillett, D. M. and Weissman, J. C. 1987. Microalgae biotechnology. *Trends in Biotechnology* 5(2), 47–53.

Bhave, R., Kuritz, T., Powell, L. and Adcock, D. 2012. Membrane-based energy efficient dewatering of microalgae in biofuels production and recovery of value added co-products. *Environmental Science and Technology* 46(10), 5599–5606.

Biller, P., Friedman, C. and Ross, A. B. 2013. Hydrothermal microwave processing of microalgae as a pre-treatment and extraction technique for bio-fuels and bio-products. *Bioresource Technology* 136, 188–195.

Borowitzka, M. A. 1998. Algae as food. In: *Microbiology of Fermented Foods* Wood, B. J. B. (Ed.) (pp. 585–602). Springer, Boston, MA.

Brady, P. V., Pohl, P. I. and Hewson, J. C. 2014. A coordination chemistry model of algal autoflocculation. *Algal Research* 5, 226–230.

Cha, K. H., Koo, S. Y., Song, D. G. and Pan, C. H. 2012. Effect of microfluidization on bioaccessibility of carotenoids from *Chlorella ellipsoidea* during simulated digestion. *Journal of Agricultural and Food Chemistry* 60(37), 9437–9442.

Chen, Y. S. and Hsiau, S. S. 2009. Cake formation and growth in cake filtration. *Powder Technology* 192(2), 217–224.

Cheng, J., Sun, J., Huang, Y., Feng, J., Zhou, J. and Cen, K. 2013. Dynamic microstructures and fractal characterization of cell wall disruption for microwave irradiation-assisted lipid extraction from wet microalgae. *Bioresource Technology* 150, 67–72.

Chinu, K., Vigneswaran, S., Erdei, L., Shon, H. K., Kandasamy, J. and Ngo, H. H. 2010. Comparison of fouling indices in assessing pre-treatment for seawater reverse osmosis. *Desalination and Water Treatment* 18(1–3), 187–191.

Chiou, Y. T., Hsieh, M. L. and Yeh, H. H. 2010. Effect of algal extracellular polymer substances on UF membrane fouling. *Desalination* 250(2), 648–652.

Chisti, Y. and Moo-Young, M. 1986. Disruption of microbial cells for intracellular products. *Enzyme and Microbial Technology* 8(4), 194–204.

Coons, J. E., Kalb, D. M., Dale, T. and Marrone, B. L. 2014. Getting to low-cost algal biofuels: a monograph on conventional and cutting-edge harvesting and extraction technologies. *Algal Research* 6, 250–270.

Corbett, L. L. and Parker, D. L. 1976. Viability of lyophilized cyanobacteria (blue-green algae). *Applied and Environmental Microbiology* 32(6), 777–780.

Cordero, B. and Voltolina, D. 1997. Viability of mass algal cultures preserved by freezing and freeze-drying. *Aquacultural Engineering* 16(4), 205–211.

Costa, R., Pereira, J. L., Gomes, J., Gonçalves, F., Hunkeler, D. and Rasteiro, M. G. 2014. The effects of acrylamide polyelectrolytes on aquatic organisms: relating toxicity to chain architecture. *Chemosphere* 112, 177–184.

da Silva, T. L. and Reis, A. 2015. Scale-up problems for the large-scale production of algae. In: *Algal Biorefinery: An Integrated Approach* Das, D. (Ed.) (pp. 125–149). Springer, Cham.

Dassey, A. J. and Theegala, C. S. 2013. Harvesting economics and strategies using centrifugation for cost effective separation of microalgae cells for biodiesel applications. *Bioresource Technology* 128, 241–245.

de Souza Leite, L., Hoffmann, M. T. and Daniel, L. A. 2019. Coagulation and dissolved air flotation as a harvesting method for microalgae cultivated in wastewater. *Journal of Water Process Engineering* 32, 100947.

Divakaran, R. and Pillai, V. S. 2002. Flocculation of algae using chitosan. *Journal of Applied Phycology* 14(5), 419–422.

Domozych, D. 2019. Algal cell walls. In *eLS*, John Wiley & Sons, Ltd. https://doi.org/10.1002/9780470015902.a0000315.pub4.

Duan, J. and Gregory, J. 2003. Coagulation by hydrolysing metal salts. *Advances in Colloid and Interface Science* 100–102, 475–502.

El Maghraby, D. M. and Fakhry, E. M. 2015. Lipid content and fatty acid composition of Mediterranean macroalgae as dynamic factors for biodiesel production. *Oceanologia* 57(1), 86–92.

Evodos 2018. Evodos Centrifuge–Company Presentation. Retrieved September 05, 2020. https://www.evodos.eu/2018/11/08/company-presentation/.

Fouad, H. A., Elhefny, R. M. and Marei, A. I. 2016. Evaluating the use of tube settlers and lamella plates in Increasing the efficiency of sedimentation tanks. *Journal of Applied Life Sciences International* 7(4), 1–8.

Gao, S., Yang, J., Tian, J., Ma, F., Tu, G. and Du, M. 2010. Electro-coagulation–flotation process for algae removal. *Journal of Hazardous Materials* 177(1–3), 336–343.

García-Vaquero, M., López-Alonso, M. and Hayes, M. 2017. Assessment of the functional properties of protein extracted from the brown seaweed *Himanthalia elongata* (Linnaeus) SF Gray. *Food Research International* 99(3), 971–978.

Garcia-Vaquero, M., Ummat, V., Tiwari, B. and Rajauria, G. 2020. Exploring ultrasound, microwave and ultrasound-microwave assisted extraction technologies to increase the extraction of bioactive compounds and antioxidants from brown macroalgae. *Marine Drugs* 18(3), 172.

Garzon-Sanabria, A. J., Davis, R. T. and Nikolov, Z. L. 2012. Harvesting *Nannochloris oculata* by inorganic electrolyte flocculation: effect of initial cell density, ionic strength, coagulant dosage, and media pH. *Bioresource Technology* 118, 418–424.

Gerardo, M. L., Hende, S. V., Vervaeren, H., Coward, T. and Skill, S. C. 2015. Harvesting of microalgae within a biorefinery approach: a review of the developments and case studies from pilot-plants. *Algal Research* 11, 248–262.

Gille, A., Trautmann, A., Posten, C. and Briviba, K. 2016. Bioaccessibility of carotenoids from *Chlorella vulgaris* and *Chlamydomonas reinhardtii*. *International Journal of Food Sciences and Nutrition* 67(5), 507–513.

Gol, R. D. 2020. Examination of microalgal biofouling in a constant-flux crossflow filtration system comparing polymeric and carbon-nanotube membranes. Master's Thesis. University of Texas at Austin.

Granados, M. R., Acién, F. G., Gómez, C., Fernández-Sevilla, J. M. and Grima, E. M. 2012. Evaluation of flocculants for the recovery of freshwater microalgae. *Bioresource Technology* 118, 102–110.

Grand View Research 2019. Algae Protein Market Size, Share & Trend Analysis Report by Product (*Spirulina*, *Chlorella*), By Source (Freshwater, Marine), By Application (Dietary Supplements, Food Products), By Region, And Segment Forecasts, 2014-2025. Report GVR-3-68038-893-0, 120 pp.

Grima, E. M., Belarbi, E. H., Fernández, F. A., Medina, A. R., and Chisti, Y. 2003. Recovery of microalgal biomass and metabolites: process options and economics. *Biotechnology Advances* 20(7--8), 491–515.

Günerken, E., D'Hondt, E., Eppink, M. H. M., Garcia-Gonzalez, L., Elst, K. and Wijffels, R. H. 2015. Cell disruption for microalgae biorefineries. *Biotechnology Advances* 33(2), 243–260.

Hadjoudja, S., Deluchat, V. and Baudu, M. 2010. Cell surface characterisation of *Microcystis aeruginosa* and *Chlorella vulgaris*. *Journal of Colloid and Interface Science* 342(2), 293–299.

Halim, R., Harun, R., Danquah, M. K. and Webley, P. A. 2012. Microalgal cell disruption for biofuel development. *Applied Energy* 91(1), 116–121.

Heasman, M., Diemar, J., O'Connor, W., Sushames, T. and Foulkes, L. 2000. Development of extended shelf-life microalgae concentrate diets harvested by centrifugation for bivalve molluscs–a summary. *Aquaculture Research* 31(8-9), 637–659.

Holdt, S. L. and Kraan, S. 2011. Bioactive compounds in seaweed: functional food applications and legislation. *Journal of Applied Phycology* 23(3), 543–597.

Holm-Hansen, O. 1964. Viability of lyophilized algae. *Canadian Journal of Botany* 42(2), 127–137.

Horiuchi, J. I., Ohba, I., Tada, K., Kobayashi, M., Kanno, T. and Kishimoto, M. 2003. Effective cell harvesting of the halotolerant microalga *Dunaliella tertiolecta* with pH control. *Journal of Bioscience and Bioengineering* 95(4), 412–415.

Hosseini, S. R. P., Tavakoli, O. and Sarrafzadeh, M. H. 2017. Experimental optimization of SC-CO2 extraction of carotenoids from *Dunaliella salina*. *The Journal of Supercritical Fluids* 121, 89–95.

Hosseinizand, H., Sokhansanj, S. and Lim, C. J. 2018. Studying the drying mechanism of microalgae *Chlorella vulgaris* and the optimum drying temperature to preserve quality characteristics. *Drying Technology* 36(9), 1049–1060.

Huang, R., Liu, Z., Yan, B., Li, Y., Li, H., Liu, D., Wang, P., Cui, F. and Shi, W. 2020. Layer-by-layer assembly of high negatively charged polycarbonate membranes with robust antifouling property for microalgae harvesting. *Journal of Membrane Science* 595, 117488.

Iqhrammullah, M., Saleha, S., Maulina, F. P. and Idroes, R. 2020. Polyurethane film prepared from ball-milled algal polyol particle and activated carbon filler for NH3-N removal. *Heliyon* 6(8), e04590.

Joannes, C., Sipaut, C. S., Dayou, J., Yasir, S. M. and Mansa, R. F. 2015. The potential of using pulsed electric field (PEF) technology as the cell disruption method to extract lipid from microalgae for biodiesel production. *International Journal of Renewable Energy Research (IJRER)* 5(2), 598–621.

Jones, J., Manning, S., Montoya, M., Keller, K. and Poenie, M. 2012. Extraction of algal lipids and their analysis by HPLC and mass spectrometry. *Journal of the American Oil Chemists' Society* 89(8), 1371–1381.

Kadam, S. U., Tiwari, B. K. and O'Donnell, C. P. 2013. Application of novel extraction technologies for bioactives from marine algae. *Journal of Agricultural and Food Chemistry* 61(20), 4667-4675.

Kapoore, R. V., Butler, T. O., Pandhal, J. and Vaidyanathan, S. 2018. Microwave-assisted extraction for microalgae: from biofuels to biorefinery. *Biology* 7(1), 18.

Kazir, M., Abuhassira, Y., Robin, A., Nahor, O., Luo, J., Israel, A., Golberg, A. and Livney, Y. D. 2019. Extraction of proteins from two marine macroalgae, *Ulva* sp. and *Gracilaria* sp., for food application, and evaluating digestibility, amino acid composition and antioxidant properties of the protein concentrates. *Food Hydrocolloids* 87, 194-203.

Kempkes, M. A. 2016. Pulsed electric fields for algal extraction and predator control. In: *Handbook of Electroporation* Miklavcic, D. (Ed.) (pp. 1-16). Springer International Publishing, NYC.

Khan, M. I., Shin, J. H. and Kim, J. D. 2018. The promising future of microalgae: current status, challenges, and optimization of a sustainable and renewable industry for biofuels, feed, and other products. *Microbial Cell Factories* 17(1), 36.

Kim, J., Stekoll, M. and Yarish, C. 2019. Opportunities, challenges and future directions of open-water seaweed aquaculture in the United States. *Phycologia* 58(5), 446 -461.

Kim, J. K., Yarish, C., Hwang, E. K., Park, M. and Kim, Y. 2017. Seaweed aquaculture: cultivation technologies, challenges and its ecosystem services. *Algae* 32(1), 1-13.

Kim, M. S. and Kwak, D. H. 2017. Effect of zeta potential on collision-attachment coefficient and removal efficiency for dissolved carbon dioxide flotation. *Environmental Engineering Science* 34(4), 272-280.

Kirschner, A., Cheng, Y.-H., Paul, D., Field, R. and Freeman, B. (2018). Fouling mechanisms in constant flux crossflow ultrafiltration. *Journal of Membrane Science*. 574. doi: 10.1016/j.memsci.2018.12.001.

Kuberkar, V. T. and Davis, R. H. 2000. Modeling of fouling reduction by secondary membranes. *Journal of Membrane Science* 168(1-2), 243-258.

Kumari, P., Kumar, M., Reddy, C. R. K. and Jha, B. 2013. Algal lipids, fatty acids and sterols. In: *Functional Ingredients from Algae for Foods and Nutraceuticals* Dominguez, H. (Ed.) (pp. 87-134). Woodhead Publishing. Cambridge, England.

Kurokawa, M., King, P. M., Wu, X., Joyce, E. M., Mason, T. J. and Yamamoto, K. 2016. Effect of sonication frequency on the disruption of algae. *Ultrasonics Sonochemistry* 31, 157-162.

Kwon, H., Lu, M., Lee, E. Y. and Lee, J. 2014. Harvesting of microalgae using flocculation combined with dissolved air flotation. *Biotechnology and Bioprocess Engineering* 19(1), 143-149.

Lee, J. Y., Yoo, C., Jun, S. Y., Ahn, C. Y. and Oh, H. M. 2010. Comparison of several methods for effective lipid extraction from microalgae. *Bioresource Technology* 101(Suppl. 1), S75-S77.

Li-Beisson, Y., Thelen, J. J., Fedosejevs, E. and Harwood, J. L. 2019. The lipid biochemistry of eukaryotic algae. *Progress in Lipid Research* 74, 31-68.

Lin, L. P. 1985. Microstructure of spray-dried and freeze-dried microalgal powders. *Food Structure* 4(2), 17.

López-Hortas, L., Gannon, L., Moreira, R., Chenlo, F., Domínguez, H. and Torres, M. D. 2018. Microwave hydrodiffusion and gravity (MHG) processing of *Laminaria ochroleuca* brown seaweed. *Journal of Cleaner Production* 197, 1108-1116.

Lorbeer, A. J., Lahnstein, J., Fincher, G. B., Su, P. and Zhang, W. 2015. Kinetics of conventional and microwave-assisted fucoidan extractions from the brown alga, *Ecklonia radiata*. *Journal of Applied Phycology* 27(5), 2079–2087.

Maeda, K., Kuramochi, H., Shinkawa, T. and Fukui, K. 2002. Solubility of two salts containing sulfate and chloride ions in water for ternary systems at 313 K. *Journal of Chemical and Engineering Data* 47(6), 1472–1475.

Maneein, S., Milledge, J. J., Nielsen, B. V. and Harvey, P. J. 2018. A review of seaweed pre-treatment methods for enhanced biofuel production by anaerobic digestion or fermentation. *Fermentation* 4(4), 100.

Marrone, B. L., Lacey, R. E., Anderson, D. B., Bonner, J., Coons, J., Dale, T., Downes, C. M., Fernando, S., Fuller, C., Goodall, B., Holladay, J. E., Kadam, K., Kalb, D., Liu, W., Mott, J. B., Nikolov, Z., Ogden, K. L., Sayre, R. T., Trewyn, B. G. and Olivares, J. A. 2018. Review of the harvesting and extraction program within the national alliance for advanced biofuels and bioproducts. *Algal Research* 33, 470 –485.

Mata, T. M., Martins, A. A. and Caetano, N. S. 2010. Microalgae for biodiesel production and other applications: a review. *Renewable and Sustainable Energy Reviews* 14(1), 217–232.

McMillan, J. R., Watson, I. A., Ali, M. and Jaafar, W. 2013. Evaluation and comparison of algal cell disruption methods: microwave, waterbath, blender, ultrasonic and laser treatment. *Applied Energy* 103, 128–134.

Meireles, M. A. A. 2013. Supercritical CO2 extraction of bioactive components from algae. In: *Functional Ingredients from Algae for Foods and Nutraceuticals* Dominguez, H. (Ed.) (pp. 561-584). Woodhead Publishing, Cambridge, England.

Milledge, J. J. and Heaven, S. 2011. Disc stack centrifugation separation and cell disruption of microalgae: a technical note. *Environment and Natural Resources Research* 1(1), 17–24.

Mimouni, V., Couzinet-Mossion, A., Ulmann, L. and Wielgosz-Collin, G. 2018. Lipids from microalgae. In: *Microalgae in Health and Disease Prevention* Levine, I. and Fleurence, J. (Eds) (pp. 109–131). Elsevier, Amsterdam.

Mohn, F. H. and Soeder, C. J. 1978. Improved technologies for harvesting and processing of microalgae and their impact on production costs. *Arch. Hydrobiol. Bech. Ergebn. Lemnol* 11, 228–253.

Montingelli, M. E., Benyounis, K. Y., Stokes, J. and Olabi, A. G. 2016. Pretreatment of macroalgal biomass for biogas production. *Energy Conversion and Management* 108, 202–209.

Ndikubwimana, T., Chang, J., Xiao, Z., Shao, W., Zeng, X., Ng, I. S. and Lu, Y. 2016. Flotation: a promising microalgae harvesting and dewatering technology for biofuels production. *Biotechnology Journal* 11(3), 315–326.

Nobles, D. R. and Manning, S. R. 2019. Extraction and characterization of lipids from macroalgae. In: *Microbial Lipid Production* Balan, V. (Ed.) (pp. 131-140). Humana Press, New York, NY.

Nogueira, D. A., da Silveira, J. M., Vidal, É. M., Ribeiro, N. T. and Veiga Burkert, C. A. 2018. Cell disruption of *Chaetoceros calcitrans* by microwave and ultrasound in lipid extraction. *International Journal of Chemical Engineering* 2018, 1-6.

Novoveská, L., Ross, M. E., Stanley, M. S., Pradelles, R., Wasiolek, V. and Sassi, J. F. 2019. Microalgal carotenoids: a review of production, current markets, regulations, and future direction. *Marine Drugs* 17(11), 640.

O'Connell, D., Savelski, M. and Slater, C. S. 2013. Life cycle assessment of dewatering routes for algae derived biodiesel processes. *Clean Technologies and Environmental Policy* 15(4), 567–577.

Oh, H. M., Lee, S. J., Park, M. H., Kim, H. S., Kim, H. C., Yoon, J. H., Kwon, G. and Yoon, B. D. 2001. Harvesting of Chlorella vulgaris using a bioflocculant from Paenibacillus sp. AM49. *Biotechnology Letters* 23(15), 1229–1234.

OriginClear 2015. The Smart Algae Harvester. https://www.originclear.com/tv-radio-news/overview-the-originoil-algae-harvester-2.

Orset, S., Leach, G. C., Morais, R. and Young, A. J. 1999. Spray-drying of the microalga *Dunaliella salina*: effects on β-carotene content and isomer composition. *Journal of Agricultural and Food Chemistry* 47(11), 4782–4790.

Papazi, A., Makridis, P. and Divanach, P. 2010. Harvesting *Chlorella minutissima* using cell coagulants. *Journal of Applied Phycology* 22(3), 349–355.

Pasquet, V., Chérouvrier, J. R., Farhat, F., Thiéry, V., Piot, J. M., Bérard, J. B., Kaas, R., Serive, B., Patrice, T., Cadoret, J. and Picot, L. 2011. Study on the microalgal pigments extraction process: performance of microwave assisted extraction. *Process Biochemistry* 46(1), 59–67.

Pereira, R. and Yarish, C. 2008. Mass production of marine macroalgae. In: *Encyclopedia of Ecology* Jørgensen, S. E. and Fath, Brian D. (Eds) (pp. 2236–2247). Academic Press, Cambridge, MA.

Pereira, R., Yarish, C. and Critchley, A. T. 2013. Seaweed aquaculture for human foods in land-based and IMTA systems. In: *Sustainable Food Production* Christou, P., Savin, R., Costa-Pierce, B. A., Misztal, I. and Whitelaw C. B. A. (Eds) (pp. 9109–9128). Springer, New York.

Phasey, J., Vandamme, D. and Fallowfield, H. J. 2017. Harvesting of algae in municipal wastewater treatment by calcium phosphate precipitation mediated by photosynthesis, sodium hydroxide and lime. *Algal Research* 27, 115–120.

Phong, W. N., Show, P. L., Ling, T. C., Juan, J. C., Ng, E. P. and Chang, J. S. 2018. Mild cell disruption methods for bio-functional proteins recovery from microalgae–recent developments and future perspectives. *Algal Research* 31, 506–516.

Plantier, S., Castaing, J. -B., Sabiri, N. -E., Massé, A., Jaouen, P. and Pontié, M. 2013. Performance of a sand filter in removal of algal bloom for SWRO pre-treatment. *Desalination and Water Treatment* 51(7–9), 1838–1846.

Poblete, R., Cortes, E., Macchiavello, J. and Bakit, J. 2018. Factors influencing solar drying performance of the red algae *Gracilaria chilensis*. *Renewable Energy* 126, 978–986.

Pomin, V. H. and Mourão, P. A. 2008. Structure, biology, evolution, and medical importance of sulfated fucans and galactans. *Glycobiology* 18(12), 1016–1027.

Postma, P. R., Pataro, G., Capitoli, M., Barbosa, M. J., Wijffels, R. H., Eppink, M. H. M., Olivieri, G. and Ferrari, G. 2016. Selective extraction of intracellular components from the microalga *Chlorella vulgaris* by combined pulsed electric field-temperature treatment. *Bioresource Technology* 203, 80–88.

Postma, P. R., Suarez-Garcia, E., Safi, C., Yonathan, K., Olivieri, G., Barbosa, M. J., Wijffels, R. H. and Eppink, M. H. M. 2017. Energy efficient bead milling of microalgae: effect of bead size on disintegration and release of proteins and carbohydrates. *Bioresource Technology* 224, 670–679.

Prakash, J., Pushparaj, B., Carlozzi, P., Torzillo, G., Montaini, E. and Materassi, R. 1997. Microalgal biomass drying by a simple solar device*. *International Journal of Solar Energy* 18(4), 303–311.

Qin, Y. 2018. Seaweed hydrocolloids as thickening, gelling, and emulsifying agents in functional food products. In: *Bioactive Seaweeds for Food Applications* Y. Qin (Ed.) (pp. 135–152). Academic Press, Cambridge, MA.

Rakesh, S., Dhar, D. W., Prasanna, R., Saxena, A. K., Saha, S., Shukla, M. and Sharma, K. 2015. Cell disruption methods for improving lipid extraction efficiency in unicellular microalgae. *Engineering in Life Sciences* 15(4), 443–447.

Raposo, M. F., De Morais, R. M. and Bernardo de Morais, A. M. 2013. Bioactivity and applications of sulphated polysaccharides from marine microalgae. *Marine Drugs* 11(1), 233–252. https://doi.org/10.3390/md11010233.

Rashid, N., Rehman, S. U. and Han, J. I. 2013. Rapid harvesting of freshwater microalgae using chitosan. *Process Biochemistry* 48(7), 1107–1110.

Rawat, I., Kumar, R. R., Mutanda, T. and Bux, F. 2011. Dual role of microalgae: phycoremediation of domestic wastewater and biomass production for sustainable biofuels production. *Applied Energy* 88(10), 3411–3424.

Renault, F., Sancey, B., Badot, P. -M. and Crini, G. 2009. Chitosan for coagulation/ flocculation processes-an eco-friendly approach. *European Polymer Journal* 45(5), 1337–1348.

Research and Markets 2020a. Carotenoids Market by Type (Astaxanthin, Beta-Carotene, Lutein, Lycopene, Canthaxanthin, and Zeaxanthin), Application (Feed, Food & Beverages, Dietary Supplements, Cosmetics, and Pharmaceuticals), Source, Formulation, and Region–Global Forecast to 2026. Report FB4411, 186 pp.

Research and Markets 2020b. Seaweed Cultivation Market by Type (Red, Brown, Green), Method of Harvesting (Aquaculture, Wild Harvesting), Form (Liquid, Powder, Flakes, Sheets), Application (Food, Feed, Agriculture, Pharmaceuticals), and Region–Global Forecast to 2025. Report FB6702, 210 pp.

Rokicka, M., Zieliński, M., Dudek, M. and Dębowski, M. 2020. Effects of ultrasonic and microwave pretreatment on lipid extraction of microalgae and methane production from the residual extracted biomass. *Bioenergy Research*, 13 1–9.

Roselet, F., Vandamme, D., Roselet, M., Muylaert, K. and Abreu, P. C. 2015. Screening of commercial natural and synthetic cationic polymers for flocculation of freshwater and marine microalgae and effects of molecular weight and charge density. *Algal Research* 10, 183–188.

Ryckebosch, E., Muylaert, K., Eeckhout, M., Ruyssen, T. and Foubert, I. 2011. Influence of drying and storage on lipid and carotenoid stability of the microalga *Phaeodactylum tricornutum*. *Journal of Agricultural and Food Chemistry* 59(20), 11063–11069.

Saadaoui, I., Cherif, M., Rasheed, R., Bounnit, T., Al Jabri, H., Sayadi, S., Hamadou, R. B. and Manning, S. R. 2020. *Mychonastes homosphaera* (Chlorophyceae): a promising feedstock for high quality feed production in the arid environment. *Algal Research* 51, 102021.

Safi, C., Charton, M., Pignolet, O., Silvestre, F., Vaca-Garcia, C. and Pontalier, P. Y. 2013. Influence of microalgae cell wall characteristics on protein extractability and determination of nitrogen-to-protein conversion factors. *Journal of Applied Phycology* 25(2), 523–529.

Safi, C., Charton, M., Ursu, A. V., Laroche, C., Zebib, B., Pontalier, P. Y. and Vaca-Garcia, C. 2014. Release of hydro-soluble microalgal proteins using mechanical and chemical treatments. *Algal Research* 3, 55–60.

Sati, H., Mitra, M., Mishra, S. and Baredar, P. 2019. Microalgal lipid extraction strategies for biodiesel production: a review. *Algal Research* 38, 101413.

Samarasinghe, N., Fernando, S., Lacey, R. and Faulkner, W. B. 2012. Algal cell rupture using high pressure homogenization as a prelude to oil extraction. *Renewable Energy* 48, 300–308.

Schlesinger, A., Eisenstadt, D., Bar-Gil, A., Carmely, H., Einbinder, S. and Gressel, J. 2012. Inexpensive non-toxic flocculation of microalgae contradicts theories; overcoming a major hurdle to bulk algal production. *Biotechnology Advances* 30(5), 1023–1030.

Sfriso, A. A., Gallo, M. and Baldi, F. 2018. Phycoerythrin productivity and diversity from five red macroalgae. *Journal of Applied Phycology* 30(4), 2523–2531.

Shi, Y., Sheng, J., Yang, F. and Hu, Q. 2007. Purification and identification of polysaccharide derived from *Chlorella pyrenoidosa*. *Food Chemistry* 103(1), 101–105.

Show, K. Y., Lee, D. J., Tay, J. H., Lee, T. M. and Chang, J. S. 2015. Microalgal drying and cell disruption-recent advances. *Bioresource Technology* 184, 258–266.

Silva, N. C., Machado, M. V. C., Brandão, R. J., Duarte, C. R. and Barrozo, M. A. S. 2019. Dehydration of microalgae *Spirulina platensis* in a rotary drum with inert bed. *Powder Technology* 351, 178–185.

Singh, A., Nigam, P. S. and Murphy, J. D. 2011. Mechanism and challenges in commercialisation of algal biofuels. *Bioresource Technology* 102(1), 26–34.

Şirin, S., Trobajo, R., Ibanez, C. and Salvadó, J. 2012. Harvesting the microalgae *Phaeodactylum tricornutum* with polyaluminum chloride, aluminium sulphate, chitosan and alkalinity-induced flocculation. *Journal of Applied Phycology* 24(5), 1067–1080.

Soeder, C. J. 1980. Massive cultivation of microalgae: results and prospects. *Hydrobiologia* 72(1–2), 197–209.

Soeder, C. J. and Pabst, W. 1975. Production, properties, preclinical and clinical testing of *Scenedesmus*. In: *The PAG Compendium* Sachs, M. Y. (Ed.) (pp. 276–23a). World Mark Press Ltd, New York.

Solana, M., Rizza, C. S. and Bertucco, A. 2014. Exploiting microalgae as a source of essential fatty acids by supercritical fluid extraction of lipids: comparison between *Scenedesmus obliquus*, *Chlorella protothecoides* and *Nannochloropsis salina*. *The Journal of Supercritical Fluids* 92, 311–318.

Soto-Sierra, L., Stoykova, P. and Nikolov, Z. L. 2018. Extraction and fractionation of microalgae-based protein products. *Algal Research* 36, 175–192.

Spiden, E. M., Yap, B. H., Hill, D. R., Kentish, S. E., Scales, P. J. and Martin, G. J. 2013. Quantitative evaluation of the ease of rupture of industrially promising microalgae by high pressure homogenization. *Bioresource Technology* 140, 165–171.

Stephens, E., Wagner, L., Ross, I. L. and Hankamer, B. 2012. Microalgal production systems: global impact of industry scale-up. In: *Microalgal Biotechnology: Integration and Economy*, Posten, C. and Walter, C. (Eds) (pp. 267–306). De Gruyter, Germany.

Sun, X. M., Ren, L. J., Zhao, Q. Y., Ji, X. J. and Huang, H. 2018. Microalgae for the production of lipid and carotenoids: a review with focus on stress regulation and adaptation. *Biotechnology for Biofuels* 11(1), 272.

Svarovsky, L. 2000. *Solid-Liquid Separation*. Elsevier, Amsterdam, The Netherlands.

Taki, K., Seki, T., Mononobe, S. and Kato, K. 2008. Zeta potential measurement on the surface of blue-green algae particles for micro-bubble process. *Water Science and Technology* 57(1), 19–25.

Taucher, J., Baer, S., Schwerna, P., Hofmann, D., Hümmer, M., Buchholz, R. and Becker, A. 2016. Cell disruption and pressurized liquid extraction of carotenoids from microalgae. *Journal of Thermodynamics and Catalysis* 7, 158–165.

Teixeira, M. R. and Rosa, M. J. 2006. Comparing dissolved air flotation and conventional sedimentation to remove cyanobacterial cells of Microcystis aeruginosa: Part I: the key operating conditions. *Separation and Purification Technology* 52(1), 84–94.

Tibbetts, S. M., Mann, J. and Dumas, A. 2017. Apparent digestibility of nutrients, energy, essential amino acids and fatty acids of juvenile Atlantic salmon (*Salmo salar* L.) diets containing whole-cell or cell-ruptured *Chlorella vulgaris* meals at five dietary inclusion levels. *Aquaculture* 481, 25–39.

Tiwari, A. and Kiran, T. 2018. Biofuels from microalgae. *Advanced Biofuels Bioenergy* 73012, 10.5772.

Toepfl, S. and Heinz, V. 2008, June. Pulsed electric fields (PEF) applications in food processing-process and equipment design and cost analysis. In *2008 IEEE 35th International Conference on Plasma Science* (pp. 1-1). IEEE.

Torres, M. D., Flórez-Fernández, N. and Domínguez, H. 2019. Integral utilization of red seaweed for bioactive production. *Marine Drugs* 17(6), 314.

Trică, B., Pârvulescu, O. C., Răducanu, C. E. and Dobre, T. 2020. Thin Layer Convective Drying Of *Cystoseira barbata* Brown Macroalga. *U.P.B. Science Bulletin*, Series B, 82(2).

Tredici, M. R., Rodolfi, L., Biondi, N., Bassi, N. and Sampietro, G. 2016. Techno-economic analysis of microalgal biomass production in a 1-ha Green Wall Panel (GWP®) plant. *Algal Research* 19, 253–263.

Uduman, N., Qi, Y., Danquah, M. K., Forde, G. M. and Hoadley, A. 2010. Dewatering of microalgal cultures: a major bottleneck to algae-based fuels. *Journal of Renewable and Sustainable Energy* 2(1), 012701.

Ummalyma, S. B., Mathew, A. K., Pandey, A. and Sukumaran, R. K. 2016. Harvesting of microalgal biomass: efficient method for flocculation through pH modulation. *Bioresource Technology* 213, 216–221.

Uribe, E., Pardo-Orellana, C. M., Vega-Gálvez, A., Ah-Hen, K. S., Pastén, A., García, V. and Aubourg, S. P. 2020. Effect of drying methods on bioactive compounds, nutritional, antioxidant, and antidiabetic potential of brown alga *Durvillaea antarctica*. *Drying Technology* 38(14), 1915–1928.

Vandamme, D., Foubert, I. and Muylaert, K. (2013). Flocculation as a low-cost method for harvesting microalgae for bulk biomass production. *Trends in Biotechnology,* Vol. 31. p. 233-239. doi: 10.1016/j.tibtech.2012.12.005

Vandamme, D. 2017. Harvesting, thickening and dewatering processes. In: *Microalgae as a Source of Bioenergy: Products, Processes and Economics* Pires, J. C., (Ed.) (pp. 202-223). Bentham Science Publishers, Sharjah, UAE .

Vandamme, D., Beuckels, A., Mathimiu, G., Foubert, I. and Muylaert, K. 2015. Reversible flocculation of microalgae using magnesium hydroxide. *Bioenergy Research* 8(2), 716-725.

Wang, H., Zhang, W., Chen, L., Wang, J. and Liu, T. 2013. The contamination and control of biological pollutants in mass cultivation of microalgae. *Bioresource Technology* 128, 745-750.

Wang, J., Zhang, M. and Fang, Z. 2019. Recent development in efficient processing technology for edible algae: a review. *Trends in Food Science and Technology* 88, 251-259.

Wang, M., Yuan, W., Jiang, X., Jing, Y. and Wang, Z. 2014a. Disruption of microalgal cells using high-frequency focused ultrasound. *Bioresource Technology* 153, 315–321.

Wang, L., Wang, X., Wu, H. and Liu, R. 2014b. Overview on biological activities and molecular characteristics of sulfated polysaccharides from marine green algae in recent years. *Marine Drugs* 12(9), 4984–5020.

White, W. L. and Wilson, P. 2015. World seaweed utilization. In: *Seaweed Sustainability* Tiwari, B. K and Troy, D. J. (Eds) (pp. 7–25). Academic Press, Amsterdam, The Netherlands.

Wilson, M. 2012. Cross flow filtration for mixed-culture algae harvesting for municipal wastewater lagoons. All Graduate Theses and Dissertations, Utah State University (p. 1162). https://digitalcommons.usu.edu/etd/1162.

Yang, F., Xiang, W., Fan, J., Wu, H., Li, T. and Long, L. 2016. High pH-induced flocculation of marine *Chlorella* sp. for biofuel production. *Journal of Applied Phycology* 28(2), 747–756.

Yu, B. 2012. CFD modeling of two-stage parallel plate sedimentation centrifuge for microalgae dewatering. Master's Thesis. Delft University of Technology.

Yuan, J., Kendall, A. and Zhang, Y. 2015. Mass balance and life cycle assessment of biodiesel from microalgae incorporated with nutrient recycling options and technology uncertainties. *GCB Bioenergy* 7(6), 1245–1259.

Zbinden, M. D. A., Sturm, B. S., Nord, R. D., Carey, W. J., Moore, D., Shinogle, H. and Stagg-Williams, S. M. 2013. Pulsed electric field (PEF) as an intensification pretreatment for greener solvent lipid extraction from microalgae. *Biotechnology and Bioengineering* 110(6), 1605–1615.

Zhang, W., Zhang, W., Zhang, X., Amendola, P., Hu, Q. and Chen, Y. 2013a. Characterization of dissolved organic matters responsible for ultrafiltration membrane fouling in algal harvesting. *Algal Research* 2(3), 223–229.

Zhang, X., Fan, L. and Roddick, F. A. 2013b. Understanding the fouling of a ceramic microfiltration membrane caused by algal organic matter released from *Microcystis aeruginosa*. *Journal of Membrane Science* 447, 362–368.

Zhao, Z., Ilyas, A., Muylaert, K. and Vankelecom, I. F. J. 2020. Optimization of patterned polysulfone membranes for microalgae harvesting. *Bioresource Technology* 309, 123367.

Zhuang, L. L., Hu, H. Y., Wu, Y. H., Wang, T. and Zhang, T. Y. 2014. A novel suspended-solid phase photobioreactor to improve biomass production and separation of microalgae. *Bioresource Technology* 153, 399–402.

Zou, S., Gu, Y., Xiao, D. and Tang, C. Y. 2011. The role of physical and chemical parameters on forward osmosis membrane fouling during algae separation. *Journal of Membrane Science* 366(1–2), 356–362.

Chapter 6

Bioprocessing of microalgal proteins and their applications in the cosmetic, nutraceutical and food industries

Jordan Wilson, Ainnatul A. Ahmad Termizi, Elvis T. Chua and Peer M. Schenk, The University of Queensland, Australia[1]

1 Introduction

The global demand for food protein is predicted to double by 2050. This raises concerns about shortages, increased greenhouse gas emissions, and other related environmental impacts, including land requirements and biodiversity, and increasing production to meet this demand, including the production of animal-based protein (Henchion et al., 2017). It is therefore of the utmost importance to develop and optimize non-animal protein-producing solutions that mitigate or remove these problems. Currently, one potentially environmental-friendly and efficient protein source is quickly emerging in the form of biomass produced from the growth and processing of microalgae.

Microalgal biomass is high in nutrients, including nitrogen and phosphorous. It is also a key source of protein; many microalgal species meet the amino acid requirements for human nutrition set by the Food and Agriculture Organization and the World Health Organization (1062007). Microalgae can produce approximately 70 times more protein per hectare than soybean and can be grown on non-arable land using fresh, brackish, or seawater (depending on the species).

1 Jordan Wilson and Ainnatul A. Ahmad Termizi contributed equally.

http://dx.doi.org/10.19103/AS.2021.0091.10

The protein content of microalgae varies across species. *Nannochloropsis* spp. (microalgae with a high production potential), for example, have a protein content, ranging from 30% to 43%. The protein, carbohydrate, and lipid content of several microalgal species are shown in Table 1.

This potential of the algal protein market can be seen in its recent growth (https://www.researchandmarkets.com/reports/4451980/global-algae-protein-market-industry-trends). Scientific understanding of and technological advances in the bioprocessing of microalgae continue to evolve, providing potential for the future. An important element is the development of refineries able to produce multiple high-value products. Subhadra and Edwards (2011)

Table 1 General composition of different microalgal species (% of dry matter)

Species	Protein	Carbohydrates	Lipids
Anabaena cylindrica	43-56	25-30	4-7
Aphanizomenon flos-aquae	62	23	3
Arthrospira maxima	60-71	13-16	6-7
Chaetoceros calcitrans	36	27	15
Chlamydomonas rheinhardii	48	17	21
Chlorella pyrenoidosa	57	26	2
Chlorella vulgaris	51-58	12-17	14-22
Diacronema vlkianum	57	32	6
Dunaliela bioculata	49	4	8
Dunaliella salina	57	32	6
Euglena gracilis	39-61	14-18	14-20
Haematococcus pluvialis	48	27	15
Isochrysis galbana	50-56	10-17	12-14
Nannochloropsis oculata	35	30	45
Nannochloropsis granulate	35.4	14.9	28.5
Nannochloropsis sp.	43	35	30
Porphyridium cruentum	28-39	40-57	9-14
Prymnesium parvum	28-45	25-33	22-38
Scenedesmus obliquus	50-56	10-17	12-14
Scenedesmus dimorphus	8-18	21-52	16-40
Scenedesmus quadricauda	47	21-52	1.9
Spirogyra sp.	6-20	33-64	11-21
Spirulina maxima	60-71	13-16	6-7
Spirulina platensis	46-63	8-14	4-9
Synechococcus sp.	63	15	11
Tetraselmis maculate	52	15	3

Source: Adapted from Becker (2007) and Koyande et al. (2019).

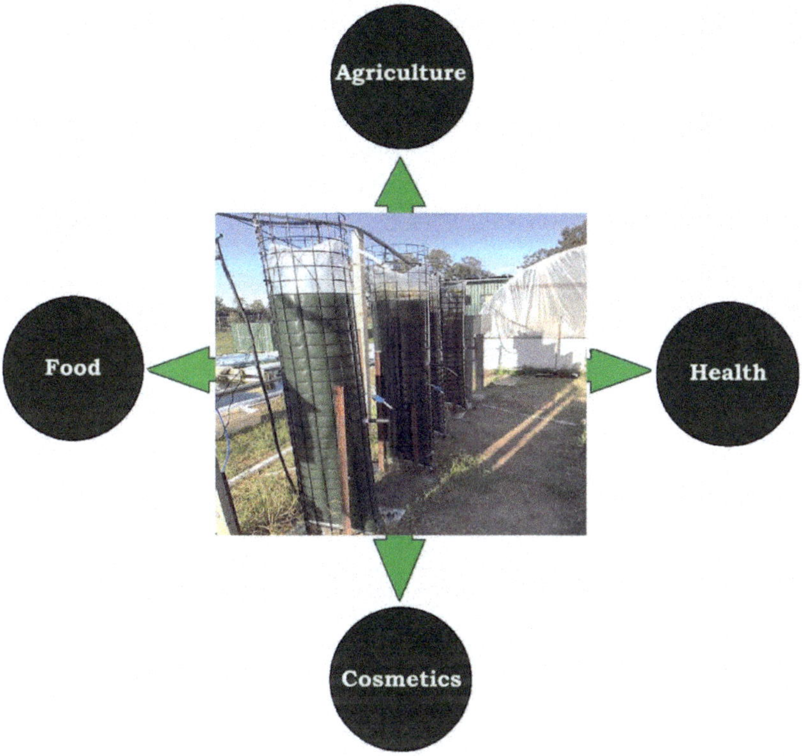

Figure 1 Multiple sectors benefit from the use of microalgae and their proteins.

identify 'clear advantages to attain high operational profit' in the sector. This is due to factors such as demand for natural food products, awareness of environmental issues, and the growth of veganism. Algal protein has a wide range of uses in the nutraceutical, cosmetic, and food industries. Figure 1 illustrates four industry sectors that benefit from the production of microalgal protein.

2 Applications of microalgal protein in the food, nutraceutical and cosmetic industries

2.1 Food

From as early as 300 AD, microalgae have been commonly incorporated into Chinese, Japanese, Korean, Chilean, Irish, African, and Indonesian diets. There is a large-scale global production of microalgae such as *Spirulina* and *Chlorella*, to produce food protein additives (e.g. for salad dressings and baked goods) and supplements (in powdered form or as pills and tablets). Protein-rich algal biomass can also be integrated into foods such as noodles, pasta, bread, biscuits, and beer (Wells et al., 2017). Table 2 lists some examples of current applications of microalgal protein products.

As an example, *Spirulina* (*Arthrospira*) has been used by NASA as a certified super-food for astronauts in space. Spira Inc., a start-up company based in Los Angeles (California), was selected by the International MoonBase Alliance and NASA Goddard to grow fresh *Spirulina* using their Grow Kit Bioreactor in the Hawaii Space Exploration Analog and Simulation (HI-SEAS). HI-SEAS, operated

Table 2 Examples of the use of protein-rich microalgae in food

Earthrise California *Spirulina*	Earthrise Nutritionals, a company based out of California, USA, operates the world's largest *Spirulina* farm. Customers are able to add *Spirulina* powder as a supplement to their food to receive multiple benefits, including antioxidant activity (the blue pigment, phycocyanin) and proteins that may increase brain, eye, and heart health and provide immune system support. Earthrise also supplies multiple delicious-looking recipes which utilize their product, including Oatmeal *Spirulina* Pancakes with 1 tsp of powdered Earthrise California *Spirulina* (Earthrise, 2020).
Electric Sky™ by Spira Inc.	Spira Inc. is another company based in California, USA, that sells *Spirulina*. The company uses its proprietary technology to extract the blue pigment **phycocyanin** from *Spirulina*, to be marketed as Electric Sky™. This beautiful cobalt-blue color can replace synthetic blue dyes in food, cosmetics, and paint (Spira Inc., 2020).
AlgaVia® whole algae protein	AlgaVia® promises a 'new vegan protein' which combines high protein richness with 'a rich collection of fiber, healthy lipids, and micronutrients'. Sold as a whole-food ingredient, this golden *Chlorella* native to freshwater in the Netherlands is harvested and processed into a fine yellow flour. One potential application is 8 g of the powder in a vegan chocolate protein beverage (Corbion Biotech Inc., 2020).
Dunaliella salina powder	Marketed as a 100% organic and natural vegan food supplement, *Dunaliella salina* powder has a characteristic bright orange color based on its high beta-carotene contents. With high amounts of magnesium, nutrients, and antioxidants, 2 tsp per serving is enough to create beautiful treats or tasty beverages. Greenfields Australia recommends a sprinkle on the top of your cereal (A.G.M. Foods Pty. Ltd., 2020).
Haematococcus pluvialis by Algatech	Algatech in Israel cultivates *Haematococcus pluvialis* in large photobioreactors. The green biomass turns red as algae run out of nutrients and astaxanthin accumulates in the red cysts. Toasted red cyst cells of *H. pluvialis* have a pleasant barbeque aroma that is suitable for vegan diets. The red motile macrozooids of these algae lack cell walls, which can also deliver astaxanthin and protein when eaten fresh.
Nannochloropsis spp.	Several companies (e.g. Ocean Orchards Australia, Ocean Pure Australia, Algae Pharm, Qponics, and Has Algae) now produce *Nannochloropsis* spp. for their high-value protein and omega-3 fatty acids to be included in the food.

by the University of Hawaii, is an isolated habitat which simulates a Mars-like environment at approximately 2500 meters above sea level. It enables astronauts to train in a Mars-like environment (HI-SEAS, 2020). As astronauts require fast-growing, nutritionally dense, and resource-efficient food that can be grown sustainably in a small space, *Spirulina*, with high protein digestibility, is an ideal candidate. The possibility of growing other microalgal species, such as *Dunaliella salina* (rich in β-carotene) and *Haematococcus pluvialis* (rich in the world's strongest antioxidant, astaxanthin), warrants further field studies to ensure that our astronauts are equipped with the best immune-boosting food on the planet, even while they are not on it.

Microalgae also have potential use as a fertilizer. Cyanobacteria's nitrogen-fixing abilities, water storage capacity, bioactive compound production, and virus protection qualities promote soil health and the growth of crops such as rice (Vaishampayan et al., 2001). One of the primary applications of microalgal protein biomass is in agriculture and aquaculture, as feed for poultry, pigs, cattle, sheep, shrimp, and multiple species of fish, including salmon, koi, and goldfish. A study by Gong et al. (2018) has assessed the apparent digestibility coefficient (ADC) of dry matter, protein, ash, and energy derived from defatted microalgal biomass of two popular microalgae: *Nannochloropsis* and *Desmodesmus*. Microalgae-containing feeds were fed to Atlantic salmon in seawater. These aquafeeds were formulated using either a cold-pelleting process or extrusion technology, with the latter significantly increasing the ADC of ash, dry matter, and protein of each species. This study demonstrated that *Nannochloropsis* sp. was more digestible than *Desmodesmus* sp. Further studies on protein digestibility are detailed in Section 4.

Another notable example of the use of microalgae in aquaculture is astaxanthin, a carotenoid derived from *H. pluvialis*, which provides fish with their pink coloration (Khoo et al., 2019). However, due to the high production costs of growing this microalga, the majority of astaxanthin marketed for fishmeal and animal feed is synthetically derived from petrochemicals, which might be linked to increased cancer risks (Butler and Golan, 2020). Further innovation in microalgal cultivation is needed to lower the cost of production. *H. pluvialis* has also been investigated for its potential as a feed ingredient for yellow perch. The results demonstrated no adverse effects on the fish when up to a quarter of their regular feed was replaced with an algal/soy protein-isolate mixture. However, an increase in replacement percentage resulted in potential nutrient imbalances and reduced benefits. Further studies are needed to optimize ideal amounts (Jiang et al., 2019).

2.2 Nutraceuticals

Algae are known to contain several bioactive peptides - short amino acid sequences, which have anti-oxidative, hypo-cholesterolemic, appetite-suppressive, and anti-microbial activities. Algal peptides have also been reported to have health benefits including hepatoprotective, UV protective, anti-osteoporotic, anti-coagulant, anti-carcinogenic, antioxidant, and anti-inflammatory activities (Bleakley and Hayes, 2017). *Spirulina* (including *Arthrospira platensis* and *Arthrospira maxima*), which is the most commercially available microalga, is currently widely used as a source of vitamins, antioxidants, and essential amino acids. Clinical studies suggest that there are no apparent toxic side effects as a result of either acute or chronic doses. *Spirulina* induces enzymatic antioxidative activity, regulates key cytokines for anti-inflammatory purposes, protects DNA and lipids from damage and eliminates free radicals (Wu et al., 2016).

Ismail et al. (2015) demonstrated significant reductions in oxidative stress, lipid peroxidation, and cholesterol, alongside increases in vitamin C, antioxidants, and superoxide dismutase (an enzyme that regulates levels of potentially harmful molecules in cells), within 30–60 days of *Spirulina* intake by patients with chronic obstructive pulmonary disease (COPD). It was recommended that studies utilizing larger sample sizes and longer durations would be helpful in determining the mechanisms by which *Spirulina* achieves these effects.

Nannochloropsis sp. is a marine microalga high in protein, essential amino acids, antioxidants, and omega-3 fatty acids with the potential to play a significant role in human nutrition (Chua and Schenk, 2017). A recent double-blind study demonstrated that it significantly lowered cholesterol (Rao et al., 2020). A 1 g/day capsule of Almega®PL – an oil derived from *Nannochloropsis* with an EPA content of > 25% – was given to volunteers with a low Omega-3 Index (OMI). Patients with the EPA-rich supplement demonstrated a ~16% increase in OMI and ~55% increase in EPA concentration. Cholesterol measurements also decreased by 25%.

2.3 Cosmetics

As the cosmetic industry has a long history of the use of synthetic and sometimes toxic chemicals, there is an increasing interest in incorporating natural ingredients into cosmetic products (Amberg and Fogarassy, 2019). Algae are a promising source because of their diversity and richness in bioactive compounds (Couteau and Coiffard, 2016). Different pigments in microalgal species make them ideal candidates for cosmetics, and the same pigments (e.g. beta-carotene, lutein, fucoxanthin, phycocyanin, etc.) can also act as antioxidants with potential anti-aging properties. Other health benefits are derived from polysaccharides, proteins, lipids, and secondary metabolites such as halogenated or phenolic compounds, nitrogen or sulfur derivatives, and terpenoids produced by algae (Ariede et al., 2017).

Compared to macroalgae, the use of microalgae in cosmetics is not new but still limited. France is currently the world's biggest market for algae application in cosmetics. A specialized facility has been developed in the west of France to provide raw marine materials, including algal extract packed in liquid form in a mixture of water and butylene glycol or propylene. Key industrial players include Aleor, Ceva, Codif International, Gelyma, and Secma (Couteau and Coiffard, 2016). Daniel Jouvance and Dior are major cosmetic manufacturers utilizing microalgae to formulate some of their cosmetic ranges (Couteau and Coiffard, 2018). *Spirulina* (*Arthrospira*) and *Chlorella* are well-established in the skin care market (Stolz and Obermayer, 2005). Algal proteins can be incorporated into basic cosmetic products such as face mask, eyeliner, or lipstick products. *Spirulina* used in facemasks is commonly marketed as a natural ingredient to support vibrant and healthy skin for both men and women, regardless of the skin type (Fig. 2).

A number of cosmetic producers make specific claims about the value of microalgal ingredients in their products. Exsymol (Monaco) has claimed that their protein-rich extract from *Spirulina* repairs the symptoms of early skin aging. Dermochlorella (St. Malo, France) claims that their product, which contains an extract from *Chlorella vulgaris*, promotes collagen synthesis that supports skin tissue regeneration and wrinkle reduction. Pentapharm (Basel, Switzerland) claims that their cosmetic ingredients including *Nannochloropsis oculata* and *D salina* exhibit skin-tightening properties and improve the skin's energy metabolism. Some major cosmetic companies in France such as LVMH and Daniel Jouvance have also invested in their own microalgae production system to incorporate this vegan ingredient in their cosmetic products (Spolaore et al., 2006).

Figure 2 *Spirulina* face mask for men and women.

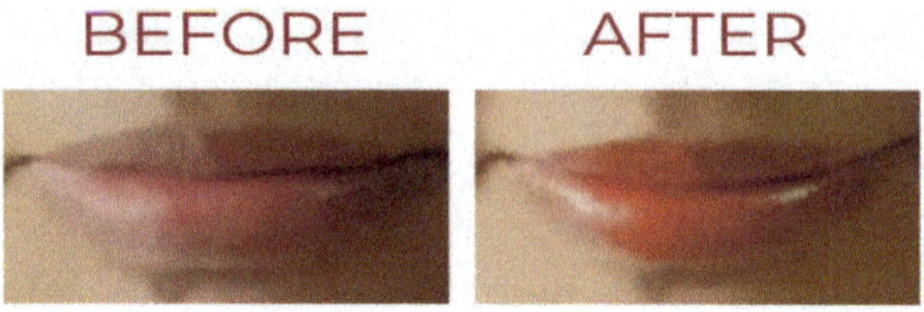

Figure 3 Edible astaxanthin lipstick derived from *Haematococcus pluvialis*, developed at the University of Queensland, Australia. The unique formulation not only provides powerful antioxidant benefits to the lips but also delivers long-lasting moisture and a natural warm pleasant color.

A study conducted by researchers from Fuji Chemical Industries Co. Ltd. assessed the cosmetic effects of astaxanthin derived from *H. pluvialis* and found significant improvements in skin wrinkle reduction, elasticity, texture, and moisture (Tominaga et al., 2009). Our own recent product development at the Algae Biotechnology Laboratory of the University of Queensland (UQ) showcased an edible astaxanthin lipstick derived from *H. pluvialis* grown at the UQ Algae Energy Farm, to provide a safe, vegan alternative to existing lipstick products in the market (Fig. 3). The market potential for microalgae application in the cosmetics industry is increasing due to changes in consumer attitudes, including awareness of climate change issues and concern about animal testing and animal-derived products. Major global producers such as Lush Cosmetics and Unilever have pledged to have 100% vegan-friendly products by 2020 as part of their corporate strategy.

3 Extraction of microalgal proteins

Microalgal species are very diverse, making the extraction of microalgal protein a complex process. Most commercial products with algal protein currently use the entire cell as an ingredient. However, the possibility of producing pure algal protein opens a path to create a multi-product microalgal biorefinery (Chua and Schenk, 2017). Both omega-3 oil and protein can, for example, be obtained from *Nannochloropsis* spp. (Chua and Schenk, 2017) and *Phaeodactylum tricornutum* (Cui et al., 2019), making microalgae farming potentially more cost-effective. This offsets challenges such as the cost of extraction and overcoming safety issues such as the use of laboratory reagents which may be harmful to humans.

Current methods of extracting proteins used in the food industry involve the use of aqueous solutions and take advantage of the physical properties of solutions such as temperature and pH. These methods usually start with the solubilization of proteins in water, followed by filtering the mixture to separate them from the solid matrices (see Fig. 4). The solubilization of proteins is influenced by various factors:

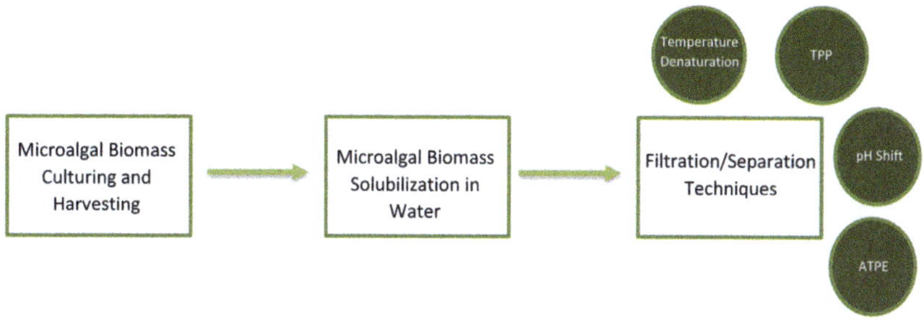

Figure 4 An overview of microalgal protein extraction. TPP - three-phase partitioning; ATPE - aqueous two-phase extraction.

- Extrinsic factors such as pH, ionic strength, temperature, and presence of solvent additives.
- Intrinsic factors, that is, the amino acid composition of the protein.

Increasing or decreasing the pH of the mixture causes amino acids to be in their ionic form, thus improving their solubility. However, the change in pH can also decrease the solubility of the proteins, especially when the pH reaches the isoelectric point of the protein, which is referred to as isoelectric precipitation (Kramer et al., 2012). In cheese-making factories, rennet or any acidic substance like vinegar or citric acid is added to lower the pH of the milk - causing coagulation. This pH-shift process has been utilized to prepare and improve the properties (e.g. emulsification) of numerous alternative plant proteins (Jiang et al., 2010; Tang et al., 2020; Wang et al., 2018). This process is attractive because of its simplicity and scalability.

Gerde et al. (2013) tested the pH-shift method to extract proteins from the *Nannochloropsis* biomass. They found that the maximum amount of protein was extracted and recovered at pH levels of 11 and 3.2, respectively. However, the total yield was only 16% when using defatted biomass. Nevertheless, the high alkaline level can also induce the dissolution of other compounds, and identifying contaminants in the protein precipitate will be crucial to streamline the process. If most of the compounds are carbohydrates, an enzyme can be added to break them into smaller sugars so that they do not precipitate at the isoelectric point.

Another well-established food protein source is soybean, specifically tofu. In tofu-making, soy proteins are denatured by exposing them to high temperatures (Liu et al., 2013a). Salts are also added to reduce the solubility of the proteins in the soymilk by reducing the interaction of water with the amino acids at the protein surfaces, causing them to separate out of the solution (Cai et al., 1997). This is commonly referred to as the salting-out effect and is used in protein purification techniques like ammonium sulfate precipitation and aqueous two-phase extraction (ATPE). Ammonium sulfate precipitation relies on the difference in the solubilities of water-soluble compounds including proteins. As the total %$(NH_4)_2SO_4$ is steadily increased, precipitates are formed at different stages, with the least water-soluble compound precipitating first (Wingfield, 2001). ATPE involves the use of two aqueous solutions which can be made of salts, alcohols, polymers, or ionic liquids (Iqbal et al., 2016; Suarez Ruiz et al., 2018). Each system has its own advantages and disadvantages. For example, the use of salt solutions in ATPE can cause protein denaturation during the purification process due to the high ionic strength of the solutions (Albertsson, 1961). This disadvantage does not occur in some polymer–polymer systems, though the high cost of some polymers can limit their use. Various factors affect the purification of compounds. A multi-factorial experimental design is usually employed to optimize the process (Iqbal et al., 2016; Liu et al., 2013b).

Numerous studies have investigated these techniques to purify microalgal proteins, especially phycobiliproteins from cyanobacteria, including *Spirulina* (Khazi et al., 2018; Pereira et al., 2020). The main advantage of extracting phycobiliproteins is their high solubility in water. The challenge is to purify them from other water-soluble proteins. Porav et al. (2020) proposed a two-step ATPE to obtain high-purity phycobiliproteins using a phosphate buffer and polyethylene glycol (PEG), resulting in analytical-purity-grade phycocyanin, reactive-purity-grade allophycocyanin, and high-purity grade phycoerythrin. Phycobiliproteins are concentrated in the upper PEG layer and need to be separated from the PEG molecules by ultrafiltration, allowing the reusability of the recovered PEG. Ruiz-Ruiz et al. (2019) proposed an engineering design with continuous ATPE extraction of C-phycocyanin from *Spirulina* maxima.

A variation to ATPE is a process referred to as three-phase partitioning (TPP), which involves the use of an alcohol (usually t-butanol and a salt solution) to force the formation of a third phase in the middle of these two aqueous solutions. Using alcohol in this process means some organic compounds, like carotenoids, can be simultaneously isolated from the microalgal biomass. Waghmare et al. (2016) showed how microalgal proteins can be extracted from *Chlorella pyrenoidosa* using TPP with t-butanol and ammonium sulfate as the main solvents. In other cases, some less hydrophilic polymers can also be used in TPP. Suarez Ruiz et al. (2020) used polypropylene glycol (PPG) and cholinium-based ionic liquids to separate different compounds from *Neochloris oleoabundans*. They optimized the process so that pigments are fractionated in the PPG layer, free glucose and proteins in the ionic liquid-rich phase, and starch and lipids form at the interface. By changing the type of polymer, proteins were collected in the interphase (Suarez Ruiz et al., 2018). The key issues include studies to optimize the different variables, scalability, toxicity, recyclability, and reusability of the reagents, to ensure the cost-effectiveness and sustainability of production.

4 Bioavailability and digestibility

The two main criteria that determine the nutritional value of protein are the amino acid profile and digestibility (Chronakis and Madsen, 2011). Protein derived from animals is generally regarded as complete protein due to a rich source of essential amino acids that humans are unable to biosynthesize. Protein derived from plants, however, is normally regarded as incomplete protein due to a lack of one or more essential amino acids (Young and Pellett, 1994). It is important for people with vegan diets to consume a variety of plant proteins from fruit, grains, legumes, and vegetables to obtain sufficient quantities of all essential amino acids (Hoffman and Falvo, 2004). Unlike animal-based proteins, plant-based proteins are harder to digest due to high concentrations of insoluble polysaccharides. However, animal-based proteins have high levels of saturated fats and cholesterol, which are closely linked to cardiovascular disease and diabetes (Bleakley and Hayes, 2017). Alternatively, as many microalgal species meet the amino acid requirements set by the Food and Agriculture Organization/World Health Organization (2007), while displaying similarity to soybean and egg (Koyande et al., 2019), they can be considered as an appropriate and alternative protein sources (Table 3).

Table 3 shows that all essential amino acids are present in most microalgae. Tryptophan can be difficult to measure for some microalgae such as *Nannochloropsis* spp. (Conde et al., 2013), but our own measurements have shown that this amino acid is present in *Nannochloropsis oceanica* at 2.1 g/100 g. Even with a complete profile of essential amino acids, low protein digestibility – linked to the rigid cellulosic cell wall in microalgae – can affect protein quality, with the exception of blue-green algae such as *Spirulina* and *D. salina* (Chronakis and Madsen, 2011). Chemical, mechanical, or enzymatic pre-treatments of microalgal biomass prior to protein extraction may therefore be necessary (Acquah et al., 2020).

Niccolai et al. (2019) tested the digestibility of 12 different microalgal species. *Spirulina* (*A platensis*), *Chlorella sorokiniana*, and *C. vulgaris* had the highest digestibility, while *Tetraselmis suecica*, *P. tricornutum*, and *Porphyridium purpureum* showed the least digestibility. Differences in cell wall composition are the likely reason: cell wall rigidity limits the action of the digestive enzymes. Cavonius et al. (2016) demonstrated that the bioaccessibility of *N. oculata* proteins can be improved by the pH-shift processing method. Kose et al. (2017) showed that the digestibility of *C. vulgaris* and *Spirulina* (*A. platensis*) can be greatly improved from 35% to 70%, and from 50% to 97%, respectively, after hydrolysis using the proteolytic enzyme, pancreatin, which effectively weakens the microalgal cell walls. *Spirulina* has thus been widely investigated as a promising alternative food ingredient. (Batista et al., 2017) found that the replacement of up to 6% of wheat flour in cookies by microalgal biomass

Table 3 Amino acid profiles of chicken breast, egg, soybean, and several microalgal species (g/100 g of dry matter)

	Source	Chicken breast	Egg	Soy-bean	*Arthrospira maxima*	*Aphanizomenon* sp.	*Chlorella* sp.	*Chlorella vulgaris*
ESSENTIAL AMINO ACIDS	His	4.5	2.4	6.0	1.8	0.9	2.4	2.0
	Ile	3.2	6.6	5.3	6.0	2.9	4.4	3.8
	Leu	6.4	8.8	7.7	8.0	5.2	9.2	8.8
	Lys	7.9	5.3	6.4	4.6	3.5	8.9	8.4
	Met	2.5	3.2	1.3	1.4	0.7	2.2	2.2
	Phe	3.2	5.8	5	4.9	2.5	5.5	5.0
	Thr	3.7	5.0	4	4.6	3.3	4.7	4.8
	Trp	NA	1.7	1.4	1.4	0.7	NA	2.1
	Val	3.5	7.2	5.3	6.5	3.2	6.1	5.5
NON-ESSENTIAL AMINO ACIDS	Ala	4.7	NA	5	6.8	4.7	8.3	7.9
	Arg	5.8	6.2	7.4	6.5	3.8	7.1	6.4
	Asn	7.8	11	1.3	8.6	4.7	9.4	9.0
	Cys-Cys	1.1	2.3	1.9	0.4	0.2	0.4	0.1
	Glu	11.2	12.6	19	12.6	7.8	12.9	11.6
	Gly	3.4	4.2	4.5	4.8	2.9	5.4	5.8
	Pro	3.2	4.2	5.3	3.9	2.9	4.8	4.8
	Ser	3.4	6.9	5.8	4.2	2.9	4.0	4.1
	Tyr	3.7	4.2	3.7	3.9	NA	4.2	3.4

Source: Adapted from Becker (2007); Koyande et al. (2019); Lorenz (1999).
NA, data not available.

did not significantly affect digestibility. Further studies are needed to test the incorporation of extracted microalgal proteins in food products and investigate how the nutritional profile of the novel product differs from non-algae control.

Further studies detailing the safety and digestibility of microalgae in mice have been conducted by Neumann et al. (2018). They concluded that algae-rich diets were suitable and beneficial with no adverse effects, particularly with regard to body weight control and healthy organ function. Freeze-dried microalgae (which had been ball-milled to limit the effect of the rigid cell wall on bioavailability) were incorporated into the regular feed of mice. Three different types of microalgae were fed in different amounts to the mice: *C. vulgaris*, *N. oceanica*, and *Phaeodactylum triconutum*. Each diet was accepted by the mice and exhibited the same caloric and proteinogenic qualities as the casein protein fed to the control group. No adverse effects were observed during the trial. Protein bioavailability and digestibility were assessed using apparent biological value (ABV), ADC, and net protein usage (NPU). Table 4 shows the results of the trial. These results indicated that each microalgae-supplemented diet did not differ in protein quality (bioavailability and digestibility) from the control diet, suggesting that up to 66% of the protein diet could be safely

Table 3 (*Continued*)

Dunaliella sp.	*Dunaliella bardawil*	*Haematococcus pluvialis*	*Nannochloropsis* sp.	*Scenedesmus* sp.	*Scenedesmus obliquus*	*Spirulina* sp.	*Spirulina platensis*
2.6	1.8	0.6	2.6	2.5	2.1	2.0	2.2
4.5	4.2	0.8	4.7	4.7	3.6	5.8	6.7
9.4	11	1.7	9.4	9.3	7.3	9.0	9.8
6.8	7.0	1.1	6.8	6.2	5.6	5.1	4.8
2.4	2.3	0.4	2.3	2.5	1.5	2.9	2.5
5.5	5.8	0.9	5.5	6.0	4.8	4.8	5.3
4.9	5.4	1.0	4.8	5.0	5.1	5.1	6.2
NA	0.7	0.3	NA	NA	0.3	NA	0.3
6.0	5.8	1.4	6.0	6.0	6.0	6.4	7.1
6.8	7.3	1.7	6.8	6.8	7.8	7.4	9.5
6.0	7.3	1.1	6.0	6.0	6.6	7.4	9.5
9.2	10	1.9	9.1	9.1	11	7.6	7.3
0.1	1.2	0.2	0.1	4.0	0.6	0.3	0.9
13.8	12.7	2.2	13.8	13.6	10.7	16.1	10.3
5.2	5.5	1.2	5.2	5.7	7.1	4.6	5.7
8.3	3.3	0.9	8.3	4.9	3.9	3.3	4.2
4.2	4.6	0.9	4.2	4.4	3.8	4.8	5.1
4.0	3.7	0.5	3.9	4.0	3.2	4.8	5.3

Table 4 Apparent biological value, apparent digestibility, and net protein usage of multiple microalgal species and casein protein

Microalgae	Diet (supplement %)	ABV	AD	NPU
Control Diet (Casein)		66.4 ± 4.8	82.7 ± 2.5	54.9 ± 4.3
Nannochloropsis oceanica	5	61.4 ± 3.6	88.8 ± 0.9	54.5 ± 3.2
	15	58.7 ± 2.2	87.6 ± 0.8	51.5 ± 2.3
	25	58.7 ± 5.2	84.4 ± 0.3	49.5 ± 4.4
Chlorella vulgaris (cultured phototrophically)	5	76.4 ± 5.5	87.3 ± 0.8	66.8 ± 5.0
	15	65.2 ± 2.9	85.6 ± 0.6	55.8 ± 2.4
	25	76.2 ± 5.4	78.3 ± 3.1	59.1 ± 3.8
Phaeodactylum tricornutum	5	69.9 ± 3.2	88.0 ± 1.2	61.3 ± 2.6
	15	77.0 ± 2.5	84.2 ± 2.9	64.9 ± 3.0
	25	69.5 ± 3.8	69.5 ± 0.7	55.2 ± 2.8

Source: Adapted from Neumann et al. (2018).
ABV, apparent biological value; AD, apparent digestibility coefficient; NPU, net protein usage.

replaced by microalgae. The study shows that microalgae possess excellent protein quality and bioavailability.

5 Conclusion

Microalgae are the foundation for a strongly emerging industry which, in the face of food scarcity, increasing environmental harm, and reduced sustainability, has the potential to develop significantly. Microalgae production is characterized by high yields, relatively low emissions, land and water usage, and ecological impacts. Microalgal protein and their numerous other high-value by-products can meet the nutritional and health needs of consumers. Research into optimizing the bioprocessing of microalgal proteins, including the sustainability of extraction techniques, will be imperative if these proteins are to be developed as pure products. Finally, the bioavailability and digestibility of the extracted proteins should also be considered, especially if these products are to be offered as potential food ingredients.

6 Where to look for further information

The following websites provide good introductions of the nutritional benefits of microalgal protein:

- https://www.veganfriendly.org.uk/health-fitness/vegan-protein-powder/algae-protein-powder/.
- https://formnutrition.com/us/inform/algae-protein/.
- https://theconversation.com/more-protein-and-good-for-the-planet-9-reasons-we-should-be-eating-microalgae-124591.

The following book provides further reading on algae as ingredients for food and nutraceuticals:

- Dominguez, H. (Ed.) 2013. *Functional Ingredients from Algae for Foods and Nutraceuticals*. Woodhead Publishing, Oxford.

The following is an informatic review article on the use of microalgae for cosmetics:

- Mourelle, M. L., Gómez, C. P. and Legido, J. L. 2017. The potential use of marine microalgae and cyanobacteria in cosmetics and thalassotherapy. *Cosmetics*, *4*(4), p.46.

The following website provides a good overview of the emerging trends and industries around the use of algae as functional food ingredients:

- https://www.greenqueen.com.hk/the-sustainable-superfood-big-players-say-microalgae-is-the-future-game-changer/.

The Algae Biomass Organization and European Algae Biomass Association are two wide-reaching societies that bring together people interested in microalgae and hold annual conferences (Algae Biomass Summit and AlgaEurope).

7 References

A.G.M. Foods Pty. Ltd. 2020. Grainfields Australia. Available at: https://agmfoods.com/ (accessed 11/11/2020).

Acquah, C., Tibbetts, S. M., Pan, S. and Udenigwe, C. 2020. Chapter 19. Nutritional quality and bioactive properties of proteins and peptides from microalgae. In: *Handbook of Microalgae-Based Processes and Products*, E. Jacob-Lopes, M. M. Maroneze, M. I. Queiroz, and L. Q. Zepka (Eds). Academic Press, pp. 493–531.

Albertsson, P. A. 1961. Fractionation of particles and macromolecules in aqueous two-phase systems. *Biochemical Pharmacology* 5(4), 351–358.

Amberg, N. and Fogarassy, C. 2019. Green consumer behavior in the cosmetics market. *Resources* 8(3), 137.

Ariede, M. B., Candido, T. M., Jacome, A. L. M., Velasco, M. V. R., de Carvalho, J. C. M. and Baby, A. R. 2017. Cosmetic attributes of algae–a review. *Algal Research* 25, 483–487.

Batista, A. P., Niccolai, A., Fradinho, P., Fragoso, S., Bursic, I., Rodolfi, L., Biondi, N., Tredici, M. R., Sousa, I. and Raymundo, A. 2017. Microalgae biomass as an alternative ingredient in cookies: sensory, physical and chemical properties, antioxidant activity and in vitro digestibility. *Algal Research* 26, 161–171.

Becker, E. W. 2007. Micro-algae as a source of protein. *Biotechnology Advances* 25(2), 207–210.

Bleakley, S. and Hayes, M. 2017. Algal proteins: extraction, application, and challenges concerning production. *Foods* 6(5), 33.

Butler, T. and Golan, Y. 2020. *Microalgae Biotechnology for Food, Health and High-Value Products*. Singapore: Springer, pp. 175–242.

Cai, T. D., Chang, K. C., Shih, M. C., Hou, H. J. and Ji, M. 1997. Comparison of bench and production scale methods for making soymilk and tofu from 13 soybean varieties. *Food Research International* 30(9), 659–668.

Cavonius, L. R., Albers, E. and Undeland, I. 2016. In vitro bioaccessibility of proteins and lipids of pH-shift processed *Nannochloropsis oculata* microalga. *Food and Function* 7(4), 2016–2024.

Chronakis, I. and Madsen, M. 2011. Algal proteins. In: *Handbook of Food Proteins*, G. O. Phillips and P. A. Williams (Eds). pp. 353–394. Woodhead Publishing.

Chua, E. T. and Schenk, P. M. 2017. A biorefinery for *Nannochloropsis*: induction, harvesting, and extraction of EPA-rich oil and high-value protein. *Bioresource Technology* 244(2), 1416–1424.

Conde, E., Balboa, E. M., Parada, M. and Falqué, E. 2013. Algal proteins, peptides and amino acids. In: *Functional Ingredients from Algae for Foods and Nutraceuticals*. Woodhead Publishing. pp. 135–180.

Corbion Biotech Inc. 2020. *algaVia*. Available at: http://www.corbion.com/algae-portfolio (accessed on 11/11/2020).

Couteau, C. and Coiffard, L. J. M. 2016. Chapter 14. Seaweed application in cosmetics. In: *Seaweed in Health and Disease Prevention*, J. Fleurence and I. Levine (Eds). pp. 423-441.

Couteau, C. and Coiffard, L. J. M. 2018. Chapter 15. Microalgal application in cosmetics. In: *Microalgae in Health and Disease Prevention*, I. A. Levine and J. Fleurence (Eds). pp. 317–321.

Cui, Y., Thomas-Hall, S. R. and Schenk, P. M. 2019. *Phaeodactylum tricornutum* microalgae as a rich source of omega-3 oil: progress in lipid induction techniques towards industry adoption. *Food Chemistry* 297, 124937.

Earthrise 2020. *Benefits of Spirulina*. Available at: https://www.earthrise.com/benefits-of-spirulina (accessed on 11/11/2020).

Gerde, J. A., Wang, T., Yao, L., Jung, S., Johnson, L. A. and Lamsal, B. 2013. Optimizing protein isolation from defatted and non-defatted *Nannochloropsis* microalgae biomass. *Algal Research* 2(2), 145–153.

Gong, Y., Guterres, H. A. D. S., Huntley, M., Sørensen, M. and Kiron, V. 2018. Digestibility of the defatted microalgae *Nannochloropsis* sp. and *Desmodesmus* sp. when fed to Atlantic salmon, Salmosalar. *Aquaculture Nutrition* 24(1), 56–64.

Henchion, M., Hayes, M., Mullen, A. M., Fenelon, M. and Tiwari, B. 2017. Future protein supply and demand: strategies and factors influencing a sustainable equilibrium. *Foods* 6(7), 53.

HI-SEAS 2020. *Hawaii Space Exploration Analog and Simulation*. Available at: https://www.hi-seas.org/ (accessed on 11/11/2020).

Hoffman, J. R. and Falvo, M. J. 2004. Protein–which is best? *Journal of Sports Science and Medicine* 3(3), 118–130.

Iqbal, M., Tao, Y., Xie, S., Zhu, Y., Chen, D., Wang, X., Huang, L., Peng, D., Sattar, A., Shabbir, M. A. B., Hussain, H. I., Ahmed, S. and Yuan, Z. 2016. Aqueous two-phase system (ATPS): an overview and advance in its applications. *Biological Procedures Online* 18(1), 18.

Ismail, M., Hossain, M. F., Tanu, A. R. and Shekhar, H. U. 2015. Effect of Spirulina intervention on oxidative stress, antioxidant status, and lipid profile in chronic obstructive pulmonary disease patients. *BioMed Research International* 2015, 486120.

Jiang, J., Xiong, Y. L. and Chen, J. 2010. pH shifting alters solubility characteristics and thermal stability of soy protein isolate and its globulin fractions in different pH, salt concentration, and temperature conditions. *Journal of Agricultural and Food Chemistry* 58(13), 8035–8042.

Jiang, M., Zhao, H. H., Zai, S. W., Shepherd, B., Wen, H. and Deng, D. F. 2019. A defatted microalgae meal (*Haematococcus pluvialis*) as a partial protein source to replace fishmeal for feeding juvenile yellow perch *Perca flavescens*. *Journal of Applied Phycology* 31(2), 1197–1205.

Khazi, M. I., Demirel, Z. and Dalay, M. C. 2018. Evaluation of growth and phycobiliprotein composition of cyanobacteria isolates cultivated in different nitrogen sources. *Journal of Applied Phycology* 30(3), 1513–1523.

Khoo, K. S., Lee, S. Y., Ooi, C. W., Fu, X., Miao, X., Ling, T. C. and Show, P. L. 2019. Recent advances in biorefinery of astaxanthin from *Haematococcus pluvialis*. *Bioresource Technology* 288, 121606.

Kose, A., Ozen, M. O., Elibol, M. and Oncel, S. S. 2017. Investigation of in vitro digestibility of dietary microalga *Chlorella vulgaris* and cyanobacterium *Spirulina platensis* as a nutritional supplement. *3 Biotech* 7(3), 170.

Koyande, A. K., Chew, K. W., Rambabu, K., Tao, Y., Chu, D.-T. and Show, P.-L. 2019. Microalgae: a potential alternative to health supplementation for humans. *Food Science and Human Wellness* 8(1), 16–24.

Kramer, R. M., Shende, V. R., Motl, N., Pace, C. N. and Scholtz, J. M. 2012. Toward a molecular understanding of protein solubility: increased negative surface charge correlates with increased solubility. *Biophysical Journal* 102(8), 1907–1915.

Liu, H.-H., Chien, J.-T. and Kuo, M.-I. 2013a. Ultra-high pressure homogenized soy flour for tofu making. *Food Hydrocolloids* 32(2), 278–285.

Liu, X., Mu, T., Sun, H., Zhang, M. and Chen, J. 2013b. Optimisation of aqueous two-phase extraction of anthocyanins from purple sweet potatoes by response surface methodology. *Food Chemistry* 141(3), 3034–3041.

Lorenz, R. 1999. A technical review of *Haematococcus algae*. *NatuRoseTM Technical Bulletin* 60, 1–12.

Neumann, U., Derwenskus, F., Gille, A., Louis, S., Schmid-Staiger, U., Briviba, K. and Cischoff, C. S. 2018. Bioavailability and safety of nutrients from the microalgae *Chlorella vulgaris, Nannochloropsis oceanica and Phaeodactylum tricornutum* in C57BL/6 Mice. *Nutrients* 10(8), 965.

Niccolai, A., Chini Zittelli, G., Rodolfi, L., Biondi, N. and Tredici, M. R. 2019. Microalgae of interest as food source: biochemical composition and digestibility. *Algal Research* 42, 101617.

Pereira, T., Barroso, S., Mendes, S. and Gil, M. M. 2020. Stability, kinetics, and application study of phycobiliprotein pigments extracted from red algae *Gracilaria gracilis*. *Journal of Food Science* 85(10), 3400–3405.

Porav, A. S., Bocăneală, M., Fălămaş, A., Bogdan, D. F., Barbu-Tudoran, L., Hegeduş, A. and Dragoş, N. 2020. Sequential aqueous two-phase system for simultaneous purification of cyanobacterial phycobiliproteins. *Bioresource Technology* 315, 123794.

Rao, A., Briskey, D., Nalley, J. O. and Ganuza, E. 2020. Omega-3 eicosapentaenoic acid (EPA) rich extract from the microalga *Nannochloropsis* decreases cholesterol in healthy individuals: a double-blind, randomized, placebo-controlled, three-month supplementation study. *Nutrients* 12(6), 869.

Ruiz-Ruiz, F., López-Guajardo, E., Vázquez-Villegas, P., del Angel-Chong, M. E., Nigam, K. D. P., Willson, R. C. and Rito-Palomares, M. 2019. Continuous aqueous two-phase extraction of microalgal C-phycocyanin using a coiled flow inverter. *Chemical Engineering and Processing–Process Intensification* 142, 107554.

Spira Inc. 2020. *Electric Sky™ Sample*. Spira Inc. Available at: https://www.spirainc.com/store/electric-sky.

Spolaore, P., Joannis-Cassan, C., Duran, E. and Isambert, A. 2006. Commercial applications of microalgae. *Journal of Bioscience and Bioengineering* 101(2), 87–96.

Stolz, P. and Obermayer, B. 2005. Manufacturing microalgae for skin care. *Cosmetics and Toiletries* 120, 99–106.

Suarez Ruiz, C. A., Emmery, D. P., Wijffels, R. H., Eppink, M. H. and van den Berg, C. 2018. Selective and mild fractionation of microalgal proteins and pigments using aqueous two-phase systems. *Journal of Chemical Technology and Biotechnology* 93(9), 2774–2783.

Suarez Ruiz, C. A., Kwaijtaal, J., Peinado, O. C., van den Berg, C., Wijffels, R. H. and Eppink, M. H. M. 2020. Multistep fractionation of microalgal biomolecules using selective aqueous two-phase systems. *ACS Sustainable Chemistry and Engineering* 8(6), 2441-2452.

Subhadra, B. G. and Edwards, M. 2011Coproduct market analysis and water footprint of simulated commercial algal biorefineries. *Applied Energy* 88(10), 3515–3523.

Tang, M., Dai, H., Ma, L., Yu, Y., Liu, T., Feng, X., Hu, W., Li, Y. and Zhang, Y. 2020. Degradation of structural proteins and their relationship with the quality of Mandarin fish (*Siniperca chuatsi*) during post-mortem storage and cooking. *International Journal of Food Science and Technology* 55(4), 1617–1628.

Tominaga, K., Hongo, N., Karato, M. and Yamashita, E. 2009. Cosmetic benefits of oral supplementation combined with topical treatment of astaxanthin on human subjects. *Food Style* 13, 25–26.

Vaishampayan, A., Sinha, R. P., Hader, D. P., Dey, T., Gupta, A. K., Bhan, U. and Rao, A. L. 2001. Cyanobacterial biofertilizers in rice agriculture. *The Botanical Review* 67(4), 453–516.

Waghmare, A. G., Salve, M. K., LeBlanc, J. G. and Arya, S. S. 2016. Concentration and characterization of microalgae proteins from *Chlorella pyrenoidosa*. *Bioresources and Bioprocessing* 3(1), 16.

Wang, Q., Jin, Y. and Xiong, Y. L. 2018. Heating-aided pH shifting modifies hemp seed protein structure, cross-linking, and emulsifying properties. *Journal of Agricultural and Food Chemistry* 66(41), 10827–10834.

Wells, M. L., Potin, P., Craigie, J. S., Raven, J. A., Merchant, S. S., Helliwell, K. E., Smith, A. G., Camire, M. E. and Brawley, S. H. 2017. Algae as nutritional and functional food sources: revisiting our understanding. *Journal of Applied Phycology* 29(2), 949–982.

Wingfield, P. 2001. Protein precipitation using ammonium sulfate. *Current Protocols in Protein Science* [Appendix(3), Appendix-3F]. https://doi.org/10.1002/0471140864.psa03fs13.

World Health Organization and United Nations University 2007. *Protein and Amino Acid Requirements in Human Nutrition* (Vol. 935). World Health Organization.

Wu, Q., Liu, L., Miron, A., Klímová, B., Wan, D. and Kuča, K. 2016. The antioxidant, immunomodulatory, and anti-inflammatory activities of *Spirulina*: an overview. *Archives of Toxicology* 90(8), 1817–1840.

Young, V. R. and Pellett, P. L. 1994. Plant proteins in relation to human protein and amino acid nutrition. *American Journal of Clinical Nutrition* 59(Suppl. 5), 1203S–1212S.

Chapter 7

Environmental impacts of seaweed cultivation: kelp farming and preservation

Jean-Baptiste Thomas, KTH Royal Institute of Technology, Sweden; José Potting, EnviroSpotting, The Netherlands and KTH Royal Institute of Technology, Sweden; and Fredrik Gröndahl, KTH Royal Institute of Technology, Sweden

1 Introduction

Seaweeds have a long history of use as food, feed and fertiliser. More recently, extracts from seaweed have also appeared in dyes, toothpastes, cosmetic creams and a range of other products. There is a fast-growing interest in seaweeds as an environmentally sustainable alternative to terrestrial plants and fossil fuels (McHugh 2003; Van den Burg et al. 2018; Van Hal et al. 2014). In Europe, a number of pilot projects and start-ups are exploring the potential of seaweed farming, preservation and processing, and novel seaweed applications that can substitute more carbon-intense or unsustainable fossil- or terrestrial biomass-based products (see other chapters in this book). The supposed environmental sustainability of seaweed farming, preservation and processing is the subject of remarkably little research. Seaweed farming is taken to include all processes, from hatchery through marine cultivation to harvesting the seaweed.

http://dx.doi.org/10.19103/AS.2021.0091.11

When talking about the environmental sustainability of seaweed, one must distinguish between two types of environmental impacts. The first relates to the environmental impacts of seaweed farming on the immediate marine environment, for example, the possible effects from the presence of infrastructure or the effects on local biodiversity. This type of impact is covered elsewhere in this book. The second type refers to the overall environmental impacts resulting from the whole seaweed supply chain, including all material and energy inputs that are used in order to produce specific seaweed products. This chapter focuses on the latter and how these impacts can be quantified by means of the life cycle assessment (LCA).

This chapter provides an overview of the environmental impacts of the supply chain for preserved seaweed. The supply chain includes the hatchery, marine infrastructure, deployment of juveniles and monitoring during cultivation (grow-out of seaweed), harvest, transport back to shore and preservation of the biomass. Not specifically addressed are the environmental impacts from downstream processing of the preserved biomass into a multitude of products, the utilisation of these products and their end-of-life disposal. However, they are discussed in light of the available literature. The environmental impacts of this further processing may or may not exceed the environmental impacts in the preceding supply chain. This will depend on the specific product and the processes involved in its manufacture.

The chapter starts with a short overview of the LCA methodology, and how it can be used to quantify the environmental impacts of seaweed supply chains. The next section introduces the 'Seafarm-LCA' taken from Thomas et al. (2021) as a case study for illustrating the environmental impacts of the preserved seaweed supply chain. Figures are based on the study published in the *ICES Journal of Marine Science*. Section 4 focuses on the overall environmental impacts of the preserved seaweed supply chain, in contrast to Sections 5–8 which focus on specific life cycle stages. Section 5 deals with spore preparation and seeding of juvenile seaweed onto the string in the hatchery, Section 6 goes into seaweed cultivation and harvesting, and Section 7 covers the preservation and storage of harvested seaweed. Each stage also includes the environmental impacts of the production of the equipment needed in that stage (i.e. the environmental impacts of cradle-to-equipment production). The chapter ends with a summary (Sections 8–9), followed by future trends in research (Section 10) and some suggestions for sources of further information for the interested reader (Section 11).

2 LCA methodology

LCA is a well-established approach to investigate the environmental performance of product systems and service systems (Baumann and Tillman

Published by Burleigh Dodds Science Publishing Limited, 2021.

2004), both referred to as product systems in the rest of this chapter. A core characteristic of systems is that they serve a function (Meadows 2015). LCA quantifies the environmental impacts per unit of *function* served by the product system, that is, per functional unit (Baumann and Tillman 2004; ISO 14044 2006), and not for a unit of product as is commonly miscommunicated.

The functional unit for a product system is typically based on its output, for example, 'one ton of dried seaweed meeting certain quality requirements', but it can sometimes also be based on system inputs (see Fig. 1). The latter is typically the case for 'waste-LCAs' comparing different types of waste treatment (Laurenti et al. 2014) but also applies to the case study LCA used for illustration in this chapter, for which the functional unit is 'one ton of freshly harvested kelp to be preserved'. This input-based functional unit enables the comparison of different methods of preserving kelp, with the rest of the compared product system being the same (Thomas et al. 2021). A further discussion of the functional unit aspects of LCAs is available in Sills et al. (2020), presented in the context of LCAs of algal biorefineries.

Many, if not most LCAs, indeed *compare* product systems, such as comparing a reference product system with an improved version of itself or, alternatively, comparing very different product systems with the same function. Comparative LCAs are typically performed to inform decision-makers, for instance within the industry, about whether to produce chemicals using one form of energy or another (Lammens et al. 2011), or to inform consumers choosing between seaweed burgers and conventional hamburgers (Van den Burg et al. 2018). Other LCAs typically focus on a single product system, usually to gain insights into environmental impact hotspots and to identify opportunities for optimisation.

A full LCA or 'cradle-to-grave' LCA covers a product system from raw material extraction, through materials production, product manufacture and use, up to and including processing the discarded product. Each of these so-called life cycle stages in a product system may (and often do) encompass more than one process. The production of woollen fabrics, for example, covers yarn spinning and fabric weaving, with several refining processes, both in between and following the spinning and weaving (Potting and Blok 1995).

While 'cradle-to-grave' LCAs encompass full product life cycles, many LCAs cover partial product life cycles. The waste-LCAs previously mentioned represent 'gate-to-grave' LCAs, where the 'gate' often is the entrance into the waste collection process (Laurenti et al. 2014). Other partial LCAs represent 'cradle-to-gate' LCAs, including the case study LCA in this chapter (see Fig. 1), where materials or substances are to be further processed after they exit the 'gate'. There are also 'gate-to-gate' LCAs, like a comparative LCA by Kouchaki-Penchah et al. (2015) for particleboard produced from agricultural and industrial waste residues.

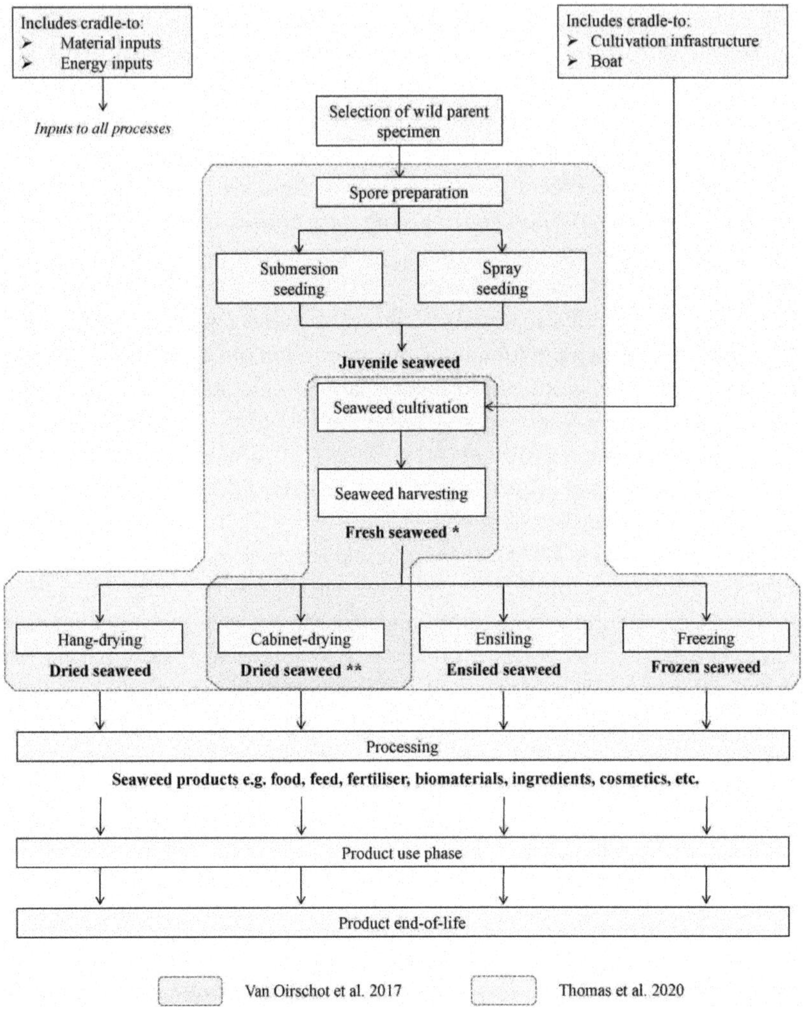

Figure 1 An overview of the 'cradle-to-gate' product system, or supply chain, for preserved brown seaweed (i.e. kelp or *Saccharina latissima*). The figure exemplifies the two related input-based (single asterisk *) and output-based (double asterisk **) functional units. Source: the figure has been modified from Thomas et al. (2021).

An LCA should cover more than one process as well as more than one environmental impact, according to the ISO 14044 (2006). Typical environmental impacts covered by LCAs are resource depletion, global warming, (stratospheric) ozone depletion, acidification, eutrophication, (tropospheric) photochemical ozone creation, human toxicity and ecotoxicity (Baumann and Tillman 2004). Life cycle studies covering just one environmental impact are not strictly LCAs, but rather footprint analyses (like carbon footprints), although they are nevertheless often referred to as LCAs. An example is a comparative 'LCA' by Taelman et al. (2015) of the cumulative exergy extraction from the natural environment of two seaweed cultivation infrastructures or a comparative 'LCA' by Alvarez-Hess et al. (2019) for the greenhouse-mitigation potential of 3-nitrooxypropanol and nitrate to cattle in cattle feeding.

The ISO-14040 series, notably the ISO 14044 (2006), provides the procedure for performing an LCA. An LCA consist of four phases:

1. Goal & scope definition, which specifies why and how an LCA is performed.
2. Inventory analysis, which quantifies all environmental and economic inputs and outputs for all processes in a product system.
3. Impact assessment, which converts all environmental inputs and outputs, from inventory analysis to a range of environmental impacts.
4. Interpretation, which evaluates the results of inventory analysis and impact assessment against the background of the defined goal & scope, in order to draw conclusions.

These are discussed below.

Goal & scope definition is the most important phase of an LCA, as it sets out how the other three phases are to be executed. In this first phase, all selected methodological choices and assumptions are primarily based on their appropriateness in answering the research questions at hand. This notably applies to defining the functional unit (in terms of which impacts will be expressed), delineation of the product system, choice of inventory methods and data for inventory analysis (e.g. allocation procedures and data requirements), impact assessment (e.g. impact categories and characterisation methods) and interpretation.

The ISO 14044 (2006) does not provide the actual methodology for performing an LCA, but there are a couple of handbooks providing detailed guidance. These handbooks are often made on behalf of a regional government. A widely used handbook is the one from the European Union, describing the so-called ILCD-methodology (ILCD 2010). Such handbooks often are of a technical character (e.g. Guinée 2002). An accessible handbook on LCA, albeit slightly outdated, is still the *Hitch Hiker's Guide to LCA* by Henrikke Baumann and Annemarie Tillman (2004). However, one can easily find a more recent alternative by searching online for 'LCA' and 'textbook'.

3 Case study: the Seafarm life cycle assessment

3.1 Choice of case study

One of the critical challenges in the LCA is a lack of available information to compile reliable and case-based inventory data across whole product life cycles. Good quality inventory analysis data is particularly difficult to obtain for seaweed product systems. This is particularly the case for the cultivation, manufacturing and subsequent life cycle stages, which are still evolving in an industrial context in Europe.

Many of the seaweed-related LCA studies in the literature focus on hypothetical production systems and limited experimental data (e.g. Langlois et al. 2012; Alvarado-Morales et al. 2013; Aitken et al. 2014; Czyrnek-Deletre et al. 2017; Van Oirschot et al. 2017; Parsons et al. 2019). Some studies also use data from pilot facilities (e.g. Taelman et al. 2015; Jung et al. 2017; Vijay Anand et al. 2018) and extrapolate potential future large-scale production systems (e.g. Seghetta et al. 2016, 2017). Inventory data are often based upon a mix of literature and personal communications, or partially based on data from case studies, experiments or established farming practices covering all processes from the hatchery through to marine cultivation to seaweed harvesting. The earliest LCA studies on seaweed product systems, such as that by Langlois et al. (2012), which itself is based on a hypothetical system, therefore form the foundation of the literature upon which many of the subsequent studies are based.

The Thomas et al. (2021) seaweed LCA study is, to our knowledge, the only one featuring inventory analysis wholly based on reliable data from a commercial operation, including detailed descriptions of kelp farm operations and a detailed inventory of activities, thus ensuring a high degree of certainty with respect to the life cycle inventory. These case data originate from the Seafarm project (www.seafarm.se) and are updated in line with optimised practices as established by Nordic Seafarm AB (formerly Koster Alg AB). The study of Thomas et al. (2021) is delimited to a cradle-to-gate perspective (where the gate is the production of preserved seaweed), to focus the assessment on those parts of the product system that are well established and from which high-quality data can be obtained. The rest of this chapter is largely based on the results from Thomas et al. (2021).

3.2 The Seafarm project

Five Swedish universities have worked together on the Seafarm project to develop a sustainable system for farming and processing seaweeds, in particular kelp (the brown seaweed *Saccharina latissima*). The end-of-project conference in June 2020 was delayed owing to the COVID-19 pandemic. Therefore, at the time of writing this chapter, the Seafarm project had not yet officially ended, but most of its research activities had been completed (Potting et al. submitted).

For research purposes, the Seafarm project was developed at a farming site on Sweden's west coast (see Fig. 2). This included a trial hatchery, cultivation, harvesting and processing operations (e.g. a biorefinery), to study both the commercial viability and the environmental impacts on the local ecosystem. The Seafarm project thus aimed to lay the foundations for a future Swedish seaweed industry. The project also broke new ground in addressing associated regulatory and licensing hurdles, which had stalled progress on a previous project (e.g. BioMara in Scotland). The project consisted of four research Focus Areas (FA) looking into specific parts of the supply chain, centred around a fifth FA. This fifth FA was strategically leading and connecting the four other FAs through a transdisciplinary sustainability assessment (Thomas 2018). The LCA by Thomas et al. (2021) formed a key part of this sustainability assessment.

The LCA by Thomas et al. (2021) serves to illustrate the environmental impacts of the supply chain for preserved seaweed (i.e. excluding downstream processing into products derived from seaweed). This section provides details of the cultivation system employed in the Seafarm project, to provide readers with a deeper understanding of how the LCA results in the case study may relate to different seaweed supply chains. The subsequent supply chain stages covered by the case study (see Fig. 1) are described in more detail in later sections. Readers should refer to Thomas et al. (2021) for full details of the precise methodology and data used in the case study LCA.

The cultivation system employed for the Seafarm project is located in the Kosterhavets National Park along the Skagerrak coast, within 5 km of the Sven Lovén Centre for Marine Sciences in Tjärnö, which is a part of the Gothenburg University. Many of the practical aspects of seaweed farming take place there. The key stages range from the cultivation, preparation, monitoring, harvesting, maintenance, the deployment of the hatchery-producing strings seeded with

Figure 2 Seafarm-cultivation in the Kosterhavets National Park, Spring 2017 (photo: Gunnar Cervin & Göran Nylund).

Published by Burleigh Dodds Science Publishing Limited, 2021.

juvenile kelp on longlines at sea and the housing of materials. The system covers 2 ha with nearly 5 km of conventional longline cultivation infrastructure, that is, longlines at circa 1 m depth kept in place by anchors and buoys (see again Fig. 1; for a complete description of the cultivation system, see Fig. 2 in Thomas et al. 2021).

In late summer, specimens of the brown seaweed *Saccharina latissima*, known as sugar kelp, are collected from the local environment (i.e. from the local gene pool), and brought back to the hatchery facility at Tjärnö. Over the following months, the kelp's reproductive cycle is triggered and juveniles settle on seeded strings (commonly also referred to as collectors). Once the juveniles have become established in late autumn, the seeded strings are deployed on longlines at sea (see Fig. 3). Over the following months, the cultivation site is monitored on a regular basis.

Most of the mature kelp is then harvested between early April and late May. The longlines are raised mechanically onto a harvesting barge. Beyond this harvest window, the kelp can become spoilt by epiphytic growth. During this harvest window, kelp individuals can more than double in size and undergo important changes in structural and biochemical composition (Thomas et al. submitted). In early April, one might expect to harvest around 5 kg m^{-1} longline of smaller, thinner kelp specimens, whereas up to 15 kg m^{-1} longline of larger, thicker kelp specimens can be harvested in May. This equates to a yield potential ranging 12.5–37.5 tons FW ha^{-1}. Early and late-harvested kelp may be more or less suitable for various end uses (Thomas et al. submitted; Vilg et al. 2015). This whole process is repeated every year, from the selection of parent specimen to harvesting.

After being harvested at sea and returned to land, kelp requires processing without delay to avoid decay and to preserve the biomass in a stable state that is practical for both transport and subsequent refining into products. At harvest, the fresh biomass contains approximately 84.9% of dry matter (Thomas et al. 2021), though this can vary to a large extent (Thomas et al. submitted). As a part of the Seafarm project, several alternative preservation approaches were trialled. Trials included hanging the biomass on wooden frames to dry outdoors (Fig. 8), drying the biomass indoors in hot air cabinets, ensilage experiments at the laboratory scale and freezing the biomass in a shipping container. Initially, only the freezing was conducted on a large scale, but later on in the project, large-scale drying methods were developed. Data in Thomas et al. (2021) are representative of larger-scale biomass preservation following scaled adaptations of these four preservation methods.

4 Overall impacts of the supply chain

Figure 4 and Table 1 give an overview of the overall environmental impacts of the seaweed supply chain, from Thomas et al. (2021), up to and including

Figure 3 Above: seeding-line of juvenile *Saccharina latissima* being deployed on a longline at the Seafarm cultivation site in autumn 2015 (photo: Gunnar Cervin & Göran Nylund). Below: harvesting the mature *Saccharina latissima* from a longline in April 2019 (photo: Jean-Baptiste Thomas).

the preservation of freshly harvested seaweed. The differences between the four preservation methods are depicted in Fig. 4, by expressing the impacts of the overall seaweed supply chain relative to the impacts of the worst-performing preservation method (freezing). Overall, the preservation of the freshly harvested seaweed (specifically using freezing or hot air cabinet drying) and the cultivation process (including the system infrastructure) are the two most important processes in terms of impacts. The freezing and hot air cabinet drying methods documented in Thomas et al. (2021) both consume a lot of energy (especially the former) in the course of the 90 days of storage used in the study (see Section 8). The impacts of the cultivation and harvest stages are dominated by the physical infrastructure at sea (see Section 6).

The impacts of the hatchery processes (spore selection and seeding) are relatively small compared to the rest of the supply chain (see Section 5). The two other preservation methods, hang-drying and ensiling, require relatively little energy and material inputs when compared to hot air cabinet drying and freezing, and so have a much reduced environmental impact per ton of preserved kelp (see Section 7). These differences mean that, in practice, overall impacts can vary by more than 50% (on average across all impact categories), depending on which preservation method is selected.

The overview in Fig. 4 also includes some environmental impact mitigation credits (bars below 0% on the y-axis) resulting from carbon and nutrient assimilation in kelp during its cultivation and the associated impact mitigation (see Section 6.3 for more details). Note that this study stops at the gate (the preserved biomass) and does not include downstream emissions when the biomass is processed into products and consumed. However, if the study were to include the processing and use of the kelp, one might argue that the assimilated nutrients and carbon would presumably be released again as emissions when kelp products are consumed. This also applies to the small environmental impact mitigation credits allocated for the co-production of biogas from ensilage effluents. These impact mitigation credits are further discussed in Section 6.2.

The overview in Fig. 4 relates to the case study LCA by Thomas et al. (2021), but other studies in the literature report similar results (Aitken et al. 2014; Taelman et al. 2015; Van Oirschot et al. 2017). These studies also identify cultivation infrastructure and processing of the biomass as impact hotspots. The parallel is not surprising in the case of Van Oirschot et al. (2017), given that the study utilises similar data from the Seafarm project. Aitken et al. (2014) is based on different case data, though the same impact assessment method (CML 2001) is used as in both Van Oirschot et al. (2017) and Thomas et al. (2021).

Although they are based on different situations, studies using other LCA impact assessment methods still come to similar results. Taelman et al. (2015), for instance, compares two cultivation infrastructures by their cumulative energy extraction from the natural environment (CEENE). They also identify cultivation infrastructure as a dominant contributor in both cases, although they both also identify boat fuel consumption as a significant contributor (a finding further discussed in Section 6). In the studies by Seghetta et al. (2016,

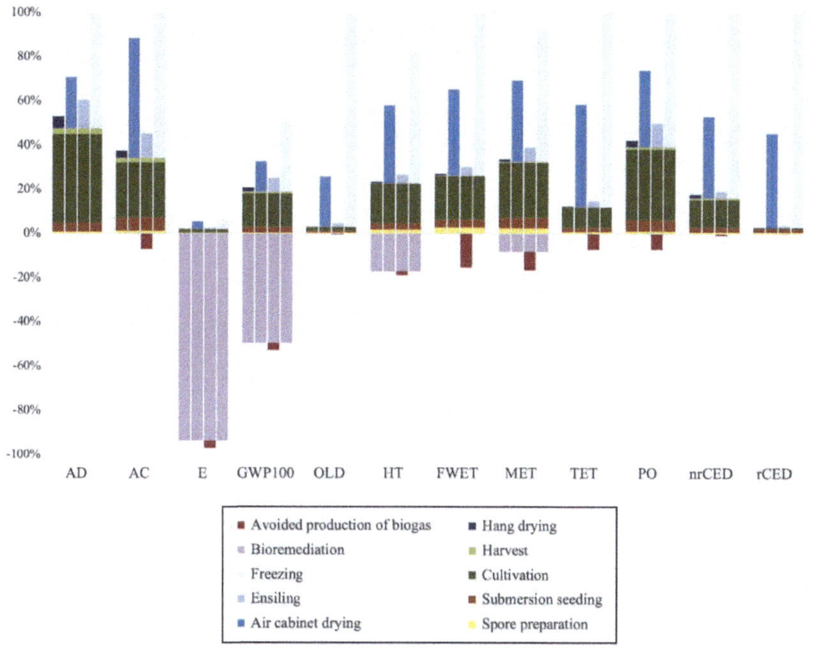

Figure 4 An overview of the environmental impacts of the preserved seaweed supply chain, highlighting the contributions from the main steps of the process chain and comparing the four ultimate preservation methods: hang drying, air-cabinet drying, ensiling and freezing. The graph also shows the impact mitigation credits accorded for bioremediative effects of kelp cultivation (i.e. carbon and nutrient assimilation), and impact mitigation credits accorded to the ensilage co-products, the ensilage effluents, which are used to for biogas production. Figure 4 is supported by Table 1, which includes the numeric values for impacts as expressed per ton fresh weight. The impacts are expressed across 10 impact categories of the CML baseline method: abiotic depletion (AD), acidification (AC), eutrophication (E), climate impact (GWP100), ozone layer depletion (OLD), human toxicity (HT), fresh water ecotoxicity (FWET), marine ecotoxicity (MET), terrestrial ecotoxicity (TET) and photochemical oxidation (PO). These are complemented by 2 additional categories showing Cumulative Energy Demand from non-renewable (nrCED) and renewable (rCED) sources. Source: figure modified from Thomas et al. (2021).

2017), the ReCiPe impact assessment method is applied and complemented by additional impact categories (e.g. USEtox). In both of those studies, the cultivation infrastructure is also identified as a critical hotspot.

To better understand the opportunities for environmental optimisation of supply chains for preserved seaweed, one must dig into each of the stages of the supply chain and look at contributions resulting from specific system components. Seaweed cultivation and downstream processing (e.g. preservation) being the two most influential processes in the product system, provide the greatest opportunities for environmentally optimising the supply

Table 1 An overview of the environmental impacts of the preserved seaweed supply chain

	AD kg Sb eq	AC kg SO_2 eq	E kg PO_4 eq	GWP100 kg CO_2 eq	OLD kg CFC-11 eq	HT kg 1,4-DB eq	FWET kg 1,4-DB eq	MET kg 1,4-DB eq	TET kg 1,4-DB eq	PO kg C2H4 eq	nrCED MJ	rCED MJ
Spore prep (1)	8.45E−03	7.58E−03	3.18E−03	1.15E+00	3.66E−07	2.97E+00	2.33E+00	2.38E+03	1.09E−02	4.33E−04	4.91E+01	2.27E+01
Subm. seed. (2A)	5.37E−02	3.85E−02	8.06E−03	7.59E+00	9.78E−07	5.26E+00	3.35E+00	4.90E+03	2.63E−02	2.03E−03	2.12E+02	6.57E+01
Spray seed. (2B)	4.17E−02	2.82E−02	6.22E−03	5.77E+00	6.40E−07	3.87E+00	2.91E+00	4.25E+03	1.54E−02	1.45E−03	1.49E+02	4.14E+01
Cultivation (3)	5.00E−01	1.63E−01	4.55E−02	4.42E+01	1.82E−06	3.11E+01	1.70E+01	2.57E+04	1.35E−01	1.28E−02	1.14E+03	2.71E+01
Harvest (4)	3.27E−02	1.36E−02	2.72E−03	2.32E+00	4.25E−07	4.69E−01	3.84E−01	5.76E+02	1.67E−03	4.94E−04	7.27E+01	4.71E−01
Bioremediation			−2.82E+00	−1.45E+02		−3.01E+01		−8.35E+03				
Hang drying (5A)	6.64E−02	2.06E−02	3.59E−03	4.97E+00	1.76E−07	1.14E+00	6.30E−01	1.07E+03	4.58E−03	1.09E−03	1.54E+02	7.58E+00
Cabinet dry. (5B)	2.90E−01	3.52E−01	1.06E−01	4.08E+01	2.69E−05	6.23E+01	3.44E+01	3.79E+04	6.82E−01	1.38E−02	3.39E+03	1.85E+03
Ensiling (5C)	1.61E−01	7.22E−02	1.90E−02	1.82E+01	1.96E−06	7.12E+00	3.64E+00	6.71E+03	4.21E−02	4.26E−03	2.73E+02	4.52E+01
AP biogas	−1.31E−05	−4.60E−02	−1.01E−01	−9.53E+00	−5.25E−07	−3.04E+00	−1.34E+01	−8.70E+03	−1.06E−01	−2.87E−03	−7.32E+01	−1.24E+01
Freezing (5D)	6.55E−01	4.28E−01	1.32E−01	9.25E+01	1.15E−04	1.06E+02	6.50E+01	6.11E+04	1.29E+00	2.42E−02	7.70E+03	4.21E+03
Total (5A)	6.61E−01	2.43E−01	−2.76E+00	−8.48E+01	3.77E−06	1.08E+01	2.37E+01	2.63E+04	1.79E−01	1.68E−02	1.63E+03	1.24E+02
Total (5B)	8.84E−01	5.75E−01	−2.66E+00	−4.90E+01	3.05E−05	7.20E+01	5.75E+01	6.31E+04	8.56E−01	2.95E−02	4.87E+03	1.96E+03
Total (5C)	7.56E−01	2.49E−01	−2.85E+00	−8.11E+01	5.03E−06	1.38E+01	1.33E+01	2.32E+04	1.11E−01	1.71E−02	1.67E+03	1.49E+02
Total (5D)	1.25E+00	6.51E−01	−2.63E+00	2.71E+00	1.18E−04	1.16E+02	8.81E+01	8.63E+04	1.47E+00	3.99E−02	9.18E+03	4.33E+03

All impacts are expressed in terms of the functional unit: per ton of fresh weight of kelp going into the preservation process (note that the functional unit is input-based and not output-based).

chain for preserved seaweed. The following sections of this chapter cover all the life cycle stages of the Thomas et al. (2021) case study in detail, starting with the hatchery process (Section 5), then the cultivation stages (Section 6) and finally the preservation stages (Section 7).

5 Environmental impacts of a kelp juvenile hatchery

The hatchery process, which provides juvenile seaweed that can be deployed at sea, involves spore preparation and seeding processes. As Fig. 4 shows, these processes contribute little to the overall environmental impacts of the preserved seaweed supply chain in the case study LCA. On average, across all impact categories, the combined hatchery processes contribute to around 5% of the impacts of the supply chain. Improving the environmental performance of hatchery processes will not, therefore, substantially reduce the overall environmental impacts of the supply chain. However, this does not mean that environmentally optimising this process is irrelevant.

In Thomas et al. (2021), the hatchery is split into two stages. First, the spore preparation process accounts for the activation of the reproductive cycle and the acquisition of a concentrated spore solution from parent specimens. Second, two alternative seeding processes, called submersion seeding and spray seeding, are compared, following the assumption that both produce seeded lines of identical quality (ready to be deployed to sea). The main difference between these processes relates to the stage of the kelp's reproductive cycle at which the settling on the strings occurs. For submersion seeding, this occurs during the formation of sporophytes, while for spray seeding, gametophytes are sprayed onto the collectors.

Figure 5 provides a comparison of the environmental impacts resulting from (a) submersion seeding and (b) spray seeding, including the impacts of all equipment, energy and materials used. On the whole, it is clear that submersion seeding performed slightly worse than spray seeding. This is due to the fact that, in practice, submersion seeding takes a little longer than spray seeding. The resulting time savings for the spray seeding approach convert to slightly reduced energy consumption and thus to slight reductions in the environmental impacts across all categories.

The energy used in the sub-processes, based on the electricity mix used in Sweden (predominantly hydro and nuclear energy), notably the energy used in temperature control (or cooling) and lighting, are the dominant factors in the environmental impacts of the hatchery. This suggests that hatcheries should, as much as possible, utilise energy-efficient temperature control and lighting systems in order to minimise environmental footprints. In terms of material contributions to impacts, the main contributors to the hatchery process were found to be the acrylic Perspex aquaria and the nylon seeding lines. Further

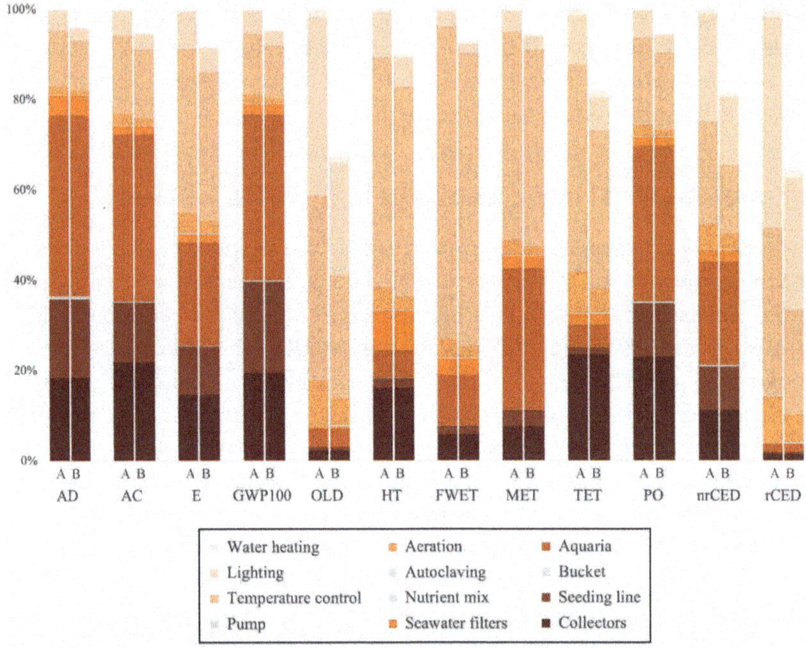

Figure 5 The breakdown of the environmental impacts of the two alternative seeding processes in the hatchery - (a) submersion seeding and (b) spray seeding - showing the relative contributions of sub-processes to highlight impact hotspots. The impacts are expressed across 10 impact categories of the CML baseline method: abiotic depletion (AD), acidification (AC), eutrophication (E), climate impact (GWP100), ozone layer depletion (OLD), human toxicity (HT), fresh water ecotoxicity (FWET), marine ecotoxicity (MET), terrestrial ecotoxicity (TET) and photochemical oxidation (PO). These are complemented by 2 additional categories showing Cumulative Energy Demand from non-renewable (nrCED) and renewable (rCED) sources.Source: the figure has been modified from Thomas et al. (2021).

reduction of impacts might therefore be made by identifying lower-impact alternatives to these materials. These findings are generally supported by LCA literature of microalgae photobioreactors, for example, in Porcelli et al. (2020) and Ketzer et al. (2017), which show that energy use, as well as the use of plastics and other materials, typically dominates the impact profiles.

The impacts of the spore preparation stage are not included in Fig. 5, as this step only results in a fraction of the impacts of the seeding methods. Spore preparation also utilises similar materials and systems as the seeding stages, although the energy requirements are much lower. As a result, any optimisation implemented in the seeding processes will also affect the spore preparation stage.

In addition to the impact hotspots and opportunities for environmental optimisation mentioned above, one additional aspect can multiply impacts from the hatchery. This is the overall failure of the hatchery to deliver high-quality seeded lines. Hatchery failure can be due to a number of reasons, for example, the contamination by epiphytes or bryozoans as the juveniles mature. The consequences can be severe in terms of life cycle impacts. A failure will result in the need to re-attempt the hatchery processes, resulting in additional energy requirements in the second attempt.

Hatchery failure can also delay the deployment of juveniles to sea by several months. Recent research has suggested that earlier deployments can help to increase the biomass yields at harvest (Handå et al. 2013; Thomas et al. submitted); thus a later deployment from a failed hatchery can also result in decreased yields. This, in turn, would translate to a relative increase in impacts, particularly if the functional unit (in terms of which the impacts are expressed) relates to productivity, for example, per ton of fresh kelp.

On the whole, therefore, though the overall contributions to environmental impacts by the hatchery stage are small, hatchery activities can still have an important effect on the overall supply chain impacts in the event of hatchery failure. The results from Thomas et al. (2021) point towards the deployment (to sea) of high-quality seeded lines in early autumn as one of the most important variables to minimise the life cycle environmental impacts of the production of preserved kelp. Optimisation strategies, therefore, should reduce energy and material requirements where possible, without compromising the production of high-quality seeded-line hatchery techniques.

6 Environmental impacts of kelp cultivation

As mentioned in Section 1, when talking about the environmental sustainability of seaweed, one must distinguish between two types of environmental impacts: the impacts on the immediate marine environment (covered elsewhere in this book), and the impacts resulting from all material and energy inputs of the system, also referred to as life cycle impacts. This section addresses the latter, that is, the life cycle environmental impacts of kelp cultivation. Section 6.1 discusses the environmental impacts of longline-cultivated kelp based on Thomas et al. (2021) and compares it to other kelp LCAs. Section 6.2 discusses the impacts of harvesting the mature kelp and discusses the key related factors that affect impacts, for instance, factors relating to fuel consumption. Section 6.3 discusses the cradle-to-gate environmental impact mitigation resulting from the assimilation of carbon and nutrients in the kelp.

6.1 Environmental impacts of longline cultivation infrastructure

The seaweed cultivation infrastructure featured in the case study LCA is a standard longline system located at a semi-sheltered site, developed as part of the Seafarm project (see Fig. 2). The seaweed cultivation infrastructure consists of 26 polyester silk longlines of 190 m, kept at a depth of 2 m by polyvinyl chloride buoys with connecting polypropylene ropes situated every 10 m along their whole length. An anchoring polyethylene buoy at the end of each longline maintains strong buoyancy via a thick and strong polypropylene rope, a low-alloy steel chain and a low-alloy steel shackle fixed to a concrete anchor on the seafloor. The 26 longlines run parallel to one another with 4 m of access corridor in-between. Another polyester silk longline runs laterally across the midpoints of each longline, to provide additional structural reinforcement. This lateral longline is also held in place by anchoring buoys connected to concrete anchors with thick ropes, chains and shackles. Further details on the life expectancies of each item can be found in Thomas et al. (2021).

Figure 6 shows a breakdown of the life cycle environmental impacts of the material and energy inputs required to build, install, operate and harvest a cultivation rig at sea. The results show distinct similarities with other studies. The ropes, and specifically the polyester silk longlines, are mainly responsible for the impacts across most impact categories, most notably the abiotic depletion and non-renewable cumulative energy-demand categories. These significant impact contributions are due to two contributing factors. The first is that collectively, the longline ropes (which are 16 mm in diameter) are the main material inputs of the system by mass (excluding the concrete anchors). The second contribution is that a considerable amount of non-renewable (fossil) resources are consumed in the production of the polyester silk ropes.

Another evident impact hotspot visible from Fig. 6, most notable for the toxicity categories, is the impact contribution of the galvanised steel chains. The important impact contribution of the chains was also identified in Van Oirschot et al. (2017), though that study features the use of stainless steel (chromium alloy) chains with impacts considerably worse than those of the galvanised steel chains used in the present case study. In addition to the ropes and chains, the combined remaining material inputs to the cultivation stage of the life cycle (buoys, anchors and shackles) account for approximately 20% of the impacts on average across all impact categories. Finally, the monitoring of the site by means of return trips using a dinghy, and the initial transport and installation of the cultivation infrastructure, together represent only a minor share of the impacts - at less than 10% on average across all impact categories.

A key environmental optimisation opportunity therefore lies in finding lower-impact alternatives to these ropes and chains. Taking into account the number of cultivation cycles, either rope alternatives manufactured from durable bio-based materials and suitable for this type of marine application could be identified, or thinner ropes might be utilised to reduce total material inputs and thus reduce impacts. To reduce the impacts from chains, research is

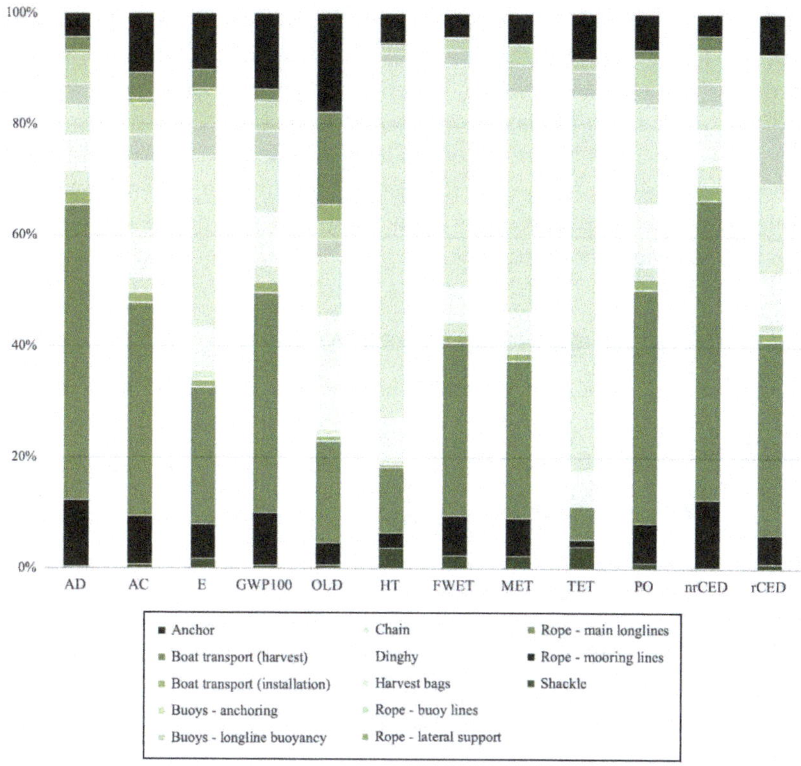

Figure 6 The breakdown of the environmental impacts of the cultivation and harvesting life cycle stages, showing the relative contributions of sub-processes to highlight impact hotspots. The impacts are expressed across 10 impact categories of the CML baseline method: abiotic depletion (AD), acidification (AC), eutrophication (E), climate impact (GWP100), ozone layer depletion (OLD), human toxicity (HT), fresh water ecotoxicity (FWET), marine ecotoxicity (MET), terrestrial ecotoxicity (TET) and photochemical oxidation (PO). These are complemented by 2 additional categories showing Cumulative Energy Demand from non-renewable (nrCED) and renewable (rCED) sources. Source data: the figure has been modified from Thomas et al. (2021).

needed to determine the feasibility of using chain-free infrastructure designs or to identify low-impact alternatives. Recycling these ropes and chains at their end-of-life may be another way to reduce their impact. Similarly, effective maintenance strategies should be developed to extend the life expectancy of these key infrastructural components.

It should be noted that most seaweed cultivation sites along the European and American coasts initially began as pilot sites. The primary concern of these pilots would have been to explore seaweed cultivation feasibility and infrastructure survivability, given local conditions. Many of those cultivation infrastructures, including the one in the case study LCA, are therefore likely to be over-engineered to ensure survivability. The design and construction of cultivation infrastructure should, of course, respect local marine conditions and other requirements. However, material and structural optimisation is also important, to optimise the economic and environmental viability of seaweed farming. Seaweed LCA literature contributes to these areas by highlighting the need for optimisation of the environmental performance of farm structures. Combining LCA approaches with survivability and cost-optimisation remain important areas for further research.

6.2 Environmental impacts of harvesting

The harvesting activities discussed in Thomas et al. (2021) are limited to the transport of the kelp from the cultivation site to a dock approximately 10 km away, and the harvest bags used to pack the seaweed as it is harvested. As can be seen from Fig. 6, together they account for less than 10% of the impacts on average across all impact categories. It should be noted, however, that this is principally due to the fact that the cultivation site is within a relatively close proximity to the dock (around 10 km). Furthermore, the low-impact contribution of the boat transport may also be because the Ecoinvent database, from which boat impacts were exported, lacks impact data for small to medium-scale boat transportation at sea. In the Thomas et al. (2021) study, transport by motorised barge was selected and adjusted by adding the estimated fuel consumption.

In other LCA studies that involve marine transport, results often show that emissions from boat transportation represent significant impact hotspots. This may be because alternatives to Ecoinvent may have been used, different types of fuel were employed or longer transport distances were involved. LCA studies that have identified marine transport as an impact hotspot highlight the need to minimise marine transport distances or to utilise more efficient vessels. This has been highlighted as an area of notable uncertainty requiring further research.

6.3 Impact mitigation by carbon and nutrient assimilation

Environmental impact categories in LCA studies can be based on a range of metrics: primary energy demand and resource depletion or emissions to the environment of specific compounds and their equivalents. Climate impacts within the CML 2001 method, for instance, are measured in terms of kilograms of CO_2-equivalent emissions, which include emissions of methane and other greenhouse gases, converted by equivalency factors to CO_2 equivalents. Similarly, eutrophication impacts within the CML 2001 method are measured in terms of kilograms of PO_4-equivalent emissions, which include emissions of nitrates, nitrogen and phosphorus converted by equivalency factors to PO_4 equivalents.

In line with this methodology, LCA studies by Seghetta et al. (2016, 2017) were designed to take into account the biogenic carbon and nutrient assimilation by kelp, and their subsequent removal from marine environments as the kelp is harvested, an approach also followed by Thomas et al. (2021). This removal from marine environments is considered to be the opposite of emissions to those environments (i.e. an assimilation or an uptake). Carbon uptake is effectively an (indirect) uptake of atmospheric CO_2 that has dissolved in marine environments as part of the marine carbon cycle (Duarte et al. 2017). Impact mitigation credits are therefore allocated for the removal of kelp from marine environments corresponding to the amounts of PO_4 and CO_2 equivalents contained in the harvested biomass.

This methodological approach applied to the Thomas et al. (2021) and Seghetta et al. (2016, 2017) studies yields similar results in both cases. In the case of Thomas et al. (2021), Fig. 4 shows that most of the carbon emissions from supply chain activities are counteracted by CO_2-equivalent uptake by the biomass. Similarly, the eutrophication mitigation potential resulting from the uptake of PO_4 equivalents far outstrips the emissions of PO_4 equivalents from supply chain activities. These results therefore suggest that the production of kelp can (potentially significantly) mitigate local eutrophication, while providing a biomass raw material for the bioeconomy that performs well from a climate perspective.

7 Environmental impacts of preservation: to dry, freeze or ensile?

As mentioned in Section 3, when kelp is harvested, it requires rapid processing to stabilise and preserve the biomass. The four alternative approaches compared in Thomas et al. (2021) are based on scaled processes that were either trialled during the Seafarm project or that have become established practices. As can be seen from Fig. 4, and particularly in Fig. 7, there is a strongly diverging contribution of the four preservation methods to the overall environmental impacts of the preserved seaweed supply chain in the case study. In each impact category, freezing dominates the impact contributions, followed closely by air cabinet drying. The ensiling and hang-drying processes were found to have far lower contributions, primarily as these require little to no energy for their operation.

One key parameter separates these four preservation methods: energy consumption. Those processes with greater energy consumption typically also have larger impact contributions. It is important to note that, once ensiled or dried, biomass does not require constant energy inputs in order to keep it

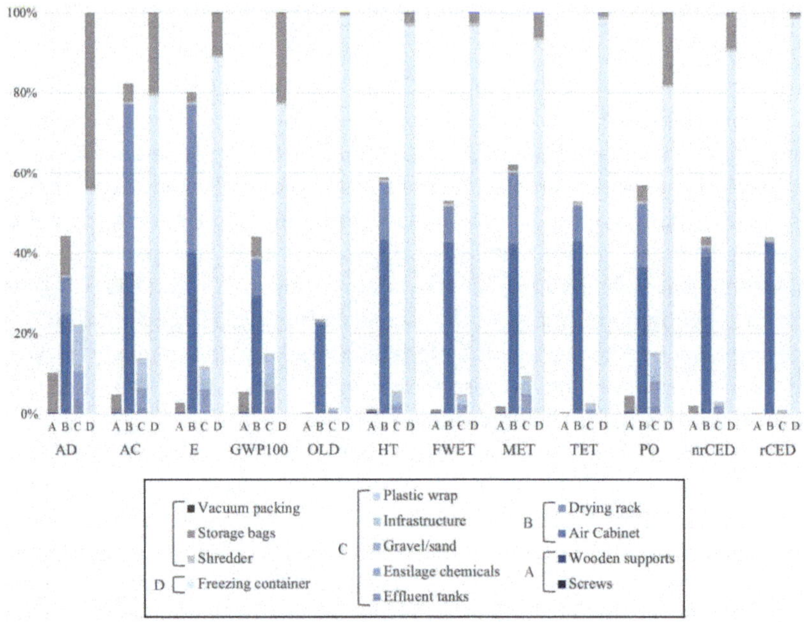

Figure 7 A comparison of the environmental impacts of the four alternative preservation methods, compared in Thomas et al. (2021): (a) hang drying, (b) air-cabinet drying, (c) ensiling and (d) freezing. The figure also presents a breakdown of the relative contributions of sub-processes for each preservation method, to highlight impact hotspots. The impacts are expressed across 10 impact categories of the CML baseline method: abiotic depletion (AD), acidification (AC), eutrophication (E), climate impact (GWP100), ozone layer depletion (OLD), human toxicity (HT), fresh water ecotoxicity (FWET), marine ecotoxicity (MET), terrestrial ecotoxicity (TET) and photochemical oxidation (PO). These are complemented by 2 additional categories showing Cumulative Energy Demand from non-renewable (nrCED) and renewable (rCED) sources. Source: the figure has been modified from Thomas et al. (2021).

preserved over time, whereas frozen biomass does require constant energy inputs to maintain cold storage. A storage time of 90 days (approximately three months) was used in Thomas et al. (2021), helping to explain why freezing has a much greater impact. This study highlights that frozen biomass should not be stored for extended periods unless in very efficient cold storage. However, whilst the production of frozen biomass had the worst performance, it also had the greatest variability.

Drying in an air cabinet was the second most energy-consuming process, and as can be seen from Fig. 7. The principal material input, stainless steel drying racks, also has a significant impact contribution in some of the impact categories, notably the acidification potential and eutrophication potential impact categories. It could be argued that this is an over-estimation, as in practice, such racks would be utilised for many purposes in the course of their life. However, they are assumed in the present study to have been exclusively built and used for this air cabinet drying method in the course of their estimated 10-year life expectancy. In summary, the drying process still dominates impacts compared to preservation methods that do not consume energy. Further research is needed to explore alternative drying methods, notably those that can be automated at scale, to identify energy-efficient methods and to explore the effects of different parameters (e.g. infrared drying or hot air drying) on biomass macromolecular and nutritional content.

The ensiling method in the study involved the use of a large concrete box built specifically for kelp ensilage. The process involves spreading kelp in the ensilage box (referred to as infrastructure in Fig. 7) using a small tractor, spreading of ensiling chemicals in the pile, and then covering the pile with a plastic sheet weighted down by sand and gravel to ensure that it is air-tight. These material inputs combined were found only to have a minor impact relative to other methods. The co-product of the ensilage, known as ensilage effluent, is also a liquid with relatively high biogas potential. Environmental impact mitigation credits were allocated in Thomas et al. (2021) to account for the production of biogas from this effluent. However, even though ensiling kelp performs well from an environmental point of view, the acidic and semi-decayed state of ensiled kelp limits its use to relatively few applications, many of which are still the subjects of research and development.

The hang-drying method described in Thomas et al. (2021) was found to have the lowest impacts. However, it may be the least scalable, since it is the most labour intensive. The main impacts can be entirely attributed to production of the wooden frames from which the kelp hang (see Fig. 8). In practice, this method is entirely weather dependent. Drying methods have therefore evolved into a two-step process, starting with an initial outdoor drying process protected by an agricultural polytunnel. Once partially dried, the kelp can be brought indoors to be finished off under more controlled conditions in heated and well-ventilated indoor spaces.

All of the preservation methods involved additional processes, either relating to shredding the biomass prior to preservation, or to packing the biomass so that it can be stored safely after preservation processes are complete. These can also be seen in Fig. 6 as the grey parts of the bars. Though these additional processes or items are nearly insignificant for most preservation methods and impact categories, plastic storage bags for freezing were found to have a relatively large impact contribution, especially in terms of abiotic depletion. This is due to the frozen biomass being relatively voluminous and thus requiring far more bags to store one ton of fresh biomass, than drying methods which involve packing a lower volume of dried biomass.

Figure 8 Above: Early trials of outdoor hang-drying in spring 2017 at the Sven Lovén Centre, Tjarnö (photo: Gunnar Cervin & Göran Nylund). Below: Polytunnel-protected outdoor drying in Spring 2019 at the Koster Alg facility on the Swedish west coast (photo: Jean-Baptiste Thomas).

8 Conclusion

The environmental performance of seaweed cultivation and processing systems is a complex puzzle. It consists of both direct local effects and effects resulting from the production and the end-of-life of material and energy inputs to the system. These latter types of impacts can be gauged by means of environmental LCA. This chapter presents an overview of the environmental performance of a kelp hatchery, cultivation and preservation system, based primarily on an LCA by Thomas et al. (2021). The knowledge gained by conducting an LCA on these types of production systems enables informed environmental optimisation of supply chain activities.

The main impact hotspots identified in both Thomas et al. (2021) and other LCA studies are a result of the physical cultivation infrastructure at sea, notably the robust and marine-grade ropes that are utilised in longline systems. Some studies also report large impact contributions from the processing of harvested

Published by Burleigh Dodds Science Publishing Limited, 2021.

biomass. Thomas et al. (2021), for example, found that freezing and air cabinet drying, both high-energy preservation methods, are important impact hotspots in the process chain. Infrastructure at sea and high-energy preservation methods are thus life cycle stages where environmental optimisation can have important impact-reduction effects.

Some studies in the LCA literature have also found that the transport of materials at sea can also contribute significantly to environmental impacts. This is notably the case in production systems that require transportation over longer distances. This is not the case in Thomas et al. (2021), where the short distance from port to farm resulted in a low-impact contribution from sea transport.

Other sub-processes, such as those of the juvenile hatchery or the alternative preservation methods of hang-drying and ensiling, were found to have lower environmental impact contributions. However, environmental optimisation strategies should still be pursued, typically based on implementing energy-efficient technologies, or of strategies that can mitigate critical risks such as hatchery failure.

Some LCA studies (notably Seghetta et al. 2016, 2017; and Thomas et al. 2021) attempt to account for the uptake of nutrients and carbon assimilated by harvested biomass from marine environments. These nutrient and carbon flows from marine environments to seaweed products (and thus back to society) attract environmental impact mitigation credits, thus (partially) offsetting impacts resulting from emissions. The result is that these product systems are found to perform very well in terms of mitigating eutrophication and from a carbon perspective.

9 Limitations of LCA

Opportunities abound for methodological development of the LCA in the coming decades. Many environmental aspects of product systems are not well covered by the LCA, notably the effects on the immediate local environment. In seaweed product systems, this could include the effects on local biodiversity, resulting from the presence of marine infrastructure. Ecosystem services are also typically not covered in the LCA. To study local impacts, alternative forms of assessment are needed. Research is ongoing to incorporate these perspectives into LCA methodology.

The spatial context of environmental impacts is also an area of LCA methodological development in need of further development. Eutrophication or oligotrophic conditions, for instance, are highly localised environmental states, each being particularly sensitive or resistant to different actions. The addition of nutrients to an oligotrophic or eutrophic lake will have very different consequences, with the former likely to shift to a eutrophic state, whereas the addition of nutrients to an already eutrophic lake will likely not cause a major regime shift. In LCA, these sorts of contexts of pre-existing environmental states and the sensitivity of these states is usually considered.

Another concern regarding LCA relates to its wider use in decision making and in reports: life cycle perspectives are often overlooked as these can be tricky to grasp by non-experts. A recent example of this is the latest Seaweed for Europe report 'Hidden champion of the ocean: seaweed as a growth engine for a sustainable European future', which omits recognition of the wide range of LCA studies on the topic. This could be considered to be symptomatic of the complexities of the LCA methods and interpretation, or perhaps of the lack of effective LCA communication strategies for research output. Regardless of the reasons, this embodies a significant challenge for future LCA developments. Efforts need to be directed to bridge the gap between LCA research and the communication of clear and useful outputs, for instance, by using the Sustainable Development Goals.

10 Future trends in seaweed-related LCA

Over the past ten years, seaweed-related LCA studies have attempted to shed light on the environmental performance of seaweed production systems, particularly within a context of seaweed being utilised as a substrate for bioenergy production. Most of these studies have been explorative in nature, based on proof-of-concept pilots to shed light on cultivation systems, and on hypothetical or experimental data for later life cycle stages (e.g. bioenergy conversion). Several studies have sought to extrapolate scales of production systems based on pilot data, to shed light on the plausible impacts and contributions to sustainability of the sector in the coming decades.

Given the limited availability of data from this emerging field, much of the focus of these LCAs has been on the earlier life cycle stages, notably the cultivation stage, in support of optimising the pilot cultivation infrastructure systems. In addition to these, data emanating from seaweed biorefinery research projects (e.g. the Danish MAB3 project) have also supported LCAs with the aim of assessing alternative pathways through biorefinery concepts. These scenario-based LCAs are also typically based on laboratory-scale experiments resulting in different combinations of end- and by-products, thus giving a sense of the potential of biorefinery activities.

Looking forward into the coming decade, European seaweed production is at a tipping point, transforming from a series of pilot projects into an emergent sector. Thomas et al. (2021) represents one of the first LCA case studies that shifts LCI focus from hypothetical and scaled production systems to a specific case-oriented pilot/commercial production system, though the scope remains on established early life cycle stages. A myriad of alternative pathways remain for the later life cycle stages, namely biorefinery processing, use-phases and end-of-life of a wide range of seaweed products. Furthermore, as seaweed farming practices scale up and move offshore, follow-up studies to Thomas et al. (2021) will be needed to document the potential of economies of scale.

As new technologies evolve, scales increase and pathway concepts materialise into commercial enterprises for specific products and processing methods, data will become available for robust LCAs to be undertaken. This will enable reliable comparisons to be made between non-renewable or fossil-based products that can be replaced by seaweed-based products (e.g. comparisons of synthetic plastics vs. alginate biopolymers). In summary, as the European seaweed industry emerges in the coming years, more LCAs will be needed to optimise the performance and guide the design principles of cultivation systems, preservation and processing methods, to ensure that the European seaweed industry delivers on its promise as a low-carbon, environmentally beneficial biomass.

11 Where to look for further information

11.1 Seaweed-related LCA articles listed in chronological order of publication

- Langlois, Juliette, Jean-François Sassi, Gwenaelle Jard, Jean-Philippe Steyer, Jean-Philippe Delgenes, and Arnaud Hélias. (2012). 'Life cycle assessment of biomethane from offshore-cultivated seaweed', *Biofuels, Bioproducts and Biorefining*, 6: 387–404.
- Alvarado-Morales, Merlin, Alessio Boldrin, Dimitar B. Karakashev, Susan L. Holdt, Irini Angelidaki, and Thomas Astrup. (2013). 'Life cycle assessment of biofuel production from brown seaweed in Nordic conditions', *Bioresource Technology*, 129: 92–99.
- Aitken, Douglas, Cristian Bulboa, Alex Godoy-Faundez, Juan L. Turrion-Gomez, and Blanca Antizar-Ladislao. (2014). 'Life cycle assessment of macroalgae cultivation and processing for biofuel production', *Journal of Cleaner Production*, 75: 45–56.
- Brockmann, Doris, Charlotte Pradinaud, Jennifer Champenois, Maud Benoit, and Arnaud Hélias. (2015). 'Environmental assessment of bioethanol from onshore grown green seaweed', *Biofuels, Bioproducts and Biorefining*, 9: 696–708.
- Cappelli, Andrea, Emanuele Gigli, Francesco Romagnoli, Silvano Simoni, Dagnija Blumberga, Massimiliano Palerno, and Elisa Guerriero. (2015). 'Co-digestion of macroalgae for biogas production: an LCA-based environmental evaluation', *Energy Procedia*, 72: 3–10.
- Taelman, Sue Ellen, Jennifer Champenois, Maeve D. Edwards, Steven De Meester, and Jo Dewulf. (2015). 'Comparative environmental life cycle

assessment of two seaweed cultivation systems in North West Europe with a focus on quantifying sea surface occupation', *Algal Research*, 11: 173–183.

- Seghetta, Michele, Xiaoru Hou, Simone Bastianoni, Anne-Belinda Bjerre, and Marianne Thomsen. (2016). 'Life cycle assessment of macroalgal biorefinery for the production of ethanol, proteins and fertilizers – a step towards a regenerative bioeconomy', *Journal of Cleaner Production*, 137: 1158–1169.
- Jung, Kyung A., Seong-Rin Lim, Yoori Kim, and Jong Moon Park. (2017). 'Opportunity and challenge of seaweed bioethanol based on life cycle CO_2 assessment', *Environmental Progress & Sustainable Energy*, 36: 200–207.
- Seghetta, M., D. Romeo, M. D'Este, M. Alvarado-Morales, I. Angelidaki, S. Bastianoni, and M. Thomsen. (2017). 'Seaweed as innovative feedstock for energy and feed – evaluating the impacts through a Life Cycle Assessment', *Journal of Cleaner Production*, 150: 1–15.
- Van Oirschot, Roel, Jean-Baptiste E. Thomas, Fredrik Gröndahl, Karen P. J. Fortuin, Willem Brandenburg, and José Potting. (2017). 'Explorative environmental life cycle assessment for system design of seaweed cultivation and drying', *Algal Research*, 27: 43–54.
- Helmes, Roel J. K., Ana M. López-Contreras, Maud Benoit, Helena Abreu, Julie Maguire, Fiona Moejes, and Sander W. K. van den Burg. (2018). 'Environmental impacts of experimental production of lactic acid for bioplastics from *Ulva* spp.', *Sustainability*, 10: 2462.
- Nishikawa, Emily, Meuris Gurgel Carlos da Silva, and Melissa Gurgel Adeodato Vieira. (2018). 'Cadmium biosorption by alginate extraction waste and process overview in Life Cycle Assessment context', *Journal of Cleaner Production*, 178: 166–175.
- Vijay Anand, K. G., K. Eswaran, and Arup Ghosh. (2018). 'Life cycle impact assessment of a seaweed product obtained from *Gracilaria edulis* – a potent plant biostimulant', *Journal of Cleaner Production*, 170: 1621–1627.
- Parsons, Sophie, Michael J. Allen, Felix Abeln, Marcelle McManus, and Christopher J. Chuck. (2019). 'Sustainability and life cycle assessment (LCA) of macroalgae-derived single cell oils', *Journal of Cleaner Production*, 232: 1272–1281.
- Heery, Eliza C., Kay Yee Lian, Lynette H. L. Loke, Hugh T. W. Tan, and Peter A. Todd. (2020). 'Evaluating seaweed farming as an eco-engineering strategy for 'blue' shoreline infrastructure', *Ecological Engineering*, 152: 105857.
- Ögmundarson, Ólafur, Sumesh Sukumara, Alexis Laurent, and Peter Fantke. (2020). 'Environmental hotspots of lactic acid production systems', *GCB Bioenergy*, 12: 19–38.
- Thomas, J. B. E., M. Sodré Ribeiro, J. Potting, G. Cervin, G. M. Nylund, J. Olsson, E. Albers, I. Undeland, H. Pavia, and F. Gröndahl. (2021). 'A comparative environmental life cycle assessment of hatchery, cultivation, and preservation of the kelp Saccharina latissima', *ICES Journal of Marine Science*.

12 References

Alvarez-Hess, P. S., Little, S. M., Moate, P. J., Jacobs, J. L., Beauchemin, K. A.. and Eckard, R. J. (2019). A partial life cycle assessment of the greenhouse gas mitigation potential of feeding 3-nitrooxypropanol and nitrate to cattle, *Agricultural Systems* 169(February), 14-23.

Baumann, H. and Tillman, A. (2004). *The Hitch Hiker's Guide to LCA. An Orientation in Life Cycle Assessment Methodology and Applications*, Sweden, Lund: Studentlitteratur Ab.

Czyrnek-Delêtre, M. M., Rocca, S., Agostini, A., Giuntoli, J. and Murphy, J. D. (2017). Life cycle assessment of seaweed biomethane, generated from seaweed sourced from integrated multi-trophic aquaculture in temperate oceanic climates, *Applied Energy* 196, 34-50.

Duarte, C. M., Wu, J., Xiao, X., Bruhn, A. and Krause-Jensen, D. (2017). Can seaweed farming play a role in climate change mitigation and adaptation? *Frontiers in Marine Science* 4(100). https://doi.org/10.3389/fmars.2017.00100.

Guinée, J. B. (2002). Handbook on life cycle assessment operational guide to the ISO standards, *The International Journal of Life Cycle Assessment* 7(5), 311-313.

Handå, A., Forbord, S., Wang, X., Broch, O. J., Dahle, S. W., Størseth, T. R., Reitan, K. I., Olsen, Y. and Skjermo, J. (2013). Seasonal- and depth-dependent growth of cultivated kelp (*Saccharina latissima*) in close proximity to salmon (*Salmo salar*) aquaculture in Norway, *Aquaculture* 414-415, 191-201.

ILCD (2010). *International Reference Life Cycle Data System (ILCD) Handbook–General Guide for Life Cycle Assessment–Detailed Guidance*, Luxembourg: Publications Office of the European Union.

ISO 14044 (2006). *ISO 14044:2006–Environmental Management–Life Cycle Assessment–Requirements and Guidelines*. ISO/TC 207/SC 5, Switzerland, Geneva: International Standard Organisation.

Ketzer, F., Skarka, J. and Rösch, C. (2017). Critical review of microalgae LCA studies for bioenergy production, *BioEnergy Research* 11(1), 95-105.

Kouchaki-Penchah, H., Sharifi, M., Mousazadeh, H. and Zarea, H. (2015). Gate to gate life cycle assessment of flat pressed particleboard production in Islamic Republic of Iran, *Journal of Cleaner Production* 112(1), 343-350

Lammens, T. M., Potting, J., Sanders, J. P. M. and De Boer, I. J. M. (2011). Environmental comparison of biobased chemicals from glutamic acid with their petrochemical equivalents, *Environmental Science and Technology* 45(19), 8521-8528.

Laurenti, A., Bakas, I., Clavreul, J., Bernstad, A., Niero, M., Gentil, E., Hauschild, M. Z. and Christensen, T. H. (2014). Review of LCA studies of solid waste management systems–part I: lessons learned and perspectives, *Waste Management* 34(3), 573-588.

McHugh, D. (2003). *A Guide to the Seaweed Industry*, Italy, Rome: Food and Agriculture Organization of the United Nations, Rome.

Meadows, D. (2015). *Thinking in Systems. A Primer*, London: Chelsea Green Publishing Co.

Porcelli, R., Dotto, F., Pezzolezi, L., Marazza, D., Greggio, N. and Righi, S. (2020). Comparative life cycle assessment of microalgae cultivation for non-energy purposes using different carbon dioxide sources, *Science of the Total Environment* 721(June), 137714.

Potting, J. and Blok, K. (1995). Life-cycle assessment of four types of floor covering, *Journal of Cleaner Production* 3(4), 201-213.

Potting, J., Thomas, J.-B. E. and Gröndahl, F. (submitted). Stakeholder participation in sustainability assessment of non-wicked problems: the case of a future seaweed industry in Sweden, *AMBIO–A Journal of the Human Environment*.

Sills, D. L., Van Doren, L. G., Beal, C. and Raynor, E. (2020). The effect of functional unit and co-product handling methods on life cycle assessment of an algal biorefinery, *Algal Research* 46, 101770.

Taelman, S. E., Champenois, J., Edwards, M. D., De Meester, S. and Dewulf, J. (2015). Comparative environmental life cycle assessment of two seaweed cultivation systems in North West Europe with a focus on quantifying sea surface occupation, *Algal Research* 11, 173–183.

Thomas, J.-B. E. (2018). Insights on the sustainability of a Swedish seaweed industry. Doctoral thesis in Industrial Ecology, Stockholm, Sweden: KTH.

Thomas, J.-B. E., Sodré-Ribeiro,M., Potting, J., Cervin, G., Nylund, G. M., Visch, W., Olsson, J., Albers, E., Undeland, I., Pavia, H.. and Gröndahl, F. (2021). A comparative environmental life cycle assessment of hatchery, cultivation and preservation of the kelp *Saccharina latissima*, *Journal of Marine Science* 78(1), 451–467.

Thomas, J.-B. E., Sterner, M., Nylund, G. M., Albers, E., Edlund, U., Undeland, I., Welander, U., Gröndahl, F. and Pavia, H. (submitted). The effect of modifying cultivation depth, deployment and harvest times on the chemical composition of *Saccharina latissima*.

Van den Burg, S., Dagevos, H. and Helmes, R. (2018). *Sustainable Seaweed Value-Chains Economics, Consumer Attitudes and Environmental Impacts*, the Netherlands, Wageningen: Wageningen University & Research.

Van Hal, J. W., Huijgen, W. J. J. and López-Contreras, A. M. (2014). Opportunities and challenges for seaweed in the biobased economy, *Trends in Biotechnology* 32(5), 231–233.

Van Oirschot, R., Thomas, J.-B. E., Gröndahl, F., Fortuin, K. P. J., Brandenburg, W. and Potting, J. (2017). Explorative environmental life cycle assessment for system design of seaweed cultivation and drying, *Algal Research* 27(November), 43–54.

Vilg, J. V., Nylund, G. M., Werner, T., Qvirist, L., Mayers, J. J., Pavia, H., Undeland, I. and Albers, E. (2015). Seasonal and spatial variation in biochemical composition of *Saccharina latissima* during a potential harvesting season for Western Sweden, *Botanica Marina* 58(6), 435–447.

Published by Burleigh Dodds Science Publishing Limited, 2021.

Part 3

Applications

Chapter 8

Nutritional and anti-methanogenic potentials of macroalgae for ruminants

Deepak Pandey, Nord University, Norway; Morteza Mansouryar, University of Copenhagen, Denmark; Margarita Novoa-Garrido, Nord University, Norway; Geir Næss, Nord University, Norway; Viswanath Kiron, Nord University, Norway; Hanne Helene Hansen, University of Copenhagen, Denmark; Mette Olaf Nielsen, Aarhus University, Denmark; and Prabhat Khanal*, Nord University, Norway*

1 Introduction

The global human population is rising rapidly and has been projected to be ~10 billion by 2050 (Holt-Giménez, 2019). This implies that by that time, an increase of 60–70% in overall food and ~78% in meat production is required (Estrada et al., 2011; Alexandratos and Bruinsma, 2012; Holt-Giménez, 2019). The livestock sector supplies ~28% of the global protein consumption, but in developed countries this contribution may be as high as 48% (Estrada et al., 2011). Thus, the livestock sector will continue to play a crucial role as a source of high-quality protein for human consumption in the future (Åby et al., 2014). However, ruminant livestock species such as cattle, goats and sheep are responsible for ~17% of the total anthropogenic enteric methane

* Corresponding authors (hhh@sund.ku.dk; prabhat.khanal@nord.no)

http://dx.doi.org/10.19103/AS.2021.0091.14

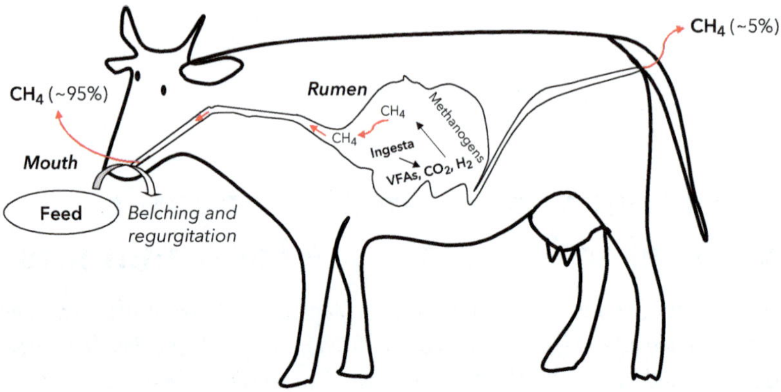

Figure 1 A general outline of methane production and emissions from ruminant animals. Enteric fermentation of ingested feeds occurs in the rumen, where the majority of CH_4 is released from the mouth by eructation. CH_4, methane; VFAs, volatile fatty acids; CO_2, carbon dioxide; H_2, hydrogen.

(CH_4) emissions, via fermentation of feeds in their forestomach (Fig. 1) (Knapp et al., 2014). Ruminants possess a unique digestive system comprised of a four-chambered stomach: rumen, reticulum, omasum and abomasum. The rumen is the residence of a large number of microorganisms, including bacteria, fungi, protozoa and archaea, and these microorganisms play a vital role in feed degradation and energy supply to the host animals (Bergman, 1990; Maia et al., 2016). Feed components, particularly carbohydrates, get partially or completely fermented in the rumen and produce volatile fatty acids (VFAs) such as acetate, propionate, butyrate, and also carbon dioxide (CO_2) and hydrogen (H_2) (Van Nevel and Demeyer, 1996) (Fig. 2). Volatile fatty acids are an important energy source for ruminants, while CO_2 and H_2 may later be reduced to CH_4 by the action of methanogenic archaea before getting eructed from animals into the environment (Bergman, 1990).

Methane is one of the major contributors of global warming and has a 28 times higher global warming potential than another greenhouse gas, CO_2 (Grossi et al., 2019). The CH_4 emission from the rumen represents a loss of up to 15% of gross energy (GE) from the feed, which could otherwise be utilized for animal growth and production (Van Nevel and Demeyer, 1996), and is, therefore, unfavorable for the animal. Enteric methanogenesis is thus both an environmental and nutritional concern, and any interruption in this process could provide nutritional benefits to the animals and result in the release of the less potent greenhouse gases CO_2 and H_2, compared to the highly potent CH_4 (Patra et al., 2017; Grossi et al., 2019). Hence, development of appropriate CH_4 abatement strategies is important to attain sustainable ruminant production systems in the future (Grossi et al., 2019).

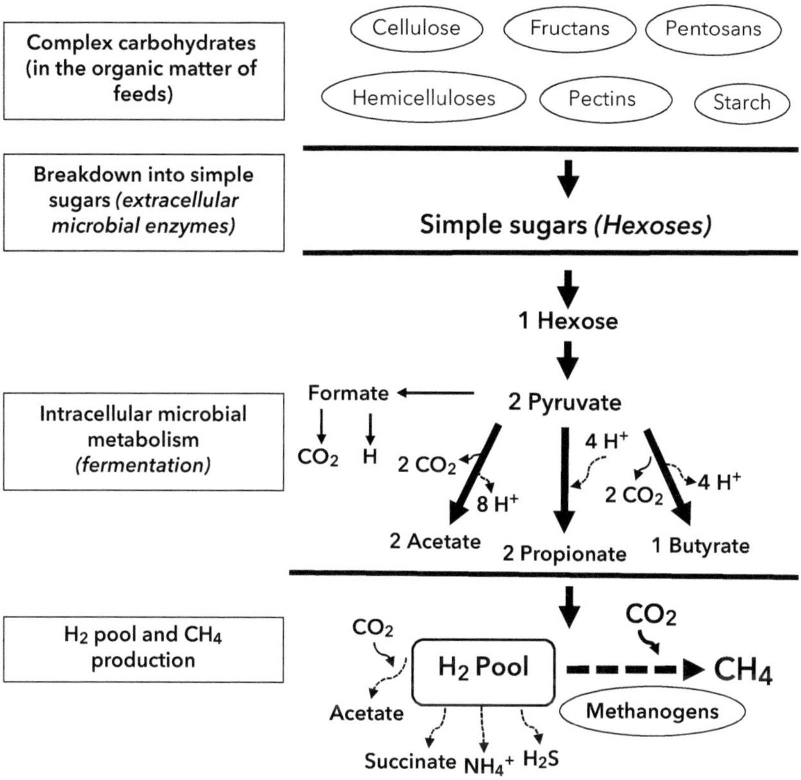

Figure 2 Major pathways involving degradation of carbohydrates, and production of volatile fatty acids and methane by microbial fermentation in the rumen. Adapted with some modifications from (McDonald et al., 2011; Haque et al., 2014; Kohn and Boston, 2000; Ungerfeld, 2013). CH_4, methane; CO_2, carbon dioxide; NH_4, ammonium; H_2S, hydrogen sulfide. Dotted lines represent pathways involving metabolic hydrogen (H) or dihydrogen (H_2).

Several CH_4 mitigation strategies have been suggested to cope with the CH_4 emissions from ruminants. These include (i) mitigation of CH_4 emissions via genetic selection (González-Recio et al., 2020), (ii) use of anti-methanogenic chemical compounds such as nitrate, chloroform and 3-nitroxy propanol (Patra et al., 2017) and (iii) dietary interventions using alternative feed ingredients and nutritional strategies (Knapp et al., 2014). Genetic selection can permanently reduce CH_4 production from an individual animal and can be inherited to the offspring (González-Recio et al., 2020). However, this approach could be technically demanding and time consuming, and convincing outcomes of genetic selection are yet to be obtained (Knapp et al., 2014; González-Recio et al., 2020). The application of anti-methanogenic chemical compounds is an effective strategy in reducing CH_4 emissions; however, their effects can be transitory and such compounds may have adverse impacts on both animal performance and the environment (Patra et al., 2017). Dietary interventions can be a relatively simple and environmentally friendly approach and can lead to no or lower negative consequences to animal health and performance (Haque, 2018; Benchaar et al., 2001; Haque et al., 2014). Despite some advantages, only a modest reduction (5–40%) in CH_4 emissions by dietary interventions has been reported (Benchaar et al., 2001; Knapp et al., 2014). In this context, identification of alternative and novel feed materials that can substantially decrease enteric CH_4 production without compromising animal health and production would be important to develop novel CH_4 mitigating strategies in the future.

Marine macroalgae (also commonly known as seaweeds) have been identified as an alternative feed resource that can largely decrease enteric CH_4 production from ruminants (Machado et al., 2016; Maia et al., 2016). Macroalgae consist of 6000–10000 diverse marine species distributed along the coastal regions worldwide, and they can be categorized into three types based on their pigmentation: brown, red and green (Tiwari and Troy, 2015; Makkar et al., 2016; Rajauria, 2015). Within the three categories of macroalgae, there are large species variations with respect to chemical composition (carbohydrates, proteins and minerals), and how digestible the organic components are in ruminant animals (Makkar et al., 2016; Dawczynski et al., 2007). Additionally, macroalgal species produce a wide range of bioactive components, such as halogenated compounds, polyphenols, complex polysaccharides and pigments (O'Sullivan et al., 2010; Charoensiddhi et al., 2017; Machado et al., 2016). Their bioactivities include antioxidative (Kannan et al., 2007; Ling et al., 2015), anti-microbial, immunomodulatory (Turner et al., 2002; Kim et al., 2018), anti-diabetic (Kang et al., 2016), anti-inflammatory, prebiotic (Cañedo-Castro et al., 2019; Reilly et al., 2008; O'Sullivan et al., 2010) and anti-methanogenic properties (Machado et al., 2014; Roque et al., 2019a). This broad range of chemical activities may enhance the future commercial application of macroalgae as multipurpose feed ingredients (Øverland et al., 2019).

Macroalgae have long been utilized as a feed ingredient for ruminant animals in different parts of the world. In many countries, including Iceland, Norway, France, Germany, Sweden, Finland, Scotland and the USA, macroalgae were used as an occasional or regular animal feed, particularly during extreme winter conditions, when the availability of other feed resources was limited (Makkar et al., 2016; Evans and Critchley, 2014; Chapman, 2012; Hansen et al., 2003; Applegate and Gray, 1995). However, there are very few published studies on the application of macroalgae as commercial and regular feed resources for ruminant animals. A brown macroalgae, *Ascophyllum nodosum*, has been reported to be used in small amounts as a feed additive for dairy cows in some organic farms in the USA (Erickson et al., 2012). North Ronaldsay (Orkney) sheep in Northern Scotland are purported to survive by grazing on different macroalgal species: *A. nodosum,*

Alaria esculenta, Fucus spp., *Laminaria* spp., *Saccharina latissima* and *Palmaria palmata* (Hansen et al., 2003; Makkar et al., 2016). However, their commercial application for farm animals on a large scale is yet to be achieved.

Macroalgal species within all three (red, brown and green) categories have been identified to have CH_4 mitigating properties both *in vitro* and *in vivo* (Machado et al., 2014; Maia et al., 2016; Belanche et al., 2016b). However, when using some macroalgal species as feed, rumen fermentation patterns and total tract digestibility may be negatively affected due to high contents of ash and complex carbohydrates of low rumen degradability (Bikker et al., 2020). This can reduce the overall animal performance, particularly when such macroalgae are fed in large amounts (Bikker et al., 2020). Hence, the implications of the anti-methanogenic properties of macroalgae must be evaluated based on their overall impacts on feed intake, digestibility and animal performance. To be able to exploit macroalgae as potential feed resources, it is essential that species of commercial relevance be extensively characterized from both a biochemical (including anti-methanogenic compounds) and a nutritional point of view as presented in Fig. 3.

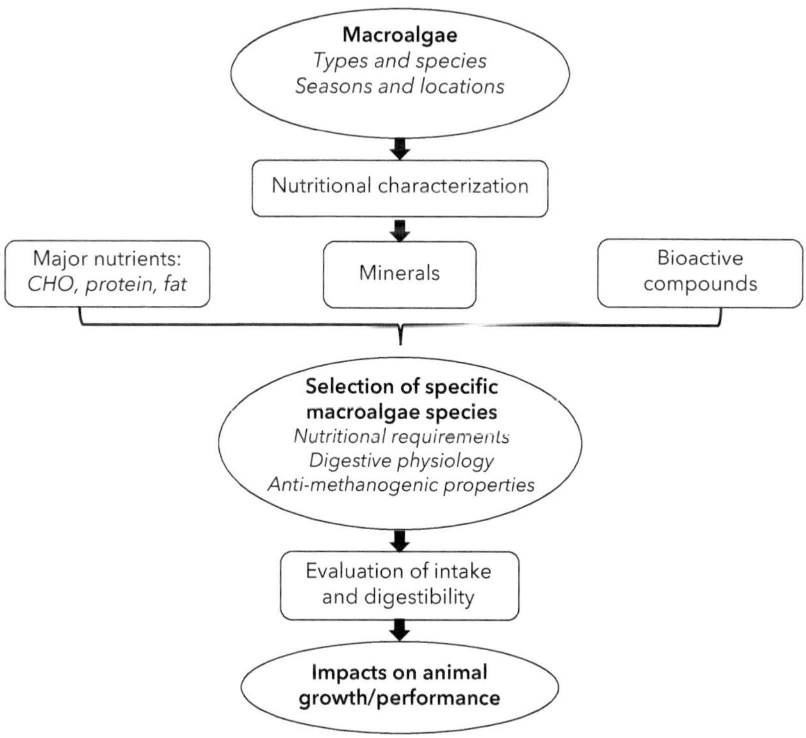

Figure 3 A flowchart for the evaluation of macroalgae as a potential ruminant feed ingredient in future. CHO, carbohydrates.

This chapter aims to evaluate the role of macroalgae as a potential anti-methanogenic ruminant feed resource. Similarly, effects of different intrinsic (macroalgal species, types) and extrinsic (growing season, post-harvest processing) factors on nutritional value as well as concentration of bioactive compounds and anti-methanogenic properties will be discussed. This will enable us to evaluate whether macroalgae can be used as anti-methanogenic dietary additive without compromising overall animal production and performance.

2 Nutritional value of macroalgae

Fresh macroalgae biomass normally contains about 70–90% water and various macro- and micro-nutrient fractions (Kılınç et al., 2013; Biancarosa et al., 2017). In this section, protein, carbohydrate, mineral and lipid contents of various macroalgal species will be described, and their potential as ruminant feeds will be evaluated. Unless otherwise stated, the contents are reported as % of dry matter (DM) to allow comparisons.

Protein: Red macroalgae species generally contain greater levels of crude protein (CP) than brown and green species. Some red species belonging to the genera *Palmaria, Pyropia* and *Porphyra* have been reported to contain 20–50% CP (Tibbetts et al., 2016; Fernández-Segovia et al., 2018; Marsham et al., 2007; Jung et al., 2016). The green macroalgae *Acrosiphonia* spp. and *Ulva* spp. also contain high levels of CP (appr. 31% and 25%, respectively) (Biancarosa et al., 2017; Peña-Rodríguez et al., 2011; Tayyab et al., 2016), whereas CP levels in most of the brown macroalgae are <15% (Dawczynski et al., 2007; Biancarosa et al., 2017). Thus, red and green macroalgae are the most relevant to consider as protein sources for animals.

Macroalgae proteins are reported to have a high quality due to their high proportion of essential amino acids (EAA) (Angell et al., 2016; Mišurcová, 2012). The red species *P. palmata, Porphyra* spp. and *Vertebrata lanosa*, the brown species *A. nodosum* and *Undaria pinnatifida* and the green species *Enteromorpha intestinalis* (*Ulva* sp.) have a higher EAA index and are thus considered to be superior compared to cereals from a nutritional point of view (Mæhre et al., 2014; Dawczynski et al., 2007; Gaillard et al., 2018). The EAA proportion in macroalgae can account for 45.7% of the total amino acids, which is similar to that of the conventional protein feed resource soybean meal (46%) and greater than fishmeal (43.4%) (Angell et al., 2016; Dawczynski et al., 2007; Biancarosa et al., 2017). Although the requirements for EAA would vary based on specific animal parameters (e.g. age, growth stage, production purpose), the EAA content of selected macroalgal species are reported to be able to fulfill the human and animal requirements (Mæhre et al., 2014). Therefore, selected macroalgae, particularly red and green species, could be considered as alternative sources of quality feed protein but their biomass yield and technologies for large-scale cultivation must be taken into account. This is, however, beyond the scope of the present chapter.

The significance of alternative proteins in ruminant nutrition depends on their digestibility. Studies regarding *in vivo* digestibility of macroalgae proteins are relatively scarce; however, *in vitro* protein digestibility (IVPD) has been explored for a number of species. The IVPD of the red macroalgae *Chondrus crispus, P. palmata, Sarcodiotheca gaudichaudii* and *Meristotheca papulosa* have been found to be ~85% of the total CP content, whereas IVPD for the brown species: *A. esculenta, A. nodosum, Fucus vesiculosus* and *S. latissima* are reported to be slightly lower (~80%) (Tibbetts et al., 2016). In ruminants, a significant part of the feed CP is degraded via microbial action in the rumen and subsequently utilized in microbial protein synthesis, including synthesis of EAA (Hvelplund and Weisbjerg, 2000). The amount of protein that passes un-degraded by the microbes to the small intestine is called rumen escape protein (REP). The bioavailability and amino acid composition of this fraction becomes particularly important, when feed protein degradability and hence microbial protein supply from the rumen is low (Hvelplund and Weisbjerg, 2000). An in situ study illustrated that 50–70% of the CP from *A. esculenta, L. digitata* and *P. palmata* is degraded in the rumen within 24 h, while for other species including *M. stellatus, Ulva* and *Pelvetia canaliculata* rumen CP degradability was substantially lower (<35%) (Tayyab et al., 2016). Hence, for many of the above mentioned macroalgal species, a large proportion (30–51% of total CP) of protein supply to the small intestine will be REP, and the intestinal digestibility of the REP becomes important for the potential amino acid supply to the animal (Tayyab et al., 2016). In the same study, degradation of CP in the small intestine was negligible for *A. esculenta* and *P. canaliculata*, while this value was similar or greater than the rumen degradability in others (*Porphyra, Palmaria, Ulva, Acrosiphonia, Mastocarpus*) (Tayyab et al., 2016). In addition, in situ total tract amino acid degradability of *Porphyra* sp. and *P. palmata*, and green macroalgae *Cladophora rupestris* and *Ulva* sp. has been found to be the highest among macroalgal species (Gaillard et al., 2018). These studies suggest that green and red macroalgae species are interesting new potential sources of rumen degradable and intestinal digestible protein for ruminants.

<u>Carbohydrates</u>: Carbohydrates are generally the most abundant organic compounds in macroalgae and may account for 25–75% of their DM (Jiménez-Escrig and Sánchez-Muniz, 2000; Rioux and Turgeon, 2015). They comprise both soluble and non-soluble carbohydrates and their relative amounts and composition vary depending upon macroalgae type and species (O'Sullivan et al., 2010). The major carbohydrates in macroalgae are unknown in terrestrial plants, and include alginate, fucoidan, mannitol and laminarin in brown; agar, carrageenan and porphyran in red; and ulvan and xylans in green species

(Cherry et al., 2019). Despite being indigestible in monogastric animals, macroalgae polysaccharides, particularly from brown species, have attracted research interest as prebiotics due to their beneficial gut impacts (O'Sullivan et al., 2010). These polysaccharides can partially or completely be fermented in the hindgut by the action of specific gut commensal bacteria producing short-chain fatty acids, and can thereby contribute to inhibit the growth of gut pathogens, such as *Clostridium* spp., *Escherichia coli* and *Salmonella* spp. (Braden et al., 2004; Seong et al., 2019). The prebiotic effect of macroalgae polysaccharides has mostly been studied in non-ruminant animals, including weanling piglets and humans (Reilly et al., 2008; Smith et al., 2011), and information about prebiotic effects for ruminant animals, containing relatively complex digestive systems, is limited. However, Tasco-14, an *A. nodosum*-based commercial additive, has been found to be effective in reducing the fecal shedding of *Escherichia coli* (O157:H7) and *Salmonella* spp. in feedlot cattle and lambs when supplemented in the diet at 2% DM basis (Braden et al., 2004; Bach et al., 2008). Further studies are needed to identify whether such reduced fecal shedding is due to the action of polysaccharides in the hind gut of cattle.

The nutritional value of macroalgae polysaccharides for ruminants depends on whether they can be degraded by the microbial population in the forestomach. Studies on the rumen microbes isolated from macroalgae-fed Ronaldsay sheep have revealed that polysaccharides from brown species, that is, alginate, fucoidan and laminarin, can variably and only partly be degraded by selective rumen microorganisms (Orpin et al., 1985; Williams et al., 2013). Only nine, out of 65, cultured isolates of rumen microorganisms were able to degrade >90% of the laminarin and 70–80% of alginates, but <20% of the fucoidans (Williams et al., 2013). The rumen microorganisms involved in the degradation of macroalgae carbohydrates include *Prevotella* spp., *Clostridium butyricum, Streptococcus bovis, Selenomonas ruminantium, Butyrivibrio fibrisolvens* and *Dasytricha ruminantium* (Orpin et al., 1985; Williams et al., 2013). However, as only a limited number of rumen microbes were included in the studies due to problems associated with microbial cultivation in artificial media, results from these *in vitro* fermentations may not be representative of the whole *in vivo* scenario of rumen degradability of macroalgal polysaccharides. Hence, further *in vivo* studies evaluating the digestibility of these polysaccharides are important to establish their nutritional value.

Minerals: Although mineral contents of macroalgae are affected by both intrinsic (macroalgae types and species) and environmental factors (culture conditions, seasons etc.), they are generally an excellent source of both macro and trace minerals. They are capable of accumulating a large quantity of minerals from seawater, and hence, the levels of various minerals including iodine, sodium, potassium, iron, chlorine and calcium in macroalgae are found to be 10–20 times higher than the levels found in terrestrial plants and fresh

Published by Burleigh Dodds Science Publishing Limited, 2021.

water algae (Pereira, 2011; Gómez-Ordónez et al., 2010; Makkar et al., 2016; Mišurcová et al., 2010). The capacity of macroalgae to concentrate minerals has been linked to their mineral-rich growing environment and the content of unique cell wall polysaccharides such as alginic acid, salts of alginate, agar and carrageenan that can absorb different inorganic ions from the seawater (Mišurcová, 2012). The ash content in macroalgal species can vary between 20% and 72% of DM (Cabrita et al., 2016; Rupérez, 2002; D'Armas et al., 2019). In general, brown and green species contain higher amounts of minerals than red species (Pereira, 2011; Cabrita et al., 2016; Fernández-Segovia et al., 2018). Due to the abundance of minerals in macroalgae, they are considered natural mineral sources for both livestock and humans; for example, they can be used for the prevention of iodine deficiency (Bańoch et al., 2010).

Minerals are important for normal functioning of different hormones and enzymes in the body (Trumbo et al., 2001; Mæhre et al., 2014). However, due to the high mineral contents such as sodium, chlorine, calcium, iron and iodine in many species (*Codium* spp., *Himanthalia elongata, Laminaria* spp., *Saccharina* spp., *Bifurcaria bifurcata* and *Ulva* spp.), an excess intake of macroalgae-based diets may result in mineral toxicity, particularly in monogastric species as they are at higher risk due to a generally lower tolerance towards excess mineral uptake than in ruminants (Bikker et al., 2020; Cabrita et al., 2016). Excessive uptake of iodine from macroalgae-based ruminant feeds can be excreted in milk or accumulated in body tissues, leading to undesirably high levels of iodine in animal products that can have adverse consequences for human health (van der Reijden et al., 2017). The maximum recommended level of iodine is 2 mg/kg feed for dairy ruminants in the European Union (EU) (Additives and Feed, 2013), due to concerns of toxic levels in ruminant products destined for human consumption. Therefore, an abundant mineral content limits the inclusion of macroalgae on a larger scale in ruminant diets, unless special precautions are undertaken while formulating diets (Bikker et al., 2020).

Macroalgae are also able to concentrate heavy metals such as arsenic, mercury and cadmium from seawater, which are known to have a range of adverse health impacts, such as cancer and renal dysfunctions (McLaughlin et al., 1999). Particularly the contents of inorganic versus organic arsenic must be considered due to the greater toxicity of the inorganic form, although the predominant form of arsenic in macroalgae is normally organic (~ 90%) (Díaz et al., 2012; Mæhre et al., 2014; Biancarosa et al., 2018). The levels of these heavy elements in 21 macroalgal species from the Norwegian coast were found to be far below the maximum tolerable levels set in the EU region (Biancarosa et al., 2018).

Lipids: Macroalgae generally contain a low level of lipids (<5%) (Makkar et al., 2016; Øverland et al., 2019). Lipids from macroalgae are considered beneficial for human health due to their bioactive properties (Mæhre et al.,

2014). However, it can be insignificant with respect to the supply of (essential) fatty acids in ruminant feeds due to their very low lipid content, and most of it is utilized by rumen microbes (Bikker et al., 2020).

3 Digestibility of macroalgae as a feed or feed ingredients

A broad range (20–97% organic matter, OM) of ruminal and post-ruminal degradability of macroalgae has been reported earlier (Table 1). The rumen degradability of brown macroalgae *A. nodosum, F. serratus,* and *F. vesiculosus* has been observed to be low (<33% of OM) when they were used as a sole ruminant diet (Greenwood et al., 1983). Moderate *in vitro* rumen DM degradability (40–65%) has been recorded for other brown (*M. pyrifera, A. esculenta, L. digitata, P. canaliculata*) as well as red (*M. stellatus, P. palmata, Porphyra* sp.) and green macroalgae (*Ulva lactuca* and *Acrosiphonia* sp.) (Gojon-Báez et al., 1998; Ventura and Castañón, 1998; Molina-Alcaide et al., 2017). However, *in vitro* rumen OM degradability for selected brown (*A. esculenta, L. digitata, L. hyperborea, Sargassum* spp., *S. latissima*) and red (*P. palmata*) species was found to be higher (80–89%) (Hansen et al., 2003; Makkar et al., 2016; Marín et al., 2009). Since the later studies were performed using rumen fluid and microbial inoculum obtained from macroalgae-eating Ronaldsay sheep (Orkney), the greater degradability could be the function of potential adaptation of rumen microbes to those particular macroalgae (Hansen et al., 2003). Hence, a gradual increase in the digestibility of macroalgae may be observed over time and therefore, animals may require an adaptation period to achieve an acceptable digestibility level. However, it is not known whether exposure to macroalgae-based diets at a young age would lead to a better feed digestibility in adults.

Feeding macroalgae at larger doses for a long duration can result in adverse health consequences such as bone and kidney dysfunctions in animals, probably due to mineral overload (Britt and Baker, 1990). Thus, long-term use of macroalgae as sole feed may not be safe, unless excess minerals and potentially toxic heavy metals are removed prior to feeding. Rinsing of macroalgae biomass with fresh cold or hot water (e.g. 40°C) for a short duration (30 min) could be effective in removing excess mineral salts from macroalgae (Magnusson et al., 2016). These types of processing may also enhance palatability of macroalgae and nutrient digestibility, though a loss of soluble nutrients can be expected (Magnusson et al., 2016). Thus, proper post-harvest processing of macroalgae biomass prior to animal feeding may minimize the risks of adverse health impacts on animals and can improve nutrient utilization.

Macroalgae can affect animal feed intake and rumen degradability of the feed, depending upon the inclusion level and the macroalgal species (Choi

Published by Burleigh Dodds Science Publishing Limited, 2021.

Table 1 Effects of macroalgae on rumen fermentation and animal performance

Macroalgae species	Study type	Animal/rumen fluid donor	Dose	Impacts on ruminant nutrition	References
Red macroalgae:					
Asparagopsis armata	*In vivo*	Dairy cow	0.5–1% OM	Reduces CH_4 but also reduces feed intake (−38%), lower milk yield (−11.6%) at 1% OM inclusion	(Roque et al., 2019b)
Asparagopsis taxiformis	*In vitro/ In vivo*	Steers/sheep	0–16.7% OM	Effectively reduces CH_4, enables ADWG at low inclusion, lowers total VFA (≥2% OM inclusion) and OMD at ≥10% inclusion	(Machado et al., 2016; Li et al., 2018; Kinley et al., 2020)
Gracilaria vermiculophyla	*In vitro*	Cow	25% DM	Reduces CH_4, no adverse effects on rumen fermentation	(Maia et al., 2016)
Gigartina sp.	*In vitro*	Cow	25% DM	Reduces CH_4, no adverse effects on rumen fermentation	(Maia et al., 2016)
Mastocarpus stellatus	*In vitro*	Sheep/dairy cow	20% DM	Lowers VFA production and has low in situ DM and crude protein degradability	(Molina-Alcaide et al., 2017)
Palmaria palmata	*In vitro/ in situ*	Sheep/dairy cow	20% DM	High DMD, supplies high level of digestible protein, high amino acid index	(Tayyab et al., 2016; de la Moneda et al., 2019; Gaillard et al., 2018)
Porphyra spp.	*In vitro/ in situ*	Sheep/dairy cow	20% DM	High DMD, provides high digestible and rumen by-pass protein (~50% of total protein content)	(Tayyab et al., 2016; de la Moneda et al., 2019; Gaillard et al., 2018)
Brown macroalgae:					
Alaria esculenta	*In vitro/ In situ*	Dairy cow, goat	20% DM	High apparent DMD and protein digestibility	(de la Moneda et al., 2019; Tayyab et al., 2016)

(*Continued*)

Table 1 Effects of macroalgae on rumen fermentation and animal performance (*Continued*)

Macroalgae species	Study type	Animal/rumen fluid donor	Dose	Impacts on ruminant nutrition	References
Ascophyllum nodosum	*In vitro/ in vivo*	Cattle, lambs, steers	2–5% DM	Alters rumen and gut microbiome, and reduces OMD, Total VFA and CH_4 production	(Wang et al., 2009a; Belanche et al., 2016b)
Fucus serratus	*In vitro*	Sheep	?	Low rumen degradability (15%)	(Greenwood et al., 1983; Makkar et al., 2016)
Fucus vesiculosus	*In vitro*	Sheep	?	Low rumen degradability (26%)	(Makkar et al., 2016; Greenwood et al., 1983)
Laminaria digitata	*In vitro*	Cow/sheep	5–20% DM	~80% *in vitro* OM digestibility, high rumen degradability, enables microbial protein synthesis, provides greater AA supply in intestine	(Belanche et al., 2016a; Hansen et al., 2003; de la Moneda et al., 2019)
Laminaria hyperborea	*In vivo*	Sheep	?	High rumen degradability, high in vitro OM digestibility (~80%)	(Hansen et al., 2003)
Laminaria ochroleuca	*In vitro*	Cow	25% DM	No adverse effects on fermentation parameters (e.g. Total VFA, CH_4)	(Maia et al., 2016)
Macrocystis pyrifera	*In situ*	Bull	?	~ 85% degradability of DM (96 hr) and high by-pass protein	(Gojon-Báez et al., 1998)
Pelvetia canaliculata	*In vitro*	Goat	Sole feed /20% DM	Low DMD, lowers VFA volume	(de la Moneda et al., 2019; Molina-Alcaide et al., 2017)
Saccharina latissima	*In vitro*	Cow/ goat	20–25% DM	Improves DMD and OMD of feed, no effects on VFA production and rumen fermentation	(de la Moneda et al., 2019; Maia et al., 2019)
Sargassum spp.	*In situ/in vivo*	Bull/sheep	10–30% DM	No effect on feed intake and digestibility, ~55–79% DMD and high protein digestibility (>85%)	(Gojon-Báez et al., 1998; Marín et al., 2009)

Undaria pinnatifida	*In vitro/ In situ*	Cow	10% DM	Improves feed digestibility and rumen environment, VFA production, provides quality and digestible protein	(Choi et al., 2019)
Green macroalgae: *Acrosiphonia* sp.	*In situ*	Dairy cow	20% DM	Provides high rumen digestible (~46%) and rumen by-pass protein (~31%)	(Tayyab et al., 2016; Molina-Alcaide et al., 2017)
Chaetomorpha sp.	*In vivo*	Sheep	20–40% DM	No adverse effects on feed digestibility and growth performance up to 30% inclusion	(Rjiba-Ktita et al., 2019)
Cladophora patentiramea	*In vitro*	Steers	16% (OM)	Reduces CH_4, feed degradability and VFA volume	(Machado et al., 2014)
Cladophora rupestris	*In situ*	Dairy cow	-	High protein content, 75% total tract AA degradability	(Gaillard et al., 2018)
Ruppia sp.	*In vivo*	Sheep	20–40% DM	Improves feed intake, no adverse effects on feed digestibility and growth performance up to 30% inclusion	(Rjiba-Ktita et al., 2019)
Ulva rigida	*In vivo/ in vitro*	Sheep/ cow	25% DM	Increases DMD of the feed, no effect on VFA and CH_4 production	(Cabrita et al., 2017; Maia et al., 2019)
Ulva sp.	*In vivo/ in vitro*	Sheep	20–40% DM	High rumen degradable and by-pass protein, no effects in feed intake and degradability for 30% DM inclusion	(Tayyab et al., 2016; Rjiba-Ktita et al., 2019)

DM, dry matter; OM, organic matter; DMD, dry matter degradability; OMD, organic matter degradability; AA, amino acids; ADWG, average daily weight gain; VFA, volatile fatty acids.

et al., 2019; Maia et al., 2019; Rjiba-Ktita et al., 2019). For instance, a low inclusion level of the anti-methanogenic red macroalga *Asparagopsis* taxiformis resulted in an improved average daily weight gain and OM degradability of the feed in cattle when included at ≤5% of OM, but the opposite effects were observed as the dose was increased to 10% OM of the total ration (Machado et al., 2016; Kinley et al., 2020). Hence, an inclusion level of <5% OM appears to be the cut off value for *A. taxiformis* in terms of maintaining the feed digestibility and fermentation parameters, such as VFA production (Machado et al., 2016; Roque et al., 2019a). Another anti-methanogenic red macroalga *Asparagopsis armata*, however, reduced feed intake, weight gain and milk yield at a relatively low inclusion level (≤1% OM) in the feed of dairy cattle (Roque et al., 2019b). This indicates that *Asparagopsis* spp. can be included in the ruminant feed at a low inclusion to achieve beneficial impacts in overall animal performance.

Other, different, red, brown and green species have also shown similar trends as *Asparagopsis* spp.; however, they can possibly be included at higher doses. Increased *in vitro* DM degradability and VFA production were observed with the edible brown macroalgae, *Undaria pinnatifida*, when it was incorporated up to 10% DM in the feed (Choi et al., 2019). Similarly, stable feed digestibility and animal performance were achieved with other brown (*A. esculenta, L. digitata, S. latissima*), red (*G. vermiculophyla, M. stellatus, P. palmata, Porphyra* sp.) and green (*Cladophora* sp. and *Ulva* spp.) macroalgae up to 20–25% of DM inclusion (de la Moneda et al., 2019; Maia et al., 2019). A few other green macroalgae (*Chaetomorpha* sp., *Ruppia* sp., and *Ulva* sp.) produced no significant negative effects on feed intake and digestibility in Barbarine sheep at up to 30% DM inclusion, but feed digestibility was reduced while the inclusion was increased to 40% (Rjiba-Ktita et al., 2019). This suggests that these macroalgal species can be incorporated to 10–30% in the ruminant rations, though more *in vivo* studies are needed to establish a beneficial inclusion level of a broad range of macroalgal species.

Macroalgae are generally low energy containing feeds due to low contents of lipid and starch, a large proportion of complex polysaccharides and relatively large content of ash (Bikker et al., 2020; Angell et al., 2016; Øverland et al., 2019). The GE contents of macroalgae, including *U. lactuca, Ulva rigida, G. vermiculophyla* and *S. latissima*, have been reported to be less (14–15.2 MJ/kg DM) than conventional ruminant feeds such as corn silage, hay silage and commercial concentrates (17.4–18.9 MJ/kg DM) (Maia et al., 2019; Ventura and Castañón, 1998). However, some brown macroalgae including *A. esculenta, A. nodosum* and *F. vesiculosus* have higher GE and digestible energy levels than the terrestrial forages such as winter rye and lichen (Applegate and Gray, 1995). Thus, although the majority of macroalgae lead to a lower energy supply compared to conventional feeds, there is a scope for their future use as feed additives in the ruminant's rations due to their high mineral contents and promising anti-methanogenic potentials as described in the following sections.

4 Anti-methanogenic properties of macroalgae

In addition to the aforementioned macro- and micro-nutrients, macroalgae are also rich sources of a wide range of bioactive components (such as pigments, tocopherols and various secondary metabolites) (Gupta and Abu-Ghannam, 2011). Macroalgae are gaining interest as anti-methanogenic feed ingredients in ruminants due to their richness of bioactive compounds, particularly halogenated and polyphenolic secondary metabolites that are able to inhibit CH_4 formation during the fermentation of feed in the forestomach (Roque et al., 2019a; Wang et al., 2008). In the following sections, the anti-methanogenic potentials of different macroalgal species will be discussed.

Red macroalgae: The potential of macroalgae to suppress enteric CH_4 formation in ruminants has been evaluated using both *in vitro* and *in vivo* studies (Table 2). The most convincing anti-methanogenic properties have been found among the red macroalgae, particularly *Asparagopsis* spp. (Machado et al., 2016; Roque et al., 2019b; Kinley et al., 2020). It was reported that a 40–98% reduction of CH_4 emission in steers could be achieved by adding as little as 0.1–0.2% (OM basis) *A. taxiformis* to a high grain diet (Kinley et al., 2020). Similarly, a 72-day feeding trial in sheep using the same macroalgae in a mixed ration (3% of the OM of the diet) containing a high proportion of fiber, resulted in an overall 80% reduction of enteric CH_4 production (Li et al., 2018). This is consistent with several *in vitro* fermentation studies, where addition of 0.5–5% OM of this macroalgae along with different substrates resulted in an ~74–99% decline in CH_4 formation over a 72-h incubation period (Roque et al., 2019a, Brooke et al., 2018; Machado et al., 2016). Another red macroalgae species belonging to the same genus, *A. armata* has also been shown to suppress the CH_4 production by ~67%, when fed to dairy cattle at 1% of OM (Roque et al., 2019b). Thus, *Asparagopsis* spp. can be an effective feed additive which can reduce enteric CH_4 production dramatically at a minimal inclusion in the ruminant diet.

The anti-methanogenic property of red macroalgae is not limited to the genus *Asparagopsis*. Two other red species *Gigartina* sp. and *Gracilaria vermiculophyla* have also demonstrated anti-methanogenic attributes in *in vitro* fermentations, but a greater amount of the macroalgae was added (16–18% OM), and the magnitude of reduction was substrate dependent (Maia et al., 2016). For example, a 60% reduction in CH_4 production was observed when *G. vermiculophyla* was supplemented to either meadow hay or corn silage, whereas *Gigartina* sp. reduced CH_4 production by 44%, but only when added

Table 2 Macroalgae species and their anti-methanogenic potential

Macroalgae species	Study type	Inclusion dose (%)	CH_4 ↓ (%)	Main active compound	Reference
Red macroalgae:					
Asparagopsis armata	*In vivo*	1% OM	67.2	Bromoform	(Roque et al., 2019b)
Asparagopsis taxiformis	*In vivo/in vitro*	0.1–16% DM/OM	74–99	Bromoform	(Kinley et al., 2020; Machado et al., 2016; Roque et al., 2019a)
Gracilaria vermiculophyla	*In vitro*	25% DM	63	?	(Maia et al., 2016)
Gigartina sp.	*In vitro*	25% DM	44	?	(Maia et al., 2016)
Brown macroalgae:					
Ascophyllum nodosum	*In vitro*	2% OM	15	Phlorotannin, polysaccharide (?)	(Belanche et al., 2016b)
Cystoseira trinodis	*In vitro*	20% OM	80	Terpenes, phlorotannin (?)	(Dubois et al., 2013)
Dictyota bartayresii	*In vitro*	2% OM	92	Phlorotannin, isoprenoids	(Machado et al., 2014)
Zonaria farlowii	*In vitro*	2% OM	11	Phenolic lipids	(Brooke et al., 2018)
Green macroalgae:					
Cladophora patentiramea	*In vitro*	16% OM	66.3	?	(Machado et al., 2014)
Oedogonium sp.	*In vitro*	75% OM	30.3	?	(Machado et al., 2016)
Ulva sp.	*In vitro*	25% DM	55	?	(Maia et al., 2016)

DM, dry matter; OM; organic matter; ↓, decrease in methane (CH_4) production; ?, unknown.

to meadow hay and not corn silage (Maia et al., 2016). However, no significant anti-methanogenic properties were detected for three other red macroalgae species studied, *M. stellatus, P. palmata* and *Porphyra* sp., when they constituted 8.4–20% fresh matter in the concentrate portion of the goat diet in an *in vitro* fermentation study (de la Moneda et al., 2019). Thus, some, but not all, red macroalgae species have strong anti-methanogenic properties, but their quantitative impact on CH_4 emission may depend on both the composition of the ruminant diet and the type and concentration of bioactive components present in the macroalgae and this requires further investigations in the future.

Brown macroalgae: Certain brown macroalgae species also have CH_4 mitigating potential, but the documentation stems primarily from *in vitro* studies. Two species *Dictyota bartayresii* and *Cystoseira trinodis* were able to reduce *in vitro* CH_4 production by 90% and 80%, respectively, when 16% of OM was added to Rhodes grass (Machado et al., 2014; Dubois et al., 2013). Other brown species, such as *A. nodosum* and *Zonaria farlowii*, have been shown to reduce CH_4 *in vitro* by 11–15% at inclusions of 2% and 5%, respectively (Brooke et al., 2018; Belanche et al., 2016b). However, no anti-methanogenic properties have been detected with 20–25% DM inclusion in feed in other *in vitro* trials, when using *L. digitata*, *L. ochroleuca*, *P. canliculata* and *S. latissima*, (Maia et al., 2016; de la Moneda et al., 2019). This suggests that only specific brown macroalgae species possess anti-methanogenic properties, and these are less powerful than those of the red species. However, further studies are needed to estimate their most effective dietary inclusion rates, and to confirm whether such outcomes are also evident *in vivo*.

Green macroalgae: Methane reduction properties have been observed in a few green macroalgae species. *Cladophora patentiramea* and the fresh water green algae *Odogonium* sp. have been shown to reduce CH_4 production by 66% and 30% *in vitro*, when added at 16% OM to decorticated cottonseed meal *in vitro* (Machado et al., 2014). With a similar inclusion rate in corn silage, another unspecified green macroalga from the genus *Ulva* illustrated a 55% suppression on enteric CH_4 production *in vitro* (Maia et al., 2016). However, the same authors later revealed that 25% DM addition of *Ulva rigida* to a mixed ration *in vitro* did not reduce CH_4 production (Maia et al., 2019). Thus, compared to brown and particularly red species, green macroalgae seem to have limited anti-methanogenic potential, which would require high levels of inclusion in the feed.

The anti-methanogenic macroalgae have also been found to affect other rumen fermentation parameters. With a concomitant reduction in the CH_4 production, they will decrease total VFA amount, feed intake and degradability when included in large amounts in the feed (Machado et al., 2014; Roque et al., 2019b). These effects were clearly observed with various macroalgal species such as *Asparagopsis* spp. (red), *C. trinodis* (brown) and *D. bartayresii* (brown) either *in vitro* or *in vivo* models, when macroalgae supplementation was gradually increased (Machado et al., 2014; Li et al., 2018; Roque et al., 2019b). Thus, it is important to include macroalgae at an optimum level so that possible negative impacts on rumen fermentation and animal performance are minimized.

4.1 Anti-methanogenic factors in macroalgae and potential mechanisms

Mechanistic insights into the anti-methanogenic properties of macroalgae are needed to identify the most efficient and safe ways of using them as feed additives to reduce enteric methane formation. Enteric CH_4 emissions can be reduced by macroalgae through two different mechanisms: (a) a direct inhibition of the methanogenic archaea themselves or rate-limiting steps in their methane formation or (b) through alteration of the rumen environment by reducing substrate availability or altering the rumen microbiota composition to disfavor the methanogens (Fig. 4).

4.1.1 Direct impacts: inhibition of methanogens and the methanogenic pathways

Macroalgae produce a number of secondary metabolites that protect them from a complex and possibly stressful seawater environment and help them to cope with various microbial infections (Li et al., 2017) and such metabolites may

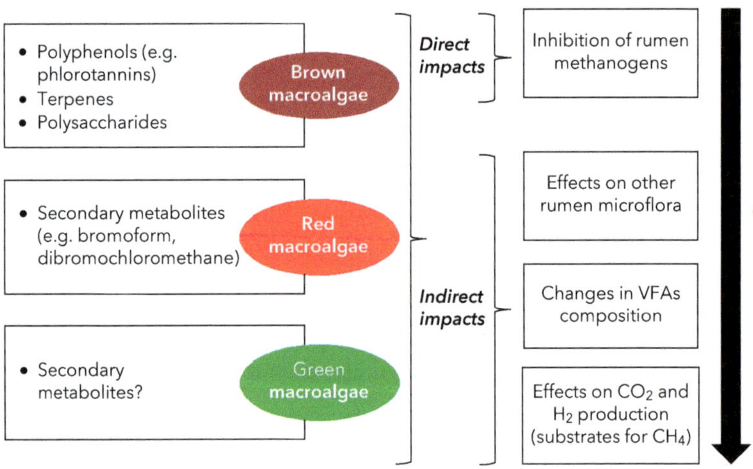

Figure 4 Potential anti-methanogenic factors of macroalgae and their mode of action in minimizing methane production in ruminants. CH_4, methane; VFAs, volatile fatty acids; CO_2, carbon dioxide; H_2, hydrogen.

also to a large extent account for the anti-methanogenic properties of some species. These compounds and their mode of action are described hereunder.

Halogenated compounds: Halogenated compounds are the aliphatic compounds containing one or two carbon atoms that are covalently linked with one or more halogen atoms (fluorine, bromine, chlorine or iodine). These compounds, such as bromoform, chloroform and bromochloromethane, irrespective of their source (synthetic or macroalgae), have shown a strong inhibitory action both *in vitro* and *in vivo* against rumen and other methanogens, significantly lowering their abundance in the rumen even at a low concentration (Paul et al., 2006; Roque et al., 2019a; Machado et al., 2018; Denman et al., 2015). Red macroalgae (e.g. *Asparagopsis* spp.) produce a high level of various brominated and chlorinated halocarbons, including bromoform, dibromochloromethane, chloroform, bromochloroacetic acid and dibromoacetic acid (Machado et al., 2018; Paul et al., 2006). They are structural analogs of CH_4 and other methanogenic intermediates and possess a higher affinity to enzymes, including corrinoid/porphinoid, which catalyzes the cobamide-dependent methyl transfer in methanogenesis (Wood et al., 1968; Yu and Smith, 2000; Roque et al., 2019b). Thus, the halogenated compounds can competitively inhibit the binding of intermediates or methane substrates into the corrinoid/porphinoid enzyme (Yu and Smith, 2000). Moreover, they are also structurally similar to CoM (a cofactor produced specifically by methanogens) which supplies the methyl group to methyl coenzyme-M reductase enzyme during the terminal reductive reaction of methanogenesis (Liu et al., 2011; Roque et al., 2019b; Li et al., 2018). Therefore, anti-methanogenic compounds from red macroalgae seem to exert their effects on CH_4 production directly by either of the two mechanisms: (a) minimizing the abundance of rumen methanogens through their anti-microbial activity or (b) interrupting their functional components such as enzymes, catalyzing the different steps of methane biosynthesis.

The anti-methanogenic property of synthetic halocarbons, such as chloroform, is dependent on the degree of chlorination, and this property can decrease over time due to the sequential reductive dechlorination during methane inhibition (Yu and Smith, 2000). In addition, methanogens have also been shown to develop resistance to synthetic anti-methanogenic compounds such as bromochloromethane when repeatedly exposed, potentially due to the adaptation of methanogens to those compounds (Patra et al., 2017). Although the rate of dechlorination and possibility of developing resistance to anti-methanogenic compounds derived from red macroalgae is unknown, two animal trials in steers and sheep have shown persistent CH_4 mitigating effect of *A. taxiformis* for 3 months (Li et al., 2018; Kinley et al., 2020). This indicates that anti-methanogenic compounds from these macroalgae might have more stable and effective CH_4 mitigation potential than synthetic halocarbons.

The excess intake of bromoform can be hazardous to human and animal health and therefore a maximum uptake level of 0.08 mg/L has been set for drinking water in the USA (EPA, 2012). In addition, synthetic aliphatic halocarbons are reported to cause ozone depletion and thus have environmental concerns (Patra et al., 2017; Roque et al., 2019b). Therefore, the possible toxicity of halogenated compounds from red macroalgae should be investigated to understand their effect on both animal health and environment.

Polyphenols: Polyphenols are a group of phenolic compounds and the concentration of these can account for up to 15% of DM in brown macroalgae (Wang et al., 2009a). The predominant form of polyphenols in brown macroalgae is phlorotannins (PT) and their anti-methanogenic properties have been described in *in vitro* studies (Hierholtzer et al., 2013; Wang et al., 2008). Though the effects of PT specific to rumen methanogenic archaea are not clear, a suppressive effect of condensed tannins (structural analogs of PT) on rumen archaea has been reported. For example, condensed tannins extracted from the terrestrial forage *Leucaena leucocephala* have exhibited a linear reduction of total rumen methanogens belonging to the orders *Methanobacteriales* and *Methanomicrobiales* with increasing doses (Tan et al., 2011). Due to the limited information available on the impacts of PT on rumen methanogens, it is too soon to evaluate whether there is a practical perspective for the use of PT in ruminants. However, because of the chemical and structural resemblances of PT and terrestrial tannins, antimicrobial activity of PT against rumen methanogens can be anticipated (Wang et al., 2008).

The mechanisms of action of PT on rumen methanogens are not known, but are described for other rumen microbes or methanogens isolated from wastewater treatment plant (Hierholtzer et al., 2013; Wang et al., 2009a). It has been revealed that PT can affect the integrity of microbial cell membrane and cell wall, via creating stress and ultimately causing cell lysis (Hierholtzer et al., 2013; Wang et al., 2009b). Other potential mechanisms of PT in relation to anti-microbial effects have been suggested to be via inactivation of extracellular enzymes and proteins necessary for growth and metabolism of microorganisms (Scalbert, 1991).

Species specific and time-dependent impacts of PT against various rumen microorganisms have been observed. For example, 500 μgmL^{-1} PT isolated from *A. nodosum* resulted in the reduction of cellulolytic rumen bacteria *Fibrobacter succinogenes* by 78%, 83% and 65% in 6, 12 and 24 h, respectively in an *in vitro* batch culture (Wang et al., 2009a). The same level of PT caused a 42% decrease in *Ruminococcus albus* without affecting the population of *Ruminococcus flavefacien* during 24 h of cultivation. In contrast, it significantly increased the number of non-cellulolytic bacteria such as *Prevotella bryantii, Ruminobacter amylophilus* and *Selenomonas ruminantium* (Wang et al., 2009a). This suggests that even within an order, various bacterial strains may be differentially affected by PT and that may also apply to the rumen methanogens. The underlying reason for this selective and differential anti-microbial property of PT is yet unknown. However, this could possibly be linked to the structure of PT, such as degree of polymerization (phloroglucinol units) and the number of reactive hydroxyl groups present (Wang et al., 2008; Hierholtzer et al., 2013). Furthermore, interspecies differences of macroalgae in the methane inhibition potential and the potency of PT from those macroalgae may also play some role in this selective action.

Polysaccharides: Macroalgae contain different kinds of polysaccharides, which are present either as structural components of the complex cell wall or as storage carbohydrates (O'Sullivan et al., 2010). Bactericidal and bacteriostatic effects of these polysaccharides have been documented against various hindgut microorganisms (Smith et al., 2011; Seong et al., 2019); however, specific information about their impacts on the rumen methanogens is yet to be evaluated. Polysaccharides from brown macroalgae (alginates, fucoidan and laminarin) can partially and selectively be fermented by specific bacteria in the rumen (such as *Prevotella* sp., *C. butyricum* and *Selenomonas* sp.) (Williams et al., 2013; Orpin et al., 1985). These polysaccharides have shown selective enrichment of beneficial gut bacteria, including *Bifidobacterium, Clostridium coccoides* and *Lactobacillus*, and a suppression of pathogenic gut microbes, including *E. coli, Salmonella* spp., *Enterococcus* and *Clostridium* spp., in monogastric animals (Charoensiddhi et al., 2017; Seong et al., 2019; Smith et al., 2011). Thus, macroalgae polysaccharides apparently have anti-microbial properties and whether such selective impacts are also evident with rumen microorganisms are not known.

Isoprenoids and terpenes: Macroalgae also produce various types of isoprenoids and terpenes, and over 200 different diterpenes have been identified in the single brown macroalgae of genus *Dictyota* (Chen et al., 2018). Isoprenoids have also been reported in other macroalgae, including *A. taxiformis* and *C. trinodis* (Brooke et al., 2018; Machado et al., 2014; Dubois et al., 2013). Although they have been suggested to contribute to the CH_4 mitigating properties, nothing is yet known about the mechanism underlying the anti-methanogenic effect of such compounds.

4.1.2 Indirect impacts: changes in the rumen environment affecting methanogenesis

In addition to the direct influences, macroalgae macroconstituents and bioactive compounds can affect numerous microorganisms in the rumen leading to changes in fermentation parameters and the overall rumen environment. The factors affected include the relative abundance and activity of non-methanogenic microorganisms, VFA production and availability of substrates for CH_4 production. The anti-methanogenic macroalgae species such as *A. taxiformis* and *A. nodosum* have been found to reduce the abundance of rumen microbes, including rumen protozoa (Roque et al., 2019a; Belanche et al., 2016b). A group of rumen methanogens (9–25% of total methanogens) live in a mutualistic relationship with protozoa and they generate a large amount of H_2 that is utilized by the methanogens to form CH_4 (Belanche et al., 2014; Newbold et al., 1995). Protozoans also get benefit from the methanogenic H_2 utilization because accumulation of H_2 in the rumen is inhibitory to their growth (Belanche et al., 2014). Thus, changes in the abundance and activity of the protozoa will result in H_2 deprivation in the rumen resulting in reduced methanogenesis (Morgavi et al., 2012).

The incorporation of macroalgae with anti-methanogenic potential in ruminant feed changes VFA production profile. They can change patterns of rumen fermentation from acetate formation towards the formation of more propionate and thus reducing acetate:propionate ratio (Machado et al., 2014; de la Moneda et al., 2019; Belanche et al., 2016b). Acetate formation in the rumen results in the release of metabolic $H_{2,}$ and this can be used by methanogens to produce CH_4 (Fig. 2). Therefore, reduced acetogenesis and increased propiogenesis are considered as factors indirectly decreasing methanogenesis (Roque et al., 2019a; Wolin and Miller, 1997). It has been noted that anti-methanogenic compounds such as PT and bromochloromethane increase the population of H_2-consuming bacteria, such as *Prevotella* spp., *F. succinogenes* and *Selenomonas* spp., in the rumen (Mitsumori et al., 2012; Denman et al., 2015; Wang et al., 2009a). The available H_2 can be re-channeled towards the formation of propionate, lactate and succinate by the action of H_2-consuming bacteria (Denman et al., 2015; Belanche et al., 2016a) which may also lead to H_2 deprivation. However, when the methanogenesis is highly inhibited (as with *A. taxiformis*), all the metabolic hydrogen produced cannot be captured by this re-channeling towards the formation of aforementioned fatty acids, and some will be eructed by animals (Martinez-Fernandez et al., 2016; Kinley et al., 2020). In fact, the actual causes and effects of macroalgae in changes of VFA, H_2 and populations of bacteria involved in the formation and consumption of these substances are yet to be clearly understood.

5 Processing and seasonal effects on anti-methanogenic properties of macroalgae

Post-harvest processing of macroalgae biomass, such as washing, drying and storage conditions may impact, not only the nutritional contents, but also the bioactive potential of harvested macroalgae (Kadam et al., 2015; Paull and Chen, 2008). The drying of *A. taxiformis* biomass at 45°C for 48 h led to a substantial reduction in the concentration of bromoform and eventual anti-methanogenic activity (Vucko et al., 2017). Similarly, oven drying at a higher temperature (80°C for 24 h) caused a significant reduction in the level of phytochemicals including polyphenols and flavonoids in *Kappaphycus alvarezzi* as compared to a lower temperature (40°C for 24 h) (Ling et al., 2015). However, the extractability of polyphenols and flavonoids could be greater when macroalgae biomass is semi-dried (e.g. 35–40°C for 2 h) as noted with the semi-dried *H. elongata* biomass (~40% increase) (Gupta et al., 2011). For other macroalgal species such as *F. vesiculosus* and *Porphyra* spp., drying methods (oven, sun or freeze drying) did not alter the amount of bioactive phytochemicals (Jiménez-Escrig et al., 2001). These results indicate that an appropriate drying/processing protocol for macroalgae biomass after harvesting may be beneficial to achieve increased levels of bioactive phytochemicals. Further studies are required to establish such optimal procedures, not in the least, in light of the high cost associated with the commercial production and transportation of macroalgae biomass.

The concentration of nutrients and bioactive components in macroalgae can vary across the seasons and geographical locations (Tayyab et al., 2016; Schiener et al., 2015). In macroalgae harvested in Norway, the level of protein and minerals have been found to be higher in spring than in autumn while polysaccharides (e.g. fucoidan and laminarin) are noted to be higher in summer (Kim, 2012; Rioux et al., 2009; Tayyab et al., 2016; Molina-Alcaide et al., 2017). On the other hand, in the same location, total polyphenol content in brown (*A. esculenta, L. digitata, P. canaliculata* and *S. latissima*), red (*P. palmata, M. stellatus* and *Porphyra* sp.) and green (*Acrosiphonia* sp., and *Cladophora rupestris*) macroalgae have been found to be around two-fold higher in autumn compared to the spring season (Molina-Alcaide et al., 2017; de la Moneda et al., 2019). In agreement with these findings, we have also found that Norwegian brown species (e.g. *A esculenta, F. vesiculosus, P. canaliculata, H. elongata, L. digitata, S. latissima*) harvested in the autumn have higher polyphenol levels compared to those harvested in the spring (Deepak et. al. unpublished data). However, a study from Scotland which included some of the aforementioned macroalgae (*A. nodosum, A. esculenta, L. digitata, L. hypeborea* and *S. latissima*) recorded a higher total polyphenol content during the summer compared to the autumn (Schiener et al., 2015). These variations are associated with the growth stage of the macroalgae and season-specific environmental factors such as temperature, light intensity and nutrient content in the seawater (Mišurcová, 2012; Parys et al., 2009). It has been mentioned that at the beginning of the spring season, vegetative growth of the macroalgae is rapid and the level of polyphenols is low during the rapid growth stage (Parys et al., 2009). Furthermore, due to the geographical differences, variations in these environmental factors may exist within the same season. Therefore, a specific harvesting period needs to be established based on compounds of interest and the spring season could be appropriate to harvest algae to achieve a maximum level of nutrients.

6 Future perspectives

Bioactive components of certain macroalgal species have the potential to be utilized as anti-methanogenic feed additives for ruminant animals to mitigate enteric CH_4 production. However, only a few species have been evaluated so far in this respect, and the most powerful anti-methanogenic compounds identified (halogenated carbons) are both ozone depleting and having documented adverse health impacts on humans (Roque et al., 2019b; Patra et al., 2017). It is therefore uncertain whether they can be approved (at least within the EU) as CH_4 mitigating instruments. Future studies should be directed towards the screening of a large number of macroalgal species to potentially identify efficient and safe compounds to be employed in climate-friendly feeding of ruminants. In this context, fractionation and/or extraction of targeted bioactive compounds (e.g. polyphenols, polysaccharides) and the characterization of their chemical and functional properties are important.

Utilization of macroalgae biomass as novel ruminant feeds on a larger scale is presently challenged by high costs associated with post-harvest processing as well as limited digestibility of several of the brown algae species that can most easily be cultivated. In addition, there are large variations between and within species in chemical composition and digestibility as well as contents of bioactive compounds depending on season, geographical location and post-harvest processing (Tiwari and Troy, 2015; Paull and Chen, 2008; Tayyab et al., 2016). This should encourage future research to develop cost-efficient techniques to increase the nutritional quality and anti-methanogenic potential of cultivable macroalgae by optimizing cultivation, harvest and post-harvest processing techniques.

7 Conclusions

The relevance of macroalgae as alternative and anti-methanogenic ruminant feeds depends upon their nutritional contents, digestibility, CH_4 mitigating potential, and influences on animal and environmental health. Red macroalgae such as *A. taxiformis and A. armata* seem to be promising anti-methanogenic feed ingredients and do not lead to significant adverse impacts on feed degradability with an inclusion rate of under 5% of OM. However, the impacts on other parameters of animal performances (e.g. feed intake, weight gain and milk yield) and rumen fermentation products, such as total VFA, should be carefully monitored. Moreover, due to a high concentration of halogenated compounds (e.g. bromoform) in those species, their potential adverse effects on human and environmental health must also be assessed. Brown macroalgae, such as

D. bartayresii, *C. trinodis* and *A. nodosum*, can be effective anti-methanogenic feed ingredients. Nevertheless, a high phlorotannin and polysaccharide content of these species can negatively impact the rumen degradability at high inclusion levels and thus optimal supplementation levels of these algae need to be carefully maintained. Green algae, including *C. patentiramea*, can also mitigate enteric CH_4 production, but the active anti-methanogenic compounds in green macroalgae are unknown.

Red macroalgae, such as *P. palmata*, *Porphyra* spp. and *Gracilaria* spp., and the green species *Acrosiphonia*, *C. rupestris*, *Ruppia* and *Ulva* can be used as a nutrient source due to their better nutritional composition and greater degradability compared to other species. Among brown macroalgae, *A. esculenta*, *Laminaria* spp., *S. latissima* and *U. pinnatifida* could be used as feed additives in up to 10–25% of DM in the ruminant feed provided that the excess mineral content is removed. Overall, macroalgal species could be an important component of future ruminant feed, but further *in vivo* studies are required to identify any potential adverse impacts on animal health and performance.

8 Where to look for further information

Further useful information about macroalgae and their applications can be obtained from the following resources:

- Sustainable use of seaweeds for food and non-food applications: Ed. Tiwari, Brijesh and Troy D. J. 2015. Seaweed sustainability-food and nonfood applications. Elsevier, https://doi.org/10.1016/C2013-0-12836-X.
- Seaweed biology - Novel insights into Ecophyisology, Ecology and Utilization. 2012. Springer. https://doi.org/10.1007/978-3-642-28451-9.
- Application of seaweeds as feed: "Seaweeds for livestock diets" Makkar et al 2016. https://www.sciencedirect.com/science/article/pii/S0377840115300274?via%3Dihub.

9 Funding

The first author is a recipient of PhD scholarship from Nord University, Norway. His research is connected to a project funded by RFF-Midt-Norge (Project number: 299081), where Nord University is the project owner, the University of Copenhagen (UCPH) and Leica AS are project partners.

10 References

Åby, B. A., Kantanen, J., Aass, L. and Meuwissen, T. (2014). Current status of livestock production in the Nordic countries and future challenges with a changing climate and human population growth. *Acta Agriculturae Scandinavica, Section A–Animal Science* 64(2): 73-97.

Additives EPo and Feed PoSuiA. (2013). Scientific opinion on the safety and efficacy of iodine compounds (E2) as feed additives for all species: calcium iodate anhydrous and potassium iodide, based on a dossier submitted by HELM AG. *EFSA Journal* 11(2): 3101.

Alexandratos, N. and Bruinsma, J. (2012). *World Agriculture Towards 2030/2050: The 2012 Revision*, 1-147. FAO.

Angell, A. R., Angell, S. F., de Nys, R. and Paul, N. A. (2016). Seaweed as a protein source for mono-gastric livestock. *Trends in Food Science and Technology* 54: 74-84.

Applegate, R. D. and Gray, P. B. (1995). Nutritional value of seaweed to ruminants. *Rangifer* 15(1): 15-18.

Bach, S. J., Wang, Y. and McAllister, T. A. (2008). Effect of feeding sun-dried seaweed (Ascophyllum nodosum) on fecal shedding of Escherichia coli O157: H7 by feedlot cattle and on growth performance of lambs. *Animal Feed Science and Technology* 142(1-2): 17-32.

Baňoch, T., Fajt, Z., Drabek, J. and Svoboda, M. (2010). Iodine and its importance in human and pigs. *Veterinářství* 60(12): 690-694.

Belanche, A., de la Fuente, G. and Newbold, C. J. (2014). Study of methanogen communities associated with different rumen protozoal populations. *FEMS Microbiology Ecology* 90(3): 663-677.

Belanche, A., Jones, E., Parveen, I. and Newbold, C. J. (2016a). A metagenomics approach to evaluate the impact of dietary supplementation with Ascophyllum nodosum or Laminaria digitata on rumen function in rusitec fermenters. *Frontiers in Microbiology* 7: 299.

Belanche, A., Ramos-Morales, E. and Newbold, C. J. (2016b). In vitro screening of natural feed additives from crustaceans, diatoms, seaweeds and plant extracts to manipulate rumen fermentation. *Journal of the Science of Food and Agriculture* 96(9): 3069-3078.

Benchaar, C., Pomar, C. and Chiquette, J. (2001). Evaluation of dietary strategies to reduce methane production in ruminants: a modelling approach. *Canadian Journal of Animal Science* 81(4): 563-574.

Bergman, E. N. (1990). Energy contributions of volatile fatty acids from the gastrointestinal tract in various species. *Physiological Reviews* 70(2): 567-590.

Biancarosa, I., Belghit, I., Bruckner, C. G., Liland, N. S., Waagbø, R., Amlund, H., Heesch, S. and Lock, E. J. (2018). Chemical characterization of 21 species of marine macroalgae common in Norwegian waters: benefits of and limitations to their potential use in food and feed. *Journal of the Science of Food and Agriculture* 98(5): 2035-2042.

Biancarosa, I., Espe, M., Bruckner, C. G., Heesch, S., Liland, N., Waagbø, R., Torstensen, B. and Lock, E. J. (2017). Amino acid composition, protein content, and nitrogen-to-protein conversion factors of 21 seaweed species from Norwegian waters. *Journal of Applied Phycology* 29(2): 1001-1009.

Bikker, P., Stokvis, L., van Krimpen, M. M., van Wikselaar, P. G. and Cone, J. W. (2020). Evaluation of seaweeds from marine waters in Northwestern Europe for application in animal nutrition. *Animal Feed Science and Technology* 263: 114460.

Braden, K. W., Blanton Jr., J. R., Allen, V. G., Pond, K. R. and Miller, M. F. (2004). Ascophyllum nodosum supplementation: a preharvest intervention for reducing Escherichia coli

O157: H7 and Salmonella spp. in feedlot steers. *Journal of Food Protection* 67(9): 1824–1828.

Britt, D. P. and Baker, J. R. (1990). Causes of death and illness in the native sheep of North Ronaldsay, Orkney I. Adult sheep. *British Veterinary Journal* 146(2): 129–142.

Brooke, C. G., Roque, B. M., Najafi, N. Gonzalez, M., Pfefferlen, A., DeAnda, V., Ginsburg, D. W., Harden, M. C., Nuzhdin, S. V. and Salwen, J. K. (2018). Evaluation of the potential of two common pacific coast macroalgae for mitigating methane emissions from ruminants. *bioRxiv*. doi: https://doi.org/10.1101/434480.

Cabrita, A. R. J., Correia, A., Rodrigues, A. R., Cortez, P. P., Vilanova, M. and Fonseca, A. J. M. (2017). Assessing in vivo digestibility and effects on immune system of sheep fed alfalfa hay supplemented with a fixed amount of Ulva rigida and Gracilaria vermiculophylla. *Journal of Applied Phycology* 29(2): 1057–1067.

Cabrita, A. R. J., Maia, M. R. G., Oliveira, H. M., Sousa-Pinto, I., Almeida, A. A., Pinto, E. and Fonseca, A. J. M. (2016). Tracing seaweeds as mineral sources for farm-animals. *Journal of Applied Phycology* 28(5): 3135–3150.

Cañedo-Castro, B., Piñón-Gimate, A., Carrillo, S., Ramos, D. and Casas-Valdez, M. (2019). Prebiotic effect of Ulva rigida meal on the intestinal integrity and serum cholesterol and triglyceride content in broilers. *Journal of Applied Phycology* 31(5): 3265-3273.

Chapman, V. (2012). *Seaweeds and Their Uses*, pp. 1–334. Springer Science & Business Media, Switzerland.

Charoensiddhi, S., Conlon, M. A., Vuaran, M. S., Franco, C. M. M. and Zhang, W. (2017). Polysaccharide and phlorotannin-enriched extracts of the brown seaweed Ecklonia radiata influence human gut microbiota and fermentation in vitro. *Journal of Applied Phycology* 29(5): 2407–2416.

Chen, J., Li, H., Zhao, Z., Xia, X., Li, B., Zhang, J. and Yan, X. (2018). Diterpenes from the marine algae of the genus Dictyota. *Marine Drugs* 16(5): 159.

Cherry, P., Yadav, S., Strain, C. R., Allsopp, P. J., McSorley, E. M., Ross, R. P. and Stanton, C. (2019). Prebiotics from seaweeds: an ocean of opportunity? *Marine Drugs* 17(6): 327.

Choi, Y. Y., Lee, S. J., Lee, Y. J.,Kim, H. S., Eom, J. S., Jo S. U. and LeeS. S. (2019). In vitro and in situ evaluation of Undaria pinnatifida as a feed ingredient for ruminants. *Journal of Applied Phycology* 1-11. 10.1007/s10811-019 01910-w.

D'Armas, H., Jaramillo, C., D'Armas, M., Echavarria, A. and Valverde, P. (2019). Proximate composition of several macroalgae from the coast of Salinas Bay, Ecuador. *Revista de Biología Tropical* 67(1): 61–68.

Dawczynski, C., Schubert, R. and Jahreis, G. (2007). Amino acids, fatty acids, and dietary fibre in edible seaweed products. *Food Chemistry* 103(3): 891–899.

de la Moneda, A., Carro, M. D., Weisbjerg, M. R., Roleda, M. Y., Lind, V., Novoa-Garrido, M. and Molina-Alcaide, E. (2019). Variability and potential of seaweeds as ingredients of ruminant diets: an in vitro study. *Animals* 9(10): 851.

Denman, S. E., Martinez Fernandez, G., Shinkai, T., Mitsumori, M. and McSweeney, C. S. (2015). Metagenomic analysis of the rumen microbial community following inhibition of methane formation by a halogenated methane analog. *Frontiers in Microbiology* 6: 1087.

Díaz, O., Tapia, Y., Muñoz, O., Montoro, R., Velez, D. and Almela, C. (2012). Total and inorganic arsenic concentrations in different species of economically important algae harvested from coastal zones of Chile. *Food and Chemical Toxicology* 50(3–4): 744–749.

Dubois, B., Tomkins, N. W., Kinley, R. D., Bai, M., Seymour, S., Paul, N. A. and Rocky, de N.(2013). Effect of tropical algae as additives on rumen in vitro gas production and fermentation characteristics. *American Journal of Plant Sciences* 4: 34–43.

EPA U. (2012). 2012 Edition of the drinking water standards and health advisories. EPA 822-S-12-001, Washington, DC.

Erickson, P. S., Marston, S. P., Gemmel, M., Deming, J., Cabral, R. G., Murphy, M. R. and Marden, J. I. (2012). Short communication: kelp taste preferences by dairy calves. *Journal of Dairy Science* 95(2): 856–858.

Estrada, A., Estrada, R. C., Meritt Jr., D., Bultrini, D. B., Klein, A., Dewenter, I. S., Tscharntke, T., Taniguchi, K., Wang, X. and Collomb, P. (2011). World livestock 2011: livestock in food security. No. FAO 363: W927.

Evans, F. D. and Critchley, A. T. (2014). Seaweeds for animal production use. *Journal of Applied Phycology* 26(2): 891–899.

Fernández-Segovia, I., Lerma-García, M. J., Fuentes, A. and Barat, J. M. (2018). Characterization of Spanish powdered seaweeds: composition, antioxidant capacity and technological properties. *Food Research International* 111: 212–219.

Gaillard, C., Bhatti, H. S., Novoa-Garrido, M., Lind, V., Roleda, M. Y. and Weisbjerg, M. R. (2018). Amino acid profiles of nine seaweed species and their in situ degradability in dairy cows. *Animal Feed Science and Technology* 241: 210–222.

Gojon-Báez, H. H., Siqueiros-Beltrones, D. A. and Hernández-Contreras, H. (1998). In situ ruminal digestibility and degradability of Macrocystis pyrifera and Sargassum spp. in bovine livestock. *Ciencias Marinas* 24(4): 463–481.

Gómez-Ordóñez, E., Alonso, E. and Rupérez, P. (2010). A simple ion chromatography method for inorganic anion analysis in edible seaweeds. *Talanta* 82(4): 1313–1317.

González-Recio, O., López-Paredes, J., Ouatahar, L., Charfeddine, N., Ugarte, E., Alenda, R. and Jiménez-Montero, J. A. (2020). Mitigation of greenhouse gases in dairy cattle via genetic selection: 2. Incorporating methane emissions into the breeding goal. *Journal of Dairy Science* 103(8): 7210–7221.

Greenwood, Y., Orpin, C. and Paterson, I. (1983). Digestibility of seaweeds in Orkney sheep. *Journal of Physiology-London* 343: 120.

Grossi, G., Goglio, P., Vitali, A. and Williams, A. G. (2019). Livestock and climate change: impact of livestock on climate and mitigation strategies. *Animal Frontiers* 9(1): 69–76.

Gupta, S. and Abu-Ghannam, N. (2011). Bioactive potential and possible health effects of edible brown seaweeds. *Trends in Food Science and Technology* 22(6): 315–326.

Gupta, S., Cox, S. and Abu-Ghannam, N. (2011). Effect of different drying temperatures on the moisture and phytochemical constituents of edible Irish brown seaweed. *LWT - Food Science and Technology* 44(5): 1266–1272.

Hansen, H. R., Hector, B. L. and Feldmann, J. (2003). A qualitative and quantitative evaluation of the seaweed diet of North Ronaldsay sheep. *Animal Feed Science and Technology* 105(1–4): 21–28.

Haque, M. N. (2018). Dietary manipulation: a sustainable way to mitigate methane emissions from ruminants. *Journal of Animal Science and Technology* 60(1): 15.

Haque, M. N., Roggenbuck, M., Khanal, P., Nielsen, M. O. and Madsen, J. (2014). Development of methane emission from lambs fed milk replacer and cream for a prolonged period. *Animal Feed Science and Technology* 198: 38–48.

Hierholtzer, A., Chatellard, L., Kierans, M., Akunna, J. C. and Collier, P. J. (2013). The impact and mode of action of phenolic compounds extracted from brown seaweed on mixed anaerobic microbial cultures. *Journal of Applied Microbiology* 114(4): 964–973.

Holt-Giménez, E. (2019). *Can We Feed the World without Destroying It?* John Wiley & Sons, Hoboken, New Jersey.

Hvelplund, T. and Weisbjerg, M. (2000). In situ techniques for the estimation of protein digestibility and postrumen availability. In: *Forage Evaluation in Ruminant Nutrition* Given, D. I., Owen, E., Axford, R. F. E. and Omed, H. M. (Eds), pp. 233–258. CABI Publishing, New York, NY.

Jiménez-Escrig, A., Jiménez-Jiménez, I., Pulido, R. and Saura-Calixto, F. (2001). Antioxidant activity of fresh and processed edible seaweeds. *Journal of the Science of Food and Agriculture* 81(5): 530–534.

Jiménez-Escrig, A. and Sánchez-Muniz, F. J. (2000). Dietary fibre from edible seaweeds: chemical structure, physicochemical properties and effects on cholesterol metabolism. *Nutrition Research* 20(4): 585–598.

Jung, S. M., Kang, S. G., Son, J. S., Jeon, J. H., Lee, H. J. and Shin, H. W. (2016). Temporal and spatial variations in the proximate composition, amino acid, and mineral content of Pyropia yezoensis. *Journal of Applied Phycology* 28(6): 3459–3467.

Kadam, S. U., Álvarez, C., Tiwari, B. K. and O'Donnell, C. P. (2015). Processing of seaweeds. In: *Seaweed Sustainability* Tiwari, B. and Troy, D. J. (Eds), pp. 61–78. Elsevier: Amsterdam, The Netherlands.

Kang, M.-C., Kang, N., Ko, S.-C., Kim, Y. and Jeon, Y. (2016). Anti-obesity effects of seaweeds of Jeju Island on the differentiation of 3T3-L1 preadipocytes and obese mice fed a high-fat diet. *Food and Chemical Toxicology* 90: 36–44.

Kannan, G., Saker, K. E., Terrill, T. H., Kouakou, B., Galipalli, S. and Gelaye, S. (2007). Effect of seaweed extract supplementation in goats exposed to simulated preslaughter stress. *Small Ruminant Research* 73(1–3): 221–227.

Kılınç, B., Cirik, S., Turan, G., Tekogul, H. and Koru, E. (2013). Seaweeds for food and industrial applications. In: *Food Industry* Muzzalupo, I. (Ed.), pp. 735–748. IntechOpen, London.

Kim, J.-Y., Kwon, Y. M., Kim, I.-S., Kim, J., Yu, D., Adhikari, B., Lee, S., Choi, I. and Cho, K. (2018). Effects of the brown seaweed Laminaria japonica supplementation on serum concentrations of IgG, triglycerides, and cholesterol, and intestinal microbiota composition in rats. *Frontiers in Nutrition* 5: 23.

Kim, K.-T. (2012). *Seasonal Variation of Seaweed Components and Novel Biological Function of Fucoidan Extracted from Brown Algae in Quebec*, Quebec, Canada. https://corpus.ulaval.ca/jspui/handle/20.500.11794/23282.

Kinley, R. D., Martinez-Fernandez, G., Matthews, M. K., de Nys, R., Magnusson, M. and Tomkins, N. W. (2020). Mitigating the carbon footprint and improving productivity of ruminant livestock agriculture using a red seaweed. *Journal of Cleaner Production* 259: 120836.

Knapp, J. R., Laur, G. L., Vadas, P. A., Weiss, W. P. and Tricarico, J. M. (2014). Invited review: enteric methane in dairy cattle production: quantifying the opportunities and impact of reducing emissions. *Journal of Dairy Science* 97(6): 3231–3261.

Kohn, R. and Boston, R. (2000). The role of thermodynamics in controlling rumen metabolism. In: *Modelling Nutrient Utilization in Farm Animals* McNamara, J. P., France, J. and Beever, D. E. (Eds), pp. 11-24. Oxfordshire, England.

Li, X., Norman, H. C., Kinley, R. D., Laurence, M., Wilmot, M., Bender, H., de Nys, R. and Tomkins, N. (2018). Asparagopsis taxiformis decreases enteric methane production from sheep. *Animal Production Science* 58(4): 681–688.

Li, Y., Fu, X., Duan, D., Liu, X., Xu, J. and Gao, X. (2017). Extraction and identification of phlorotannins from the brown alga, Sargassum fusiforme (Harvey) Setchell. *Marine Drugs* 15(2): 49.

Ling, A. L. M., Yasir, S., Matanjun, P. and Abu Bakar, M. F. (2015). Effect of different drying techniques on the phytochemical content and antioxidant activity of Kappaphycus alvarezii. *Journal of Applied Phycology* 27(4): 1717–1723.

Liu, H., Wang, J., Wang, A. and Chen, J. (2011). Chemical inhibitors of methanogenesis and putative applications. *Applied Microbiology and Biotechnology* 89(5): 1333–1340.

Machado, L., Magnusson, M., Paul, N. A., de Nys, R. and Tomkins, N. (2014). Effects of marine and freshwater macroalgae on in vitro total gas and methane production. *PLoS ONE* 9(1): e85289.

Machado, L., Magnusson, M., Paul, N. A., Kinley, R., de Nys, R. and Tomkins, N. (2016). Dose-response effects of Asparagopsis taxiformis and Oedogonium sp. on in vitro fermentation and methane production. *Journal of Applied Phycology* 28(2): 1443–1452.

Machado, L., Tomkins, N., Magnusson, M., Midgley, D. J., de Nys, R. and Rosewarne, C. P. (2018). In vitro response of rumen microbiota to the antimethanogenic red macroalga Asparagopsis taxiformis. *Microbial Ecology* 75(3): 811–818.

Mæhre, H. K., Malde, M. K., Eilertsen, K. E. and Elvevoll, E. O. (2014). Characterization of protein, lipid and mineral contents in common Norwegian seaweeds and evaluation of their potential as food and feed. *Journal of the Science of Food and Agriculture* 94(15): 3281–3290.

Magnusson, M., Carl, C., Mata, L., de Nys, R. and Paul, N. A. (2016). Seaweed salt from Ulva: a novel first step in a cascading biorefinery model. *Algal Research* 16: 308–316.

Maia, M. R. G., Fonseca, A. J. M., Cortez, P. P. and Cabrita, A. R. J. (2019). In vitro evaluation of macroalgae as unconventional ingredients in ruminant animal feeds. *Algal Research* 40: 101481.

Maia, M. R. G., Fonseca, A. J. M., Oliveira, H. M., Mendonça, C. and Cabrita, A. R. J. (2016). The potential role of seaweeds in the natural manipulation of rumen fermentation and methane production. *Scientific Reports* 6(1): 32321.

Makkar, H. P. S., Tran, G., Heuzé, V., Giger-Reverdin, S., Lessire, M., Lebas, F. and Ankers, P. (2016). Seaweeds for livestock diets: a review. *Animal Feed Science and Technology* 212: 1–17.

Marín, A., Casas-Valdez, M., Carrillo, S., Hernández, H., Monroy, A., Sanginés, L. and Pérez-Gil, F. (2009). The marine algae Sargassum spp.(Sargassaceae) as feed for sheep in tropical and subtropical regions. *Revista de biología tropical* 57(4): 1271–1281.

Marsham, S., Scott, G. W. and Tobin, M. L. (2007). Comparison of nutritive chemistry of a range of temperate seaweeds. *Food Chemistry* 100(4): 1331–1336.

Martinez-Fernandez, G., Denman, S. E., Yang, C., Cheung, J., Mitsumori M. and McSweeney C. S. (2016). Methane inhibition alters the microbial community, hydrogen flow, and fermentation response in the rumen of cattle. *Frontiers in Microbiology* 7: 1122.

McDonald, P., Edwards, R., Greenhalgh, J., Morgan, C., Sinclair, L. and Wilkinson R. (2011). *Animal Nutrition* (7th edn.). Pearson Education Limited, London, UK.

McLaughlin, M. J., Parker, D. R. and Clarke, J. M. (1999). Metals and micronutrients-food safety issues. *Field Crops Research* 60(1–2): 143–163.

Mišurcová, L. (2012). *Chemical Composition of Seaweeds*. In: *Handbook of Marine Macroalgae: Biotechnology and Applied Phycology* Se-Kwon Kimpp (Ed.), pp. 171–192. Wiley Online Library, Hoboken, New Jersey.

Mišurcová, L., Kráčmar, S., Klejdus, B. and Vacek, J. (2010). Nitrogen content, dietary fiber, and digestibility in algal food products. *Czech Journal of Food Sciences* 28: 27-35.
Mitsumori, M., Shinkai, T., Takenaka, A., Enishi, O., Higuchi, K., Kobayashi, Y., Nonaka, I., Asanuma, N., Denman, S. E. and McSweeney, C. S. (2012). Responses in digestion, rumen fermentation and microbial populations to inhibition of methane formation by a halogenated methane analogue. *British Journal of Nutrition* 108(3): 482-491.
Molina-Alcaide, E., Carro, M. D., Roleda, M. Y., Weisbjerg, M. R., Lind, V. and Novoa-Garrido, M. (2017). In vitro ruminal fermentation and methane production of different seaweed species. *Animal Feed Science and Technology* 228: 1-12.
Morgavi, D. P., Martin, C., Jouany, J. P. and Ranilla, M. J. (2012). Rumen protozoa and methanogenesis: not a simple cause-effect relationship. *British Journal of Nutrition* 107(3): 388-397.
Newbold, C. J., Lassalas, B. and Jouany, J. P. (1995). The importance of methanogens associated with ciliate protozoa in ruminal methane production in vitro. *Letters in Applied Microbiology* 21(4): 230-234.
Orpin, C. G., Greenwood, Y., Hall, F. J. and Paterson, I. W. (1985). The rumen microbiology of seaweed digestion in Orkney sheep. *Journal of Applied Bacteriology* 58(6): 585-596.
O'Sullivan, L., Murphy, B., McLoughlin, P., Duggan, P., Lawlor, P. G., Hughes, H. and Gardiner, G. E. (2010). Prebiotics from marine macroalgae for human and animal health applications. *Marine Drugs* 8(7): 2038-2064.
Øverland, M., Mydland, L. T. and Skrede, A. (2019). Marine macroalgae as sources of protein and bioactive compounds in feed for monogastric animals. *Journal of the Science of Food and Agriculture* 99(1): 13-24.
Parys, S., Kehraus, S., Pete, R., Küpper, F. C., Glombitza, K. and König, G. M. (2009). Seasonal variation of Polyphenolics in Ascophyllum nodosum (Phaeophyceae). *European Journal of Phycology* 44(3): 331-338.
Patra, A., Park, T., Kim, M. and Yu, Z. (2017). Rumen methanogens and mitigation of methane emission by anti-methanogenic compounds and substances. *Journal of Animal Science and Biotechnology* 8(1): 13.
Paul, N. A., de Nys, R. and Steinberg, P. (2006). Chemical defence against bacteria in the red alga Asparagopsis armata: linking structure with function. *Marine Ecology Progress Series* 306: 87-101.
Paull, R. E. and Chen, N. J. (2008). Postharvest handling and storage of the edible red seaweed Gracilaria. *Postharvest Biology and Technology* 48(2): 302-308.
Peña-Rodríguez, A., Mawhinney, T. P., Ricque-Marie, D. and Cruz-Suárez, L. E. (2011). Chemical composition of cultivated seaweed Ulva clathrata (Roth) C. Agardh. *Food Chemistry* 129(2): 491-498.
Pereira, L. (2011). A review of the nutrient composition of selected edible seaweeds. In: *Seaweed: Ecology, Nutrient Composition and Medicinal Uses* Pomin, V. H. (Ed.), pp. 15-47. Nova Science Publishers, Inc., Coimbra.
Rajauria, G. (2015). Seaweeds: a sustainable feed source for livestock and aquaculture. In: *Seaweed Sustainability* Tiwari, B. and Troy, D. J. (Eds), pp. 389-420. Elsevier, Amsterdam, The Netherlands.
Reilly, P., O'doherty, J. V., Pierce, K. M., Callan, J. J., O'Sullivan, J. T. and Sweeney, T. (2008). The effects of seaweed extract inclusion on gut morphology, selected intestinal microbiota, nutrient digestibility, volatile fatty acid concentrations and the immune status of the weaned pig. *Animal* 2(10): 1465-1473.

Rioux, L.-E. and Turgeon, S. L. (2015). Seaweed carbohydrates. In: *Seaweed Sustainability* Tiwari, B. and Troy, D J. (Eds), pp. 141–192. Elsevier, Amsterdam, The Netherlands.

Rioux, L.-E., Turgeon, S. L. and Beaulieu, M. (2009). Effect of season on the composition of bioactive polysaccharides from the brown seaweed Saccharina longicruris. *Phytochemistry* 70(8): 1069–1075.

Rjiba-Ktita, S., Chermiti, A., Valdés, C. and López, S. (2019). Digestibility, nitrogen balance and weight gain in sheep fed with diets supplemented with different seaweeds. *Journal of Applied Phycology*: 1-9.

Roque, B. M., Brooke, C. G., Ladau, J., Polley, T., Marsh, L. J., Najafi, N., Pandey, P., Singh, L., Kinley, R., Salwen, J. K., Eloe-Fadrosh, E., Kebreab, E. and Hess, M. (2019a). Effect of the macroalgae Asparagopsis taxiformis on methane production and rumen microbiome assemblage. *Animal Microbiome* 1(1): 3.

Roque, B. M., Salwen, J. K., Kinley, R. and Kebreab, E. (2019b). Inclusion of Asparagopsis armata in lactating dairy cows' diet reduces enteric methane emission by over 50 percent. *Journal of Cleaner Production* 234: 132–138.

Rupérez, P. (2002). Mineral content of edible marine seaweeds. *Food Chemistry* 79(1): 23–26.

Scalbert, A. (1991). Antimicrobial properties of tannins. *Phytochemistry* 30(12): 3875–3883.

Schiener, P., Black, K. D., Stanley, M. S. and Green, D. H. (2015). The seasonal variation in the chemical composition of the kelp species Laminaria digitata, Laminaria Hyperborea, Saccharina latissima and Alaria esculenta. *Journal of Applied Phycology* 27(1): 363–373.

Seong, H., Bae, J.-H., Seo, J. S., Kim, S., Kim, T. and Han, N. S. (2019). Comparative analysis of prebiotic effects of seaweed polysaccharides laminaran, porphyran, and ulvan using in vitro human fecal fermentation. *Journal of Functional Foods* 57: 408–416.

Smith, A. G., O'doherty, J. V., Reilly, P., Ryan, M. T., Bahar, B. and Sweeney, T. (2011). The effects of laminarin derived from Laminaria digitata on measurements of gut health: selected bacterial populations, intestinal fermentation, mucin gene expression and cytokine gene expression in the pig. *British Journal of Nutrition* 105(5): 669–677.

Tan, H. Y., Sieo, C. C., Abdullah, N., Liang, J. B., Huang, X. D. and Ho, Y. W. (2011). Effects of condensed tannins from Leucaena on methane production, rumen fermentation and populations of methanogens and protozoa in vitro. *Animal Feed Science and Technology* 169(3–4): 185–193.

Tayyab, U., Novoa-Garrido, M., Roleda, M. Y., Lind, V. and Weisbjerg, M. R. (2016). Ruminal and intestinal protein degradability of various seaweed species measured in situ in dairy cows. *Animal Feed Science and Technology* 213: 44–54.

Tibbetts, S. M., Milley, J. E. and Lall, S. P. (2016). Nutritional quality of some wild and cultivated seaweeds: nutrient composition, total phenolic content and in vitro digestibility. *Journal of Applied Phycology* 28(6): 3575–3585.

Tiwari, B. K. and Troy, D. J. (2015). Seaweed sustainability-food and nonfood applications. In: *Seaweed Sustainability* Tiwari, B. and Troy, D. J. (Eds), pp. 1–472. Elsevier, Amsterdam, The Netherlands.

Trumbo, P., Yates, A. A., Schlicker, S. and Poos, M. (2001). Dietary reference intakes: vitamin A, vitamin K, arsenic, boron, chromium, copper, iodine, iron, manganese, molybdenum, nickel, silicon, vanadium, and zinc. *Journal of the American Dietetic Association* 101(3): 294–301.

Turner, J. L., Dritz, S. S., Higgins, J. J. and Minton, J. E. (2002). Effects of Ascophyllum nodosum extract on growth performance and immune function of young pigs challenged with Salmonella typhimurium. *Journal of Animal Science* 80(7): 1947-1953.

Ungerfeld, E. M. (2013). A theoretical comparison between two ruminal electron sinks. *Frontiers in Microbiology* 4: 319.

van der Reijden, O. L., Zimmermann, M. B. and Galetti, V. (2017). Iodine in dairy milk: sources, concentrations and importance to human health. *Best Practice and Research. Clinical Endocrinology and Metabolism* 31(4): 385-395.

Van Nevel, C. J. and Demeyer, D. I. (1996). Control of rumen methanogenesis. *Environmental Monitoring and Assessment* 42(1-2): 73-97.

Ventura, M. R. and Castañón, J. I. R. (1998). The nutritive value of seaweed (Ulva Lactuca) for goats. *Small Ruminant Research* 29(3): 325-327.

Vucko, M. J., Magnusson, M., Kinley, R. D., Villart, C. and de Nys, R. (2017). The effects of processing on the in vitro antimethanogenic capacity and concentration of secondary metabolites of Asparagopsis taxiformis. *Journal of Applied Phycology* 29(3): 1577-1586.

Wang, Y., Alexander, T. W. and McAllister, T. A. (2009a). In vitro effects of phlorotannins from Ascophyllum nodosum (brown seaweed) on rumen bacterial populations and fermentation. *Journal of the Science of Food and Agriculture* 89(13): 2252-2260.

Wang, Y., Xu, Z., Bach, S. J. and McAllister, T. A. (2008). Effects of phlorotannins from Ascophyllum nodosum (brown seaweed) on in vitro ruminal digestion of mixed forage or barley grain. *Animal Feed Science and Technology* 145(1-4): 375-395.

Wang, Y., Xu, Z., Bach, S. J. and McAllister, T. A. (2009b). Sensitivity of Escherichia coli to seaweed (Ascophyllum nodosum) phlorotannins and terrestrial tannins. *Asian-Australasian Journal of Animal Sciences* 22(2): 238-245.

Williams, A. G., Withers, S. and Sutherland, A. D. (2013). The potential of bacteria isolated from ruminal contents of seaweed-eating North Ronaldsay sheep to hydrolyse seaweed components and produce methane by anaerobic digestion in vitro. *Microbial Biotechnology* 6(1): 45-52.

Wolin, M. and Miller, T. (1997). Microbe-microbe interactions. In: *The Rumen Microbial Ecosystem* Hobson, P. N. and Stewart, C. S. (Eds), pp. 467-488. Chapman and Hall, London.

Wood, J. M., Kennedy, F. S. and Wolfe, R. S. (1968). Reaction of multihalogenated hydrocarbons with free and bound reduced vitamin B12. *Biochemistry* 7(5): 1707-1713.

Yu, Z. and Smith, G. B. (2000). Inhibition of methanogenesis by C1-and C2-polychlorinated aliphatic hydrocarbons. *Environmental Toxicology and Chemistry* 19(9): 2212-2217.

Chapter 9

Developing seaweed/macroalgae as feed for pigs

Marta López-Alonso, Universidade de Santiago de Compostela, Spain; Marco García-Vaquero, University College Dublin, Ireland; and Marta Miranda, Universidade de Santiago de Compostela, Spain

1 Introduction

The global human population is projected to reach 9.7 billion by 2050. It has been estimated that overall food production will have to double by 2050 to meet the increased demand (Aiking, 2014). Animal farming systems will have to increase production by an estimated 70%, increasing the demand for animal feed by up to 235% (Herman and Schmidt, 2016). In meeting the increasing demand, intensive production systems have traditionally focused on maximising production outputs while minimising production costs (Jones et al., 2012). High-intensity farming has been characterised by genetically selecting animals for fast growth and high meat content, kept at high stocking densities and fed nutrient-dense plant-based feeds (Swanson, 1995). Potential consequences have been the greater levels of stress for animals and their vulnerability to disease, leading to the increased prophylactic use of antibiotics in some cases (Van Boeckel et al., 2015).

It has been increasingly recognised that some aspects of these production systems are no longer sustainable. Researchers have begun to investigate alternative feed sources that can be more sustainably produced than

http://dx.doi.org/10.19103/AS.2021.0091.15

conventional, land-based crops used for feed, and which also have the potential to boost animals' natural immunity to disease whilst meeting their nutritional requirements, thus reducing the reliance on antibiotics (García-Vaquero, 2018; Miranda et al., 2017; Morais et al., 2020).

Macroalgae or seaweeds comprise a diverse group with more than 10 000 different species described to date (Collins et al., 2016). Approximately 5% of them are currently exploited for feed applications (Michalak and Chojnacka, 2014). They are able to adapt to changing and extreme marine environmental conditions by producing secondary metabolites including lipids, polysaccharides and minerals, to protect the biomass from cell damage caused by these marine stressors (Garcia-Vaquero et al., 2017). These metabolites make them a promising potential feed ingredient.

The use of macroalgae as a source of protein and other essential nutrients (such as polyunsaturated fatty acids, phenolic compounds and minerals) in animal nutrition are promising dietary strategies to complement traditional feed sources (Øverland et al., 2019). The incorporation of macroalgal extracts containing carbohydrates such as laminarin and fucoidan has, for example, shown promising results as an alternative to antibiotics in piglets, either directly by feed supplementation and/or indirectly via maternal dietary supplementation (Sweeney and O'Doherty, 2016). This chapter summarises the nutritional properties of macroalgae for use in animal feed, both as a source of protein and as antimicrobial compounds in swine production systems.

2 Challenges in using macroalgae for feed applications

The use of macroalgae to feed livestock has been documented for thousands of years since at least the time of Ancient Greece (Makkar et al., 2016). In the nineteenth and early twentieth centuries, seaweed was used in the coastal regions of France, the Scottish islands and Scandinavia, mainly to feed ruminants but also other livestock species, including pigs (Chapman, 2012; Sauvageau, 1920). The use of brown algae like *Ascophyllum nodosum* or *Fucus vesiculosus* in pig nutrition has been documented since the beginning of the twentieth century. Macroalgae were traditionally mixed with cereal meals to fatten pigs in countries in Northern Europe (Chapman, 2012; Evans and Critchley, 2014). The interest in this feedstock increased during periods of food shortage such as World War I, but usage declined in the first half of the twentieth century due to the perceived poor nutritional quality of the biomass (Makkar et al., 2016). The low levels of protein and energy available for metabolism, and the high mineral content of brown seaweeds led to their replacement by other protein ingredients such as fishmeal and soybean meal in formulated feed for monogastric animals (Øverland et al., 2019). It was also found that high amounts of brown seaweed could be harmful to health: the inclusion of 10% of *A. Nodosum* in pig feed over several weeks, for example, has been related to weight loss (Jones et al., 1979). To maximise the potential benefits of the inclusion of seaweeds in feeds for improved pig production and health, seaweeds can only be used at low concentrations in feed (1–2%) rather than as a major source of macronutrients (Morais et al., 2020).

As previously mentioned, the high amount of ash in macroalgae has hindered the inclusion of intact or dried macroalgae in animal diets, as feeds high in minerals have resulted in diarrhoea and decreased animal performance in pigs and poultry (Koreleski et al., 2010; Wilkinson, 1992). There has been a recent focus on refining biomass to reduce unwanted mineral content in favour of the more beneficial compounds.

When extracting polysaccharides, their molecular weight, monosaccharide composition and sulphate contents will change depending on the extraction and purification methods used, affecting the biological properties and effects of these molecules (Garcia-Vaquero et al., 2017). Ale and Meyer (2013) attributed the lack of official approval of polysaccharides or their derived fractions for pharmaceutical, dermatological, nutraceutical or other commercial applications to the lack of standardised extraction methodologies.

Recent research has focused on identifying the optimum extraction parameters such as temperature, time and pH (Garcia-Vaquero et al., 2019; Yuan et al., 2018). Yuan et al. (2018) extracted sulphated polysaccharides from the green macroalga *Ulva prolifera* using microwave-assisted hydrothermal extraction. The sulphur content increased with the temperature and acid concentration of the solvent. Antioxidant and pancreatic lipase inhibition properties were also influenced by extraction conditions due to structural modifications of the active molecules (Yuan et al., 2018). Garcia-Vaquero et al. (2019) used hydrothermal-assisted extraction and found significant variation in the levels of fucoidan, laminarin and antioxidant activity under differing conditions. The authors determined the optimum process parameters for a combined maximised extraction of these compounds from *Laminaria hyperborea* (120°C, 80.9 min and 12.02 volume of solvent-to-seaweed ratio). Other techniques, such as the combination of ultrasound and microwave technologies, have also been explored for brown macroalga *A. nodosum*, achieving increased yields of fucoidan (Garcia-Vaquero et al., 2020).

Techniques such as enzyme-assisted extraction have also attracted attention due to their high efficiency and mild extraction conditions that ensure less modification in the chemical structure of the compounds of interest (Michalak and Chojnacka, 2014). Enzymes have been used to extract multiple compounds including proteins, phenols, carotenoids and lipids (Billakanti et al., 2013; Wang et al., 2010). However, the use of these methods has been hindered by the high price of enzymes and the cost of scaling up production (Michalak and Chojnacka, 2014).

Improvements in the extraction of beneficial compounds have been one factor in renewed scientific interest in macroalgae as a source of multiple high-value compounds including proteins, polysaccharides, lipids, pigments and minerals with a wide variety of antioxidant, anti-bacterial, anti-tumour and other properties (Holdt and Kraan, 2011). This has made them promising ingredients for a wide variety of applications, for example, as nutraceuticals or functional foods, pharmaceuticals, cosmetics and in animal feed.

3 Composition of macroalgae

Based on their pigment composition, macroalgae can be classified as:

- brown macroalgae (Phaeophyta);
- red macroalgae (Rhodophyta); and
- green macroalgae (Chlorophyta).

Macroalgae are rich in carbohydrates, with medium or high amounts of proteins, low levels of lipids and variable mineral content (Dominguez and Loret, 2019; Kraan, 2013). The composition of macroalgae varies greatly depending on the macroalgal class, species, parts of the macroalgae sampled as well as the season and location (Øverland et al., 2019). The variable composition of macroalgae (water, ash, crude protein, crude lipids and polysaccharides) compiled from the recent scientific literature are summarised in Table 1.

As seen in Table 1, the moisture content of the macroalgal biomass is high, accounting in some cases for up to 94% of the weight of the collected fresh biomass. This means that once the biomass is collected, the common process is drying (freeze-drying or oven drying) to preserve the biomass, before applying further processing or technological treatments to obtain macroalgal

Table 1 Summary of the composition ranges described for Phaeophyta (brown macroalgae), Chlorophyta (green macroalgae) and Rhodophyta (red macroalgae) in the recent scientific literature. The content of this table was modified from the work of Øverland et al. (2019)

Proximate composition	Phaeophyta	Chlorophyta	Rhodophyta
Water (% water in wet biomass)	61-94	78-92	72-91
Ash (% DW basis)	15-45	11-55	12-42.2
Polysaccharides (% DW basis)	38-61	15-65	36-66
Crude protein (% DW basis)	2.4-16.8	3.2-35.2	6.4-37.6
Crude lipids (% DW basis)	0.3-9.6	0.3-2.8	0.2-12.9

DW, dry weight.

ingredients (Garcia-Vaquero et al., 2020; Rioux et al., 2010; Yuan et al., 2018). The following sections review in more detail the composition of macroalgae: protein, carbohydrates, lipids, minerals and other compounds. This is partly because variations in these compounds are a key challenge in the effective use of macroalgae in feed, and partly because their accurate characterisation is essential to the use of bioactive compounds with beneficial effects in pig nutrition.

3.1 Protein content in macroalgae

Protein content varies significantly, depending on multiple factors including species. Comparing protein accumulation between studies is difficult due to the different approaches used to estimate protein content in macroalgae. The most commonly used method is based on the quantification of the overall nitrogen content, estimating the value of crude protein by using multiple conversion factors. Due to the high amounts of non-protein nitrogen in macroalgae, several conversion factors have been calculated (Biancarosa et al., 2017; Makkar et al., 2016). The conversion factor used in each study affects the comparisons between studies.

In general, protein concentrations in brown macroalgae are lower (1-16%) compared to green (11-26%) and red (11-33%) macroalgae that have protein levels comparable to other traditionally used protein-rich products (Garcia-Vaquero and Hayes, 2016). The protein concentration is also influenced by season. The average protein content of *Laminaria digitata*, for example, is 6.8%, with the highest levels accumulated during the first quarter of the year (Schiener et al., 2015). Interspecies and intraspecies variations in protein content are shown in Table 2 (García-Vaquero, 2018). This table summarises the variable protein contents (1.1-26.7%) of selected brown, red and green macroalgae.

As shown in Table 2, the reported protein content of brown macroalgae varied between 1.1% (Nielsen et al., 2016) and 16.1% (Mols-Mortensen et al., 2017). Huge variations in the protein content have also been reported for red macroalgal species, with levels varying from 9.2 % (Chan and Matanjun, 2017) to 25.2% (Lozano et al., 2016). The protein levels described in green macroalgae varied between 11.2 % (Biancarosa et al., 2017) and 26.7% (Silva et al., 2015). A significant variation in the levels of protein has also been described within these macroalgal groups depending on the species and climatological conditions affecting the biomass. Studying compounds in European kelp varieties (*Saccharina latissima* and *L digitata*) collected in Denmark, Nielsen et al. (2016) reported the highest biomass production and protein content in seaweed from highly saline water, while low salinity resulted in higher amounts of fermentable sugars and pigments. Manns et al. (2017) reported that the variation in composition was mainly related to season, species and location. The

Table 2 Protein content in selected brown, red and green macroalgae species, originally published by García-Vaquero (2018) and reproduced with permission from Springer Nature

Macroalgae (sp.)	Protein content (% dry weight)	References
Phaeophyta (brown macroalgae)		
Fucus sp.	3.9–4.0	(Biancarosa et al., 2017)
Ascophyllum nodosum	3.0	(Biancarosa et al., 2017)
	6.8	(Schiener et al., 2015)
Laminaria sp.	1.5–5.1	(Nielsen et al., 2016)
	6.6	(Biancarosa et al., 2017)
	2.4–12.7	(Manns et al., 2017)
Saccharina latissima	1.1–7.5	(Nielsen et al., 2016)
	4.8–14.1	(Manns et al., 2017)
	4.0–16.1	(Mols-Mortensen et al., 2017)
	10.6	(Stévant et al., 2017)
Eisenia arborea	9.2–12.5	(Landa-Cansigno et al., 2017)
Alaria esculenta	10.6	(Stévant et al., 2017)
Rhodophyta (red macroalgae)		
Porphyra sp.	13.5–20.6	(Biancarosa et al., 2017)
Gracilaria sp.	9.2	(Chan and Matanjun, 2017)
	12.6	(Chan and Matanjun, 2017)
Chondrus crispus	11.0	(Biancarosa et al., 2017)
Palmaria palmata	10.6	(Biancarosa et al., 2017)
	10.9	(Schiener et al., 2017)
Pyropia columbina	25.2	(Lozano et al., 2016)
Chlorophyta (green macroalgae)		
Ulva sp.	26.7	(Silva et al., 2015)
	13.6	(Angell et al., 2017)
	11.2–15	(Biancarosa et al., 2017)
	22.5	(Bikker et al., 2016)
Cladophora rupestris	12.0	(Biancarosa et al., 2017)

environmental variables studied by the authors included temperature, salinity, phosphate, nitrate and ammonia. The authors reported high levels of protein (15–20% w/w) in seaweed samples collected during February and March from multiple locations, while the lowest amount of protein (2.3–8.4% w/w) was reported in the biomass collected during July and August (Manns et al., 2017). Research on *Saccharina* spp. from the Faroe Islands by Mols-Mortensen et al. (2017) found significantly higher protein content in biomass collected in April–May compared to July–August, while no significant differences were found between locations (Mols-Mortensen et al., 2017).

In contrast to carbohydrates, protein levels are highest during the winter and low during the summer (Adams et al., 2011; Fleurence, 1999; Schiener et al., 2015). It has been suggested that high levels of protein support the build-up of nitrogen reserves for rapid growth during the summer months (Chapman and Craigie, 1977). Schiener et al. (2015) reported the average contents of protein at 6.9 ± 1.1%, 6.8 ± 1.3%, 7.1 ± 1.7% and 11.0 ± 1.4% in *L. digitata*, *L. hyperborea*, *S. latissima* and *A. esculenta*, respectively. The contents of protein varied significantly through the year, with high levels of protein during the first quarter of the year for all macroalgal species, while the lowest protein contents were found in the third quarter of the year.

Amino acids combine to make proteins. Glutamic and aspartic acids are the most abundant amino acids in most macroalgae, with generally low levels of methionine (Holdt and Kraan, 2011). Brown macroalgae are considered rich sources of threonine, valine, leucine, lysine, glycine and alanine (Holdt and Kraan, 2011). Macroalgae usually contain high levels of glutamic acid, present both in its bound and free forms, contributing to the characteristic umami taste of algae (Mæhre et al., 2014). Lourenço et al. (2002) studied the amino acid and protein composition of 19 tropical seaweed species. The authors reported that green algae had low percentages of aspartic and glutamic acids, while red algae showed high percentages of lysine and arginine, and brown macroalgae had higher amounts of methionine. The authors also reported significant variations in the concentration of individual amino acids depending on macroalgal species. The highest concentration of glutamic acid (17.6% of total amino acids) was found in the brown macroalga *Sargassum vulgare*, while the lowest amount (10.7%) was reported in the green macroalga *Caulerpa fastigiate*. The amino acid content varied between species, location and season of collection. Manns et al. (2017) reported that the brown macroalgae *L. digitata* and *S. latissima* contained mainly glutamic acid, aspartic acid and alanine. Individual amino acid contents vary significantly between spring and summer, with levels more dependent on location than species. Glutamic acid represented over 25% of the total amino acid content of *L. digitata* collected in March, while these levels decreased to 12.9% in biomass collected in August. The highest levels of aspartic acid were found in August (14.5%) compared to March (9.5%).

The high levels of protein in macroalgae make it possible to produce bioactive peptides or cryptides. Bioactive peptides are sequences of 2–30 amino acids with no activity within the parent protein. After their release by several enzymatic hydrolysis or fermentation processes, these peptides have been linked to a wide range of health benefits in food and feed. Several bioactive peptides have been identified from *Undaria pinnatifida* (Suetsuna and Nakano, 2000). Some have been commercialised, such as the Wakame peptide jelly (Riken Vitamin Co., Ltd., Tokyo, Japan) and Nori peptide S (Shirako Co., Ltd., Tokyo, Japan) (Fukami, 2010).

3.2 Carbohydrates in macroalgae

Carbohydrates are one of the major components of macroalgae, with contents ranging from 4% to 76% depending on species, location and season (Holdt and Kraan, 2011). Macroalgal carbohydrates include a wide range of compounds (Lafarga et al., 2020). The major polysaccharides include alginate, carrageenan, and other phycocolloids such as agar, commonly used as stabilisers, thickeners and emulsifiers in food production. Red macroalgae mainly contain floridean starch, cellulose and other minor compounds such as mannans, xylans and sulphated galactans. Other carbohydrates, such as glucans (laminarin) and fucoidan, are mainly present in brown macroalgae, while green macroalgae are the main source of energy-reserve carbohydrates such as ulvans (Lafarga et al., 2020; Stiger-Pouvreau et al., 2016). As noted, the levels depend on factors such as species and season. Fucoidan levels in *F. vesiculosus,* for example, were higher than in *U. pinnatifida* (Holdt and Kraan, 2011). The highest amount of fucoidan in cultured *L. japonica* occurred when the biomass matured, rather than during the early stage of development (Honya et al., 1999).

Macroalgal carbohydrates such as fucoidan and laminarin have been associated with improved piglet performance and gut health (Gahan et al., 2009; McDonnell et al., 2010), and have been considered as alternatives to antibiotics and zinc oxide (Sweeney and O'Doherty, 2016). These polysaccharides demonstrate a range of anti-inflammatory, antioxidant, anti-coagulant, anti-viral and anti-tumour properties *in vitro* and *in vivo* (Garcia-Vaquero et al., 2017), with the potential to benefit human health through improving the antioxidant content of animal-derived products (Moroney et al., 2012, 2015). Dietary supplementation of extracts from brown seaweed *(L. digitata*) containing laminarin and fucoidan also improved the quality and shelf life of pig meat by lowering lipid oxidation (Moroney et al., 2012, 2015). Since polysaccharides such as alginates, fucoidan, and laminarin are not digested in the upper digestive tract of the animals, they can be considered as a source of dietary fibre for animals (Garcia-Vaquero et al., 2017). Lahaye (1991) estimated total dietary fibre ranging from 32.7% to 74.6% (on a DW basis) and water-soluble fractions ranging from 51.6% to 85%.

3.3 Lipids and other compounds in macroalgae

In general, macroalgae have a low lipid content, normally ranging between 0.2% and 12.9% as seen in Table 1. However, macroalgal lipids are rich in polyunsaturated fatty acids (PUFA), such as omega 6 and 3 (n6 and n3), with health benefits for both humans and animals. The relative abundance of a particular PUFA depends on the species. Red macroalgae normally have high concentrations of eicosapentaenoic acid (C20:5, n3) that can reach levels of up to half of the total fatty acids of the macroalgae; while brown macroalgae have in general high concentrations of oleic acid (C18:1, n9), linoleic acid (C18:2, n6) and α-linolenic acid (C18:3, n3) (Dawczynski et al., 2007).

Macroalgae are also rich in other phytochemicals including phenolic compounds, pigments and vitamins. As previously mentioned, the main macroalgal classes are established on the basis of the pigmentation of the biomass. Brown macroalgae are rich in fucoxanthin, while the colour of red macroalgae is associated with an abundance of phycobilins and other pigments such as chlorophylls (a and b) and carotenes. Xanthophylls are the main compounds influencing the pigmentation of green macroalgae (Øverland et al., 2019). Macroalgae also contain a wide range of phenolic compounds (i.e. phlorotannins and phloroglucinols) with potent antioxidant properties (Rajauria, 2018). Roleda et al. (2019) reported differences in phenolic compound levels depending on the species and season of collection. The levels of polyphenols in *Palmaria palmata* were significantly higher in Spring, compared to Summer and Autumn (Roleda et al., 2019).

3.4 Mineral contents in macroalgae

The amount of ash in the macroalgal biomass, ranging from 11% to 55% (Table 1), hinders the utilisation of full or intact dried macroalgae as animal feed. The incorporation of high levels of minerals in the feed of monogastric species, including pigs and poultry, resulted in diarrhoea and decreased animal performance (Koreleski et al., 2010; Wilkinson, 1992). There has been a recent study on the extraction of high-value compounds, concentrating them while decreasing the amount of minerals present in the raw biomass. Several innovative technologies, including ultrasound and microwaves, have been explored to obtain macroalgal ingredients or extracts rich in biologically active compounds, such as polysaccharides (i.e. fucoidan and laminarin), for feed applications (Garcia-Vaquero et al., 2017; Yuan et al., 2018).

Macroalgae can accumulate variable levels of minerals including essential trace elements (Ca, Co, Cr, Cu, Fe, I, Mg, Mn, Mo, Ni, P, Se and Zn), providing a source of essential minerals in animal feed (Rey-Crespo et al., 2014). The mineral levels vary by species due to varying metal-binding capacities (Güven et al., 1995). The cell wall polysaccharides produced by brown macroalgae have a high ability to absorb and retain metals. As a result, alginates produced by brown macroalgae have the strongest metal-binding capacity, followed by carrageenans and agar, predominantly produced by red macroalgae (Güven et al., 1995). Brown macroalgae have been described as being exceptionally high in iodine, compared to red and green macroalgae (Biancarosa et al., 2018; Mæhre et al., 2014).

Biancarosa et al. (2018) analysed the mineral profile of 21 macroalgal species collected from the coasts of Norway, and reported generally high iodine contents, varying from 22 mg/kg to 10 000 mg/kg dried weight (DW), with the highest accumulation reported in the brown macroalgae *L digitata*. Seaweed supplements in pigs are considered a way to increase the iodine content of meat (Dierick et al., 2009; He et al., 2002). Dierick et al. (2009) demonstrated that feeding pigs with 2% of *A. nodosum* increased the concentration of iodine in the tissue by 2.7 to 6.8, depending on the tissue.

Macroalgae are also known to accumulate toxic metals such as As, Cd and Hg. Yamada et al. (2007) reported different concentrations of Cd and Pb in *U pinnatifida* related to industrial waste. A number of regulators are addressing content of toxic metals in macroalgae when used as animal feed. In the European Union, calcareous marine algae must contain less than 10 mg As/kg and 15 mg Pb/kg relative to a feed, with moisture content of 12% (Commission Regulation 574/2011) (Regulation, 2011). A feedstuff containing macroalgae must contain less than 40 mg As/kg, with levels of the most toxic As species (inorganic As) of less than 2 mg/kg (Commission Regulation 574/2011) (Regulation, 2011). European legislation specifically mentions levels of As in the brown macroalga *Hizikia fusiforme*, as a result of previous reports of high levels (Besada et al., 2009; Rose et al., 2007).

Most studies analysing the mineral content of macroalgae have focused on analysing total As concentrations (Biancarosa et al., 2018; Khan et al., 2015; Ronan et al., 2017), even though As may be present in its organic form with little or no toxicity (Contam, 2009). Rose et al. (2007) reported that the level of inorganic As was less than 7% of total As. A recent study determining the mineral profile of three brown macroalgae (*L digitata, L hyperborea* and *A. nodosum*) collected in Ireland reported low levels of toxic metals (Cd, Hg and Pb), while levels of total As were high (49–64 mg/kg DW macroalgae) compared with previous reports (Garcia-Vaquero et al., 2021).

4 Biological functions and health-promoting effects of macroalgae and macroalgal-derived extracts in pig nutrition

Based on the properties of macroalgae, particularly brown seaweeds, a large body of recent research has been conducted in pigs. Table 3 summarises the experimental work conducted in the last decade on the use of macroalgae or macroalgal extracts in pigs.

There is limited information on the effects of dietary supplementation with full macroalgal biomass. Choi et al. (2017) described the beneficial effects on growth performance, gut microflora and intestinal morphology in weaning pigs receiving intact *Ecklonia cava* that could be attributed to the high levels of fucoidan (ca. 112 ± 6 g/Kg) in this brown macroalgae. However, Michiels et al. (2012) reported that the addition of *A. nodosum* had no effect on the

Table 3 Effects of macroalgae supplementation on the performance and gut health in weaned piglets. Supplemented diets were tested against a control diet

		End-points*						
Compounds/Substances tested (doses)	Duration (days)	Growth performance	Nutrient digestibility	Gut microbiota	Gut architecture	Antioxidant activity	Immune system modulation	References
Macroalgal extracts								
Laminarin (300 mg/kg) and/ or fucoidan (240 mg/kg) extracts	Weaning to 8 d			+	+		+	(Walsh et al., 2013a)
Brown seaweed (alginates oligosaccharides (50–200 mg/kg)	Weaning to 14 d	+	+			+	+	(Wan et al., 2017)
Brown seaweed (alginic acid oligosaccharides (100 mg/kg)	Weaning to 14 d				+	+	+	(Wan et al., 2018)
Fucoidan extract (44%), (125 and 250 mg/kg)	Weaning to 14 d	-	+					(Rattigan et al., 2019)
Laminarin extract (65%) (100, 200 and 300 mg/kg)	Weaning to 14 d	+		+				(Rattigan et al., 2020)
Laminarin extract (65%) (300 mg/kg)	Weaning to 14 d	+		+	+			(Vigors et al., 2020)
Brown seaweed (alginic acid oligosaccharides (100 mg/kg)	Weaning to 21 d	+	+	+	+	+	+	(Wan et al., 2016)
Laminarin (300 mg/kg) and/ or fucoidan (240 mg/kg) extracts	Weaning to 21 d	+		+				(McDonnell et al., 2010)

(Continued)

Table 3 (*Continued*)

Compounds/Substances tested (doses)	Duration (days)	End-points*						References
		Growth performance	Nutrient digestibility	Gut microbiota	Gut architecture	Antioxidant activity	Immune system modulation	
Laminarin (300 mg/kg) and/or fucoidan (236 mg/kg) extracts	Weaning to 21 d	+		+				(O'Doherty et al., 2010)
Laminarin (300 mg/kg) and/or fucoidan (240 mg/kg) extracts	Weaning to 21 d	+	+					(McAlpine et al., 2012)
Seaweed extract containing laminarin (314 mg/kg) and fucoidan (249 mg/kg)	Weaning to 25 d	+	+	+				(Dillon et al., 2010)
Laminarin (300 mg/kg) and/or fucoidan (240 mg/kg) extracts	Weaning to 32 d	+	+	+				(Heim et al., 2014)
Laminarin (150–300 mg/kg) and/or fucoidan (240 mg/kg) extracts	Weaning to 35 d	+	+	+				(Walsh et al., 2013b)
Fucoidan extract (44%), (250 mg/kg) and laminarin extract (65%) (300 mg/kg)	Weaning to 35 d	-	+	+	+			(Vigors et al., 2021)
Laminarin (300 mg/kg) and/or fucoidan (240 mg/kg) extracts	Weaning to 40 d	+	+	+				(O'Shea et al., 2014)

Commercial seaweed extract (OceanFeed Swine®) (5 g/kg)	Weaning to 160 d	+		+				(Ruiz et al., 2018)
Seaweed extract containing laminarin (sows 1 g/d; piglets 300 mg/kg) and fucoidan (sows 0.8 g/d; piglets 240 mg/kg)	107 d gestation-weaning (28 d)	+						(Draper et al., 2016)
Seaweed extract (10 g/d) containing laminarin (1 g) and fucoidan (0.8 g)	107 d gestation-weaning (28 d)				+		+	(Heim et al., 2015a)
Seaweed extract containing laminarin (280 mg/kg) and fucoidan (244 mg/kg)	107 d gestation-weaning (26 d)	+	–	+	–		+	(Leonard et al., 2011)
Seaweed extract (10 g/d) containing laminarin (1 g) and/or fucoidan (0.8 g)	107 d gestation-weaning (24 d)	+			+		+	(Heim et al., 2015b)
Whole macroalgae								
Ecklonia cava (0. 5–1.5 g/kg)	Weaning to 28 d	+	–	+	+			(Choi et al., 2017)
Ascophyllum nodosum (2.5–10 g/kg)	Weaning to 28 d	–		–				(Michiels et al., 2012)

*Key of outputs: + (significant positive); – (significant negative) effect of the treatment compared to the control group; empty cell, not indicated or not determined.

performance or gut health and plasma oxidative status of piglets, possibly due to levels of bioactive compounds being too low to exert any prebiotic effect.

As shown in Table 3, many studies involve the use of macroalgal extracts in feed. The extracts studied mainly include mixtures of carbohydrates, such as laminarin and/or fucoidan, given as feed supplements to piglets during the post-weaning period (from weaning up to 40 days), since the weaning phase is a critical period due to the high incidence of enteric pathologies. Supplements have also been given to sows at the end of the gestation period, and the effects of this maternal supplementation on piglets were then analysed (Heim et al., 2015a,b; Leonard et al., 2011). Other studies have focused on the long-term effects of dietary supplementation fed to animals from weaning to slaughter (Draper et al., 2016; Ruiz et al., 2018).

4.1 Growth performance and digestion

The studies shown in Table 3 reported the overall positive effects on productive performance with the inclusion of macroalgal extracts in the diet. The average daily gain (ADG) of piglets fed brown seaweeds and/or macroalgal-derived ingredients was in all cases higher (ca. 5% to 40%) compared to that of piglets fed a control diet. Studies evaluating animals at slaughter found the effects of supplementation on ADG to be limited, with improvements of < 5 % compared to those receiving a control diet (Draper et al., 2016; Heim et al., 2015b). The positive effects of macroalgal extracts on growth seem to be related to the prebiotic effects of seaweed polysaccharides on piglet gut function and health (Corino et al., 2019). Overall, the inclusion of macroalgal extracts had positive effects on the digestibility of nitrogen (up to 8% increase), gross energy (up to 10% increase), fibre (up to 73% increase) and ash (up to 80% increase) in most of the studies. The improvement in nutrient digestibility seems to be related to the effect of carbohydrates and antioxidants on gut microbiota and on the villous architecture, with an increase in absorptive capacity and nutrient transporters (Sweeney and O'Doherty, 2016), and also to the beneficial effect on the intestinal mucosal cells and volatile fatty acid production (i.e. butyric acid) (Corino et al., 2019).

4.2 Gut function

Macroalgal supplements also have significant effects on gut microbiota (Table 3). Algal supplements stimulate the growth of *Lactobacilli* and reduce the population of Enterobacteriaceae such as *Escherichia coli*. Dierick et al. (2010) observed that 1% of *A. nodosum* added to feed reduced *E. coli*, while increasing the Lactobacilli/*E. coli* ratio, leading to a lower susceptibility of the animals to intestinal disorders.

Several studies have also demonstrated the positive influence of extracts from brown seaweed on the morphological features of the small intestine of suckling piglets, with increased villus height in the ileum. These positive effects have been attributed to an upregulation in the expression of the tight junction proteins, occludin (OCLN) and zonula occludens 1 (ZO-1) in the small intestine (Wan et al., 2016), together with enhanced antioxidant capacity, decreased mast cell degranulation and prevention of the release of pro-inflammatory cytokines via restraining the TLR4/NF-kB and NOD1/NF-kB signalling pathways (Wan et al., 2016, 2017, 2018). Maternal dietary supplementation with macroalgal-derived polysaccharides also down-regulated the gene expression of pro-inflammatory cytokines involved in *E. coli*-caused diarrhoea in piglets at 48 h after birth and weaning (Heim et al., 2015a).

A number of recent studies have focused on the effects of feeding fucoidan and laminarin for 14 days post-weaning (Rattigan et al., 2019; Vigors et al., 2020). Rattigan et al. (2020) found that laminarin supplementation at 300 ppm provided the most beneficial effects, with improvements in animal performance and positive effects in small intestinal morphology, microbial populations and gene expression. When analysing the microbiome of the laminarin-supplemented group, the changes observed included an increased abundance of *Prevotella* and reductions in *Enterobacteriaceae* (Vigors et al., 2020). Laminarin supplementation had a positive influence on intestinal health through alterations in the gastrointestinal microbiome and increased butyrate production (Vigors et al., 2021). However, there was no additional benefit on performance with either laminarin or fucoidan supplementation up to day 35 post-weaning.

4.3 Immune function and antioxidant capacity

Seaweed extracts also seem to play an immunomodulatory role. Leonard et al. (2011) reported that the dietary supplementation of sows from 109 days of gestation until weaning with extracts from the brown macroalga *Laminaria spp.* increased levels of immunoglobulins (Ig) G and A in the colostrum of sows by 19% and 25%, respectively, as well as the levels of IgG in the serum of piglets by 10%. Feeding seaweed extracts during lactation suppressed pro-inflammatory IL-1a mRNA expression in the ileum of pigs 11 days after weaning. Dietary seaweed extract supplementation post-weaning also induced up-regulation in colonic MUC2 mRNA expression in pigs 11 days post-weaning. These results demonstrate that feed supplementation post-weaning can improve gut health and growth performance in starter pigs.

Other macroalgal polysaccharides such as alginic acid supplemented at 100 mg/kg feed over 21 days post-weaning also increased IgG and IgA concentrations in piglet serum by 20% and 53%, respectively (Wan et al., 2016).

Alginic acid supplementation also increased superoxide dismutase, catalase and total antioxidant activities, and decreased the concentration of malonic dialdehyde in the serum of supplemented animals. The positive effects of alginic acid on the growth performance, antioxidant capacity, immunity and intestinal development in weaned pigs suggest that alginic acid could serve as a useful bioactive feed additive.

Other studies have evaluated the effect of dietary supplementation with laminarin and fucoidan on antioxidant activity (Moroney et al., 2012, 2015; Rajauria et al., 2016). Dietary supplementation of laminarin and fucoidan (0.45–0.9 g/kg) in pigs 3 weeks pre-slaughter resulted in enhanced meat due to deposition of marine-derived bioactive antioxidant components in muscle, reduced saturated fatty acids and lowered lipid oxidation in muscle (Moroney et al., 2015). This antioxidant response was attributed to mechanisms such as the reduction of saturated fatty acids and decreased lipid oxidation in the muscle, though it is still unclear if the free radical scavenging abilities of the extract are responsible for the antioxidant activity observed in the muscle. The enhanced meat quality effected by seaweed polysaccharides may be mediated through the health-promoting effects of gut-associated immunity. The improved fatty acid profiles and enhanced lipid stability of pork meat had no impact on the tenderness, flavour or other sensory properties of the meat. This suggests that dietary supplementation of seaweed extracts containing laminarin and fucoidan in pigs could result in an enhanced meat product. Rajauria et al. (2016) found an improvement in meat quality from the addition of a seaweed extract containing laminarin, due to decreased lipid oxidation. Improved antioxidant capacity and colour stability, together with the reduced lipid peroxidation in the muscle longissimus dorsi, suggest that feed supplementation with macroalgal polysaccharides may protect animals from oxidative stress-induced diseases and improve meat quality.

5 Conclusion and future trends

So far, beneficial effects have been mainly found when adding macroalgae at levels of ≤ 10% of the total concentration in animal feed. More studies of the biochemical profile of seaweed macro and micronutrients, and particularly seaweed bioactive metabolites are needed to properly calculate optimum rates of inclusion of the macroalgae in feed, both to optimise the beneficial effects and to avoid the negative or toxic effects of other compounds (i.e. As, Cd and Hg toxic metals). Wild seaweed biomass varies enormously in nutritional value because of seasonal geographical variations and varying risks of bioaccumulation of heavy metals. Regular monitoring is essential to provide a reliable source of safe animal feed supplementation. A large body of research has been conducted in the last decade on the supplementation of pigs' feed with macroalgal extracts, with beneficial effects related mainly to the presence of polysaccharides (laminarin and/or fucoidan) extracted from brown macroalgae. These studies demonstrate the potential of macroalgal compounds as feed supplements and their application in pig nutrition and production. Despite these promising results, particularly when used as substitutes for in-feed antibiotics, further research is needed to elucidate the effects of individual compounds present in macroalgal extracts. The production of extracts following optimised and standardised protocols, and the characterisation of the molecules within these extracts, will allow the establishment of clear structure–function relationships. This will allow industrial exploitation of macroalgal extracts for improved intestinal physiology, morphology, microbiology and immune response of post-weaning piglets, leading to industrial products with proven health-enhancement benefits that can be applied routinely in the herds.

6 Where to look for further information

As seen in the current chapter, macroalgae can be an excellent source of multiple compounds that can be used in animal feed for multiple purposes. However, the incorporation of macroalgae in animal feed will ultimately depend on the composition of the original feedstock. An introduction to the wide variation of the macroalgal biomass with respect to the species, season and year of collection can be accessed in Garcia-Vaquero et al. (2021). 'Seasonal variation of the proximate composition, mineral content, fatty acid profiles and other phytochemical constituents of selected brown macroalgae' (see full details in the reference list). This publication may serve as a reference guide for the inclusion of macroalgae in animal feed and the selective collection of this feedstock depending on the intended use in animals' diet. Moreover, one of the main organisations in Europe promoting the knowledge on algal biomass is the European Algae Biomass Association (EABA) (https://www.eaba-association.org/en/aboutus). This group include key information related to algal biomass and main key stakeholders of this industry in Europe.

Macroalgae, despite being an excellent source of multiple beneficial compounds, may also contain high levels of minerals and other minor compounds (heavy metals) that may limit the utilisation of this biomass either for practical purposes or legal issues related to the presence of these contaminants in the biomass. An extensive legislative framework on the use of macroalgae as animal feed in Europe can be accessed from Garcia-Vaquero and Hayes (2016) 'Red and green macroalgae for fish and animal feed and human functional food development' (see in the reference list). Furthermore, the use of macroalgae in animal feed is regulated in Europe by the European Food Safety Authority (EFSA) within the 'feed additives' panel. The readers of the chapter are referred to the EFSA and, in particular, to the publications of this panel to get the latest information related to the use of any fed additive (https://www.efsa.europa.eu/en/topics/topic/feed-additives). Individual countries may also provide their own guidance or interpretation of these European guidelines. As an example, the Food Safety Authority of Ireland (FSAI) published the report 'Safety Considerations of Seaweed and Seaweed-derived Foods Available on the Irish Market' (ISBN: 978-1-910348-42-0) and established recommendations for the seaweed sector in Ireland in addition to the aforementioned European regulations.

As the field of animal nutrition specialises, further studies will be needed not only including macroalgae or macroalgal extracts in animal feed, but targeting specific macroalgal molecules in animal feed to establish a clear association between the health benefits appreciated in the animals and particular macroalgal compounds that could be industrially exploited. In order to achieve that, multidisciplinary projects, including the optimisation of the extraction of macroalgal compounds as well as their *in vitro* and *in vivo* evaluation, should be further explored. Currently, the European project BIOCARB-4-FOOD (17RDSUSFOOD2ERA-NET1, https://www.biocarb4food.eu/) focuses on targeting the extraction and utilisation of macroalgal carbohydrates for multiple industrial applications. Other promising projects following also a multidisciplinary approach include the 'Enhance microalgae' (https://www.enhancemicroalgae.eu/). This European project offers a comprehensive approach to the study of algae for the generation of high-value compounds, starting from cultivation to extraction of these compounds from algae using novel technologies, as well as the final potential applications of these compounds in multiple industries.

7 Acknowledgements

This chapter did not receive any specific grant from funding agencies in the public, commercial or not-for-profit sectors.

8 References

Adams, J. M. M., Toop, T. A., Donnison, I. S. and Gallagher, J. A. (2011). Seasonal variation in Laminaria digitata and its impact on biochemical conversion routes to biofuels. *Bioresource Technology* 102(21): 9976–9984. DOI: 10.1016/j.biortech.2011.08.032.

Aiking, H. (2014). Protein production: planet, profit, plus people? *The American Journal of Clinical Nutrition* 100(Suppl. 1): 483S–489S. DOI: 10.3945/ajcn.113.071209.

Ale, M. T. and Meyer, A. S. (2013). Fucoidans from brown seaweeds: an update on structures, extraction techniques and use of enzymes as tools for structural elucidation. *RSC Advances*. Royal Society of Chemistry 3(22): 8131–8141. DOI: 10.1039/C3RA23373A.

Angell, A. R., Paul, N. A. and de Nys, R. (2017). A comparison of protocols for isolating and concentrating protein from the green seaweed Ulva ohnoi. *Journal of Applied Phycology* 29(2): 1011–1026. DOI: 10.1007/s10811-016-0972-7.

Besada, V., Andrade, J. M., Schultze, F. and González, J. J. (2009). Heavy metals in edible seaweeds commercialised for human consumption. *Journal of Marine Systems* 75(1-2): 305-313. DOI: 10.1016/j.jmarsys.2008.10.010.

Biancarosa, I., Espe, M., Bruckner, C. G., Heesch, S., Liland, N., Waagbø, R., Torstensen, B. and Lock, E. J. (2017). Amino acid composition, protein content, and nitrogen-to-protein conversion factors of 21 seaweed species from Norwegian waters. *Journal of Applied Phycology* 29(2): 1001-1009. DOI: 10.1007/s10811-016-0984-3.

Biancarosa, I., Belghit, I., Bruckner, C. G., Liland, N. S., Waagbø, R., Amlund, H., Heesch, S. and Lock, E. J. (2018). Chemical characterization of 21 species of marine macroalgae common in Norwegian waters: benefits of and limitations to their potential use in food and feed. *Journal of the Science of Food and Agriculture*. John Wiley & Sons, Ltd 98(5): 2035-2042. DOI: 10.1002/jsfa.8798.

Bikker, P., van Krimpen, M. M., van Wikselaar, P., Houweling-Tan, B., Scaccia, N., van Hal, J. W., Huijgen, W. J. J., Cone, J. W. and López-Contreras, A. M. (2016). Biorefinery of the green seaweed Ulva Lactuca to produce animal feed, chemicals and biofuels. *Journal of Applied Phycology*. Springer Netherlands 28(6): 3511-3525. DOI: 10.1007/s10811-016-0842-3.

Billakanti, J. M., Catchpole, O. J., Fenton, T. A., Mitchell, K. A. and MacKenzie, A. D. (2013). Enzyme-assisted extraction of fucoxanthin and lipids containing polyunsaturated fatty acids from Undaria pinnatifida using dimethyl ether and ethanol. *Process Biochemistry* 48(12): 1999-2008. DOI: 10.1016/j.procbio.2013.09.015.

Chan, P. T. and Matanjun, P. (2017). Chemical composition and physicochemical properties of tropical red seaweed, Gracilaria changii. *Food Chemistry* 221: 302-310. DOI: 10.1016/j.foodchem.2016.10.066.

Chapman, A. R. O. and Craigie, J. S. (1977). Seasonal growth in Laminaria longicruris: relations with dissolved inorganic nutrients and internal reserves of nitrogen. *Marine Biology* 40(3): 197-205. DOI: 10.1007/BF00390875.

Chapman, V. (2012). *Seaweeds and Their Uses*. New York: Springer Science & Business Media.

Choi, Y., Hosseindoust, A., Goel, A., Choi, Y., Hosseindoust, A., Goel, A., Lee, S., Jha, P. K., Kwon, I. K., & Chae, B. J. (2017). Effects of Ecklonia cava as fucoidan-rich algae on growth performance, nutrient digestibility, intestinal morphology and caecal microflora in weanling pigs. *Asian-Australasian Journal of Animal Sciences*. Asian-Australasian Association of Animal Production Societies (AAAP) and Korean Society of Animal Science and Technology (KSAST) 30(1): 64-70. DOI: 10.5713/ajas.16.0102.

Collins, K. G., Fitzgerald, G. F., Stanton, C. and Ross, R. P. (2016). Looking beyond the terrestrial: the potential of seaweed derived bioactives to treat non-communicable diseases. *Marine Drugs*. MDPI AG 14(3). DOI: 10.3390/md14030060.

Contam EP on C in the FC (2009) Scientific opinion on arsenic in food. *EFSA Journal* Wiley-Blackwell Publishing Ltd 7(10). DOI: 10.2903/j.efsa.2009.1351.

Corino, C., Modina, S. C., Di Giancamillo, A., Chiapparini, S. and Rossi, R. (2019). Seaweeds in pig nutrition. *Animals: An Open Access Journal from MDPI*. MDPI AG 9(12): 1126. DOI: 10.3390/ani9121126.

Dawczynski, C., Schubert, R. and Jahreis, G. (2007). Amino acids, fatty acids, and dietary fibre in edible seaweed products. *Food Chemistry* 103(3): 891-899. DOI: 10.1016/j.foodchem.2006.09.041.

Dierick, N., Ovyn, A. and De Smet, S. (2009). Effect of feeding intact brown seaweed Ascophyllum nodosum on some digestive parameters and on iodine content in edible tissues in pigs. *Journal of the Science of Food and Agriculture* 89(4): 584–594. DOI: 10.1002/jsfa.3480.

Dierick, N., Ovyn, A. and De Smet, S. (2010). In vitro assessment of the effect of intact marine brown macro-algae Ascophyllum nodosum on the gut flora of piglets. *Livestock Science* 133(1–3): 154–156. DOI: 10.1016/j.livsci.2010.06.051.

Dillon, S., Sweeney, T., Figat, S., Callan, J. J. and O'Doherty, J. V. (2010). The effects of lactose inclusion and seaweed extract on performance, nutrient digestibility and microbial populations in newly weaned piglets. *Livestock Science* 134(1–3): 205-207. DOI: 10.1016/j.livsci.2010.06.142.

Dominguez, H. and Loret, E. P. (2019). Ulva Lactuca, A source of troubles and potential riches. *Marine Drugs*. MDPI AG 17(6): 357. DOI: 10.3390/md17060357.

Draper, J., Walsh, A. M., McDonnell, M. and O'Doherty, J. V. (2016). Maternally offered seaweed extracts improves the performance and health status of the postweaned pig1. *Journal of Animal Science* 94(suppl_3): 391–394. DOI: 10.2527/jas.2015-9776.

Evans, F. D. and Critchley, A. T. (2014). Seaweeds for animal production use. *Journal of Applied Phycology* 26(2): 891–899. DOI: 10.1007/s10811-013-0162-9.

Fleurence, J. (1999). Seaweed proteins. *Trends in Food Science and Technology* 10(1): 25–28. DOI: 10.1016/S0924-2244(99)00015-1.

Fukami, H. (2010). *Functional Foods and Biotechnology in Japan*. DOI: 10.1201/9781420087123.

Gahan, D. A., Lynch, M. B., Callan, J. J., O'Sullivan, J. T. and O'Doherty, J. V. (2009). Performance of weanling piglets offered low-, medium- or high-lactose diets supplemented with a seaweed extract from Laminaria spp. *Animal* 3(1): 24–31. DOI: 10.1017/S1751731108003017.

Garcia-Vaquero, M. and Hayes, M. (2016). Red and green macroalgae for fish and animal feed and human functional food development. *Food Reviews International*. Taylor & Francis 32(1): 15–45. DOI: 10.1080/87559129.2015.1041184.

Garcia-Vaquero, M., Rajauria, G., O'Doherty, J. V. and Sweeney, T. (2017). Polysaccharides from macroalgae: recent advances, innovative technologies and challenges in extraction and purification. *Food Research International* 99(3): 1011–1020. DOI: 10.1016/j.foodres.2016.11.016.

García-Vaquero, M. (2018). Seaweed proteins and applications in animal feed. In: *Novel Proteins for Food, Pharmaceuticals and Agriculture*. Wiley Online Books, pp. 139-161. DOI: 10.1002/9781119385332.ch7.

Garcia-Vaquero, M., O'Doherty, J. V., Tiwari, B. K., Sweeney, T. and Rajauria, G. (2019). Enhancing the extraction of polysaccharides and antioxidants from macroalgae using sequential hydrothermal-assisted extraction followed by ultrasound and thermal technologies. *Marine Drugs*. MDPI AG 17(8): 457. DOI: 10.3390/md17080457.

Garcia-Vaquero, M., Ummat, V., Tiwari, B. and Rajauria, G. (2020). Exploring ultrasound, microwave and ultrasound–microwave assisted extraction technologies to increase the extraction of bioactive compounds and antioxidants from brown macroalgae. *Marine Drugs*. MDPI AG 18(3): 172. DOI: 10.3390/md18030172.

Garcia-Vaquero, M., Rajauria, G., Miranda, M., Sweeney, T., Lopez-Alonso, M. and O'Doherty, J. (2021). Seasonal variation of the proximate composition, mineral content, fatty acid profiles and other phytochemical constituents of selected brown macroalgae. *Marine Drugs* 19(4) . DOI: 10.3390/md19040204.

Güven, K. C., Akyüz, K. and Yurdun, T. (1995). Selectivity of heavy metal binding by algal polysaccharides. *Toxicological and Environmental Chemistry*. Taylor & Francis 47(1-2): 65–70. DOI: 10.1080/02772249509358127.

He, M. L., Hollwich, W. and Rambeck, W. A. (2002). Supplementation of algae to the diet of pigs: a new possibility to improve the iodine content in the meat. *Journal of Animal Physiology and Animal Nutrition* 86(3-4): 97–104. DOI: 10.1046/j.1439-0396.2002.00363.x.

Heim, G., Walsh, A. M., Sweeney, T., Doyle, D. N., O'Shea, C. J., Ryan, M. T. and O'Doherty, J. V. (2014). Effect of seaweed-derived laminarin and fucoidan and zinc oxide on gut morphology, nutrient transporters, nutrient digestibility, growth performance and selected microbial populations in weaned pigs. *British Journal of Nutrition*. Cambridge University Press 111(9): 1577–1585. DOI: 10.1017/S0007114513004224.

Heim, G., O'Doherty, J. V., O'Shea, C. J., Doyle, D. N., Egan, A. M., Thornton, K. and Sweeney, T. (2015a). Maternal supplementation of seaweed-derived polysaccharides improves intestinal health and immune status of suckling piglets. *Journal of Nutritional Science*. Cambridge University Press 4: e27–e27. DOI: 10.1017/jns.2015.16.

Heim, G., Sweeney, T., O'Shea, C. J., Doyle, D. N. and O'Doherty, J. V. (2015b). Effect of maternal dietary supplementation of laminarin and fucoidan, independently or in combination, on pig growth performance and aspects of intestinal health. *Animal Feed Science and Technology* 204: 28–41. DOI: 10.1016/j.anifeedsci.2015.02.007.

Herman, E. M. and Schmidt, M. A. (2016). The potential for engineering enhanced functional-feed soybeans for sustainable aquaculture feed. *Frontiers in Plant Science* 7: 440. Available at: https://www.frontiersin.org/article/10.3389/fpls.2016.00440.

Holdt, S. L. and Kraan, S. (2011). Bioactive compounds in seaweed: functional food applications and legislation. *Journal of Applied Phycology* 23(3): 543–597. DOI: 10.1007/s10811-010-9632-5.

Honya, M., Mori, H., Anzai, M., Araki, Y. and Nisizawa, K. (1999). Monthly changes in the content of fucans, their constituent sugars and sulphate in cultured *Laminaria japonica*. In: *Proceedings of the Sixteenth International Seaweed Symposium*, pp. 411–416. DOI: 10.1007/978-94-011-4449-0_49.

Jones, R. T., Blunden, G. and Probert, A. J. (1979). Effects of dietary Ascophyllum nodosum on blood parameters of rats and pigs. *Botanica Marina* 22: 393–402.

Jones, C. K., Gabler, N. K., Main, R. G. and Patience, J. F. (2012). Characterizing growth and carcass composition differences in pigs with varying weaning weights and postweaning performance1. *Journal of Animal Science* 90(11): 4072–4080. DOI: 10.2527/jas.2011-4793.

Khan, N., Ryu, K. Y., Choi, J. Y., Nho, E. Y., Habte, G., Choi, H., Kim, M. H., Park, K. S. and Kim, K. S. (2015). Determination of toxic heavy metals and speciation of arsenic in seaweeds from South Korea. *Food Chemistry* 169: 464–470. DOI: 10.1016/j.foodchem.2014.08.020.

Koreleski, J., Świątkiewicz, S. and Arczewska, A. (2010). The effect of dietary potassium and sodium on performance, carcass traits, and nitrogen balance and excreta moisture in broiler chicken. *Journal of Animal and Feed Sciences* 19(2): 244-256. DOI: 10.22358/jafs/66285/2010.

Kraan, S. (2013). 6 - Pigments and minor compounds in algae. In: Domínguez HBT-FI from A for F and N (Ed.) *Functional Ingredients from Algae for Foods and Nutraceuticals. Woodhead Publishing Series in Food Science, Technology and Nutrition*. Woodhead Publishing, pp. 205–251. DOI: 10.1533/9780857098689.1.205.

Lafarga, T., Acién-Fernández, F. G. and Garcia-Vaquero, M. (2020). Bioactive peptides and carbohydrates from seaweed for food applications: natural occurrence, isolation, purification, and identification. *Algal Research* 48. DOI: 10.1016/j.algal.2020.101909: 101909.

Lahaye, M. (1991). Marine algae as sources of fibres: determination of soluble and insoluble dietary fibre contents in some 'sea vegetables'. *Journal of the Science of Food and Agriculture*. John Wiley & Sons, Ltd 54(4): 587–594. DOI: 10.1002/jsfa.2740540410.

Landa-Cansigno, C., Hernández-Carmona, G., Arvizu-Higuera, D. L., Muñoz-Ochoa, M. and Hernández-Guerrero, C. J. (2017). Bimonthly variation in the chemical composition and biological activity of the brown seaweed Eisenia arborea (Laminariales: Ochrophyta) from Bahía Magdalena, Baja California Sur, Mexico. *Journal of Applied Phycology* 29(5): 2605–2615. DOI: 10.1007/s10811-017-1195-2.

Leonard, S. G., Sweeney, T., Bahar, B., Lynch, B. P. and O'Doherty, J. V. (2011). Effects of dietary seaweed extract supplementation in sows and post-weaned pigs on performance, intestinal morphology, intestinal microflora and immune status. *British Journal of Nutrition*. Cambridge University Press 106(5): 688–699. DOI: 10.1017/S0007114511000997.

Lourenço, S. O., Barbarino, E., De-Paula, J. C., Pereira, L. OdS. and Marquez, U. M. L. (2002). Amino acid composition, protein content and calculation of nitrogen-to-protein conversion factors for 19 tropical seaweeds. *Phycological Research*. John Wiley & Sons, Ltd 50(3): 233–241. DOI: 10.1111/j.1440-1835.2002.tb00156.x.

Lozano, I., Wacyk, J. M., Carrasco, J. and Cortez-San Martín, M. A. (2016). Red macroalgae Pyropia columbina and Gracilaria chilensis: sustainable feed additive in the Salmo salar diet and the evaluation of potential antiviral activity against infectious salmon anemia virus. *Journal of Applied Phycology* 28(2): 1343–1351. DOI: 10.1007/s10811-015-0648-8.

Mæhre, H. K., Malde, M. K., Eilertsen, K. E. and Elvevoll, E. O. (2014). Characterization of protein, lipid and mineral contents in common Norwegian seaweeds and evaluation of their potential as food and feed. *Journal of the Science of Food and Agriculture*. John Wiley & Sons, Ltd 94(15): 3281–3290. DOI: 10.1002/jsfa.6681.

Makkar, H. P. S., Tran, G., Heuzé, V., Giger-Reverdin, S., Lessire, M., Lebas, F. and Ankers, P. (2016). Seaweeds for livestock diets: a review. *Animal Feed Science and Technology* 212: 1–17. DOI: 10.1016/j.anifeedsci.2015.09.018.

Manns, D., Nielsen, M. M., Bruhn, A., Saake, B. and Meyer, A. S. (2017). Compositional variations of brown seaweeds Laminaria digitata and Saccharina latissima in Danish waters. *Journal of Applied Phycology* 29(3): 1493–1506. DOI: 10.1007/s10811-017-1056-z.

McAlpine, P., O'Shea, C. J., Varley, P. F., Flynn, B. and O'Doherty, J. V. (2012). The effect of seaweed extract as an alternative to zinc oxide diets on growth performance, nutrient digestibility, and fecal score of weaned piglets. *Journal of Animal Science* 90 (Suppl. 4): 224–226. DOI: 10.2527/jas.53956.

McDonnell, P., Figat, S. and O'Doherty, J. V. (2010). The effect of dietary laminarin and fucoidan in the diet of the weanling piglet on performance, selected faecal microbial populations and volatile fatty acid concentrations. *Animal* Cambridge University Press 4(4): 579–585. DOI: 10.1017/S1751731109991376.

Michalak, I. and Chojnacka, K. (2014). Algal extracts: technology and advances. *Engineering in Life Sciences*. John Wiley & Sons, Ltd 14(6): 581–591. DOI: 10.1002/elsc.201400139.

Michiels, J., Skrivanova, E., Missotten, J., Ovyn, A., Mrazek, J., De Smet, S. and Dierick, N. (2012). Intact brown seaweed (Ascophyllum nodosum) in diets of weaned piglets: effects on performance, gut bacteria and morphology and plasma oxidative status. *Journal of Animal Physiology and Animal Nutrition*. John Wiley & Sons, Ltd 96(6): 1101–1111. DOI: 10.1111/j.1439-0396.2011.01227.x.

Miranda, M., Lopez-Alonso, M. and Garcia-Vaquero, M. (2017). Macroalgae for functional feed development: applications in aquaculture, ruminant and swine feed industries. In: *Seaweeds: Biodiversity, Environmental Chemistry and Ecological Impacts*, pp. 133–153. Available at: https://www.scopus.com/inward/record.uri?eid=2-s2.0-85048822636&partnerID=40&md5=97e44f36e4ee7081f3c8bb790bae7eba.

Mols-Mortensen, A., Ortind, E. áG., Jacobsen, C. and Holdt, S. L. (2017). Variation in growth, yield and protein concentration in Saccharina latissima (Laminariales, Phaeophyceae) cultivated with different wave and current exposures in the Faroe Islands. *Journal of Applied Phycology* 29(5): 2277–2286. DOI: 10.1007/s10811-017-1169-4.

Morais, T., Inácio, A., Coutinho, T., Ministro, M., Cotas, J., Pereira, L. and Bahcevandziev, K. (2020). Seaweed potential in the animal feed: a review. *Journal of Marine Science and Engineering*. MDPI AG 8(8). DOI: 10.3390/JMSE8080559.

Moroney, N. C., O'Grady, M. N., O'Doherty, J. V. and Kerry, J. P. (2012). Addition of seaweed (Laminaria digitata) extracts containing laminarin and fucoidan to porcine diets: influence on the quality and shelf-life of fresh pork. *Meat Science* 92(4): 423–429. DOI: 10.1016/j.meatsci.2012.05.005.

Moroney, N. C., O'Grady, M. N., Robertson, R. C., Stanton, C., O'Doherty, J. V. and Kerry, J. P. (2015). Influence of level and duration of feeding polysaccharide (laminarin and fucoidan) extracts from brown seaweed (Laminaria digitata) on quality indices of fresh pork. *Meat Science* 99: 132–141. DOI: 10.1016/j.meatsci.2014.08.016.

Nielsen, M. M., Manns, D., D'Este, M., Krause-Jensen, D., Rasmussen, M. B., Larsen, M. M., Alvarado-Morales, M., Angelidaki, I. and Bruhn, A. (2016). Variation in biochemical composition of Saccharina latissima and Laminaria digitata along an estuarine salinity gradient in inner Danish waters. *Algal Research* 13: 235–245. DOI: 10.1016/j.algal.2015.12.003.

O'Doherty, J. V., McDonnell, P. and Figat, S. (2010). The effect of dietary laminarin and fucoidan in the diet of the weanling piglet on performance and selected faecal microbial populations. *Livestock Science* 134(1–3): 208–210. DOI: 10.1016/j.livsci.2010.06.143.

O'Shea, C. J., McAlpine, P., Sweeney, T., Varley, P. F. and O'Doherty, J. V. (2014). Effect of the interaction of seaweed extracts containing laminarin and fucoidan with zinc oxide on the growth performance, digestibility and faecal characteristics of growing piglets. *British Journal of Nutrition*. Cambridge University Press 111(5): 798–807. DOI: 10.1017/S0007114513003280.

Øverland, M., Mydland, L. T. and Skrede, A. (2019). Marine macroalgae as sources of protein and bioactive compounds in feed for monogastric animals. *Journal of the Science of Food and Agriculture*. John Wiley & Sons, Ltd 99(1): 13–24. DOI: 10.1002/jsfa.9143.

Rajauria, G. (2018). Optimization and validation of reverse phase HPLC method for qualitative and quantitative assessment of polyphenols in seaweed. *Journal of Pharmaceutical and Biomedical Analysis* 148: 230–237. DOI: 10.1016/j.jpba.2017.10.002.

Rajauria, G., Draper, J., McDonnell, M. and O'Doherty, J. V. (2016). Effect of dietary seaweed extracts, galactooligosaccharide and vitamin E supplementation on meat quality parameters in finisher pigs. *Innovative Food Science and Emerging Technologies* 37: 269–275. DOI: 10.1016/j.ifset.2016.09.007.

Rattigan, R., Sweeney, T., Vigors, S., Thornton, K., Rajauria, G. and O'Doherty, A. J. V. (2019). The effect of increasing inclusion levels of a fucoidan-rich extract derived from Ascophyllum nodosum on growth performance and aspects of intestinal health of pigs post-weaning. *Marine Drugs* 17(12) . DOI: 10.3390/md17120680.

Rattigan, R., Sweeney, T., Maher, S., Thornton, K., Rajauria, G. and O'Doherty, J. V. (2020). Laminarin-rich extract improves growth performance, small intestinal morphology, gene expression of nutrient transporters and the large intestinal microbial composition of piglets during the critical post-weaning period. *British Journal of Nutrition*. Cambridge University Press 123(3): 255–263. DOI: 10.1017/S0007114519002678.

Regulation EC (2011) Commission Regulation (EU) No 574/2011 of 16 June 2011 amending Annex I to Directive 2002/32/EC of the European Parliament and of the Council as regards maximum levels for nitrite, melamine. Ambrosia spp. and carry-over of certain coccidiostats and histom. *Official Journal of the European Union, L* 159: 7–24.

Rey-Crespo, F., López-Alonso, M. and Miranda, M. (2014). The use of seaweed from the Galician coast as a mineral supplement in organic dairy cattle. *Animal* 8(4): 580–586. DOI: 10.1017/S1751731113002474.

Rioux, L. E., Turgeon, S. L. and Beaulieu, M. (2010). Structural characterization of laminaran and galactofucan extracted from the brown seaweed Saccharina longicruris. *Phytochemistry* 71(13): 1586–1595. DOI: 10.1016/j.phytochem.2010.05.021.

Roleda, M. Y., Marfaing, H., Desnica, N., Jónsdóttir, R., Skjermo, J., Rebours, C. and Nitschke, U. (2019). Variations in polyphenol and heavy metal contents of wild-harvested and cultivated seaweed bulk biomass: health risk assessment and implication for food applications. *Food Control* 95: 121–134. DOI: 10.1016/j.foodcont.2018.07.031.

Ronan, J. M., Stengel, D. B., Raab, A., Feldmann, J., O'Hea, L., Bralatei, E. and McGovern, E. (2017). High proportions of inorganic arsenic in Laminaria digitata but not in Ascophyllum nodosum samples from Ireland. *Chemosphere* 186: 17–23. DOI: 10.1016/j.chemosphere.2017.07.076.

Rose, M., Lewis, J., Langford, N., Baxter, M., Origgi, S., Barber, M., MacBain, H. and Thomas, K. (2007). Arsenic in seaweed–forms, concentration and dietary exposure. *Food and Chemical Toxicology* . Elsevier 45(7): 1263–1267.

Ruiz, Á. R., Gadicke, P., Andrades, S. M. and Cubillos, R. (2018). Supplementing nursery pig feed with seaweed extracts increases final body weight of pigs. *Austral Journal of Veterinary Sciences*. Universidad Austral de Chile. Facultad de Ciencias Veterinarias 50(2): 83–87. DOI: 10.4067/S0719-81322018000200083.

Sauvageau, C. (Camille) (1920). *Utilisation Des Algues Marines*. Paris: O. Doin. Available at: https://www.biodiversitylibrary.org/item/122691.

Schiener, P., Black, K. D., Stanley, M. S. and Green, D. H. (2015). The seasonal variation in the chemical composition of the kelp species Laminaria digitata, Laminaria Hyperborea, Saccharina latissima and Alaria esculenta. *Journal of Applied Phycology* 27(1): 363–373. DOI: 10.1007/s10811-014-0327-1.

Schiener, P., Zhao, S., Theodoridou, K., Carey, M., Mooney-McAuley, K. and Greenwell, C. (2017). The nutritional aspects of biorefined Saccharina latissima, Ascophyllum nodosum and Palmaria palmata. *Biomass Conversion and Biorefinery* 7(2): 221–235. DOI: 10.1007/s13399-016-0227-5.

Silva, D. M., Valente, L. M. P., Sousa-Pinto, I., Pereira, R., Pires, M. A., Seixas, F. and Rema, P. (2015). Evaluation of IMTA-produced seaweeds (Gracilaria, Porphyra, and Ulva) as dietary ingredients in Nile tilapia, Oreochromis niloticus L., juveniles. Effects on growth performance and gut histology. *Journal of Applied Phycology* 27(4): 1671–1680. DOI: 10.1007/s10811-014-0453-9.

Stévant, P., Marfaing, H., Rustad, T., Sandbakken, I., Fleurence, J. and Chapman, A. (2017). Nutritional value of the kelps Alaria esculenta and Saccharina latissima and effects of short-term storage on biomass quality. *Journal of Applied Phycology* 29(5): 2417–2426. DOI: 10.1007/s10811-017-1126-2.

Stiger-Pouvreau, V., Bourgougnon, N. and Deslandes, E. (2016). Carbohydrates from seaweeds. In: Fleurence, J. (Translator) and Levine IBT-S in H and DP (eds) *Seaweed in Health and Disease Prevention*. San Diego: Academic Press, pp. 223–274. DOI: 10.1016/B978-0-12-802772-1.00008-7.

Suetsuna, K. and Nakano, T. (2000). Identification of an antihypertensive peptide from peptic digest of wakame (Undaria pinnatifida). *The Journal of Nutritional Biochemistry* 11(9): 450–454. DOI: 10.1016/S0955-2863(00)00110-8.

Swanson, J. C. (1995). Farm animal well-being and intensive production systems2. *Journal of Animal Science* 73(9): 2744–2751. DOI: 10.2527/1995.7392744x.

Sweeney, T. and O'Doherty, J. V. (2016). Marine macroalgal extracts to maintain gut homeostasis in the weaning piglet. *Domestic Animal Endocrinology* 56 (Suppl.): S84–S89. DOI: 10.1016/j.domaniend.2016.02.002.

Van Boeckel, T. P., Brower, C., Gilbert, M., Grenfell, B. T., Levin, S. A., Robinson, T. P., Teillant, A. and Laxminarayan, R. (2015). Global trends in antimicrobial use in food animals. *Proceedings of the National Academy of Sciences of the United States of America*. National Academy of Sciences 112(18): 5649-5654. DOI: 10.1073/pnas.1503141112.

Vigors, S., O'Doherty, J. V., Rattigan, R., McDonnell, M. J., Rajauria, G. and Sweeney, T. (2020). Effect of a laminarin rich macroalgal extract on the caecal and colonic microbiota in the post-weaned pig. *Marine Drugs* 18(3). DOI: 10.3390/md18030157.

Vigors, S., O'Doherty, J., Rattigan, R. and Sweeney, T. (2021). Effect of supplementing seaweed extracts to pigs until d35 post-weaning on performance and aspects of intestinal health. *Marine Drugs* 19(4) . DOI: 10.3390/md19040183.

Walsh, A. M., Sweeney, T., O'Shea, C. J., Doyle, D. N. and O'Doherty, J. V. (2013a). Effect of dietary laminarin and fucoidan on selected microbiota, intestinal morphology and immune status of the newly weaned pig. *British Journal of Nutrition*. Cambridge University Press 110(9): 1630–1638. DOI: 10.1017/S0007114513000834.

Walsh, A. M., Sweeney, T., O'Shea, C. J., Doyle, D. N. and O'Doherty, J. V. O. (2013b). Effect of supplementing varying inclusion levels of laminarin and fucoidan on growth performance, digestibility of diet components, selected faecal microbial populations and volatile fatty acid concentrations in weaned pigs. *Animal Feed Science and Technology* 183(3-4): 151–159. DOI: 10.1016/j.anifeedsci.2013.04.013.

Wan, J., Jiang, F., Xu, Q., Chen, D. and He, J. (2016). Alginic acid oligosaccharide accelerates weaned pig growth through regulating antioxidant capacity, immunity and intestinal development. *RSC Advances*. The Royal Society of Chemistry 6(90): 87026–87035. DOI: 10.1039/C6RA18135J.

Wan, J., Zhang, J., Chen, D., Yu, B. and He, J. (2017). Effects of alginate oligosaccharide on the growth performance, antioxidant capacity and intestinal digestion-absorption function in weaned pigs. *Animal Feed Science and Technology* 234: 118–127. DOI: 10.1016/j.anifeedsci.2017.09.006.

Wan, J., Zhang, J., Chen, D., Yu, B., Huang, Z., Mao, X., Zheng, P., Yu, J. and He, J. (2018). Alginate oligosaccharide enhances intestinal integrity of weaned pigs through altering intestinal inflammatory responses and antioxidant status. *RSC Advances*. The Royal Society of Chemistry 8(24): 13482–13492. DOI: 10.1039/C8RA01943F.

Wang, T., Jónsdóttir, R., Kristinsson, H. G., Hreggvidsson, G. O., Jónsson, J. Ó, Thorkelsson, G. and Ólafsdóttir, G. (2010). Enzyme-enhanced extraction of antioxidant ingredients from red algae Palmaria palmata. *LWT–Food Science and Technology* 43(9): 1387–1393. DOI: 10.1016/j.lwt.2010.05.010.

Wilkinson, M. (1992). *Seaweed Resources in Europe: Uses and Potential*. Guiry, M. D. and Blunden, G. (Eds). Chichester: John Wiley & Sons Ltd, 1991. xi+432pp. Price: £65.00. ISBN 0 471 92947 6. Aquatic Conservation: Marine and Freshwater Ecosystems 2(2). John Wiley & Sons, Ltd: 209–210. DOI: 10.1002/aqc.3270020206.

Yamada, M., Yamamoto, K., Ushihara, Y. and Kawai, H. (2007). Variation in metal concentrations in the brown alga Undaria pinnatifida in Osaka Bay, Japan. *Phycological Research*. John Wiley & Sons, Ltd 55(3): 222–230. DOI: 10.1111/j.1440-1835.2007.00465.x.

Yuan, Y., Zhang, J., Fan, J., Clark, J., Shen, P., Li, Y. and Zhang, C. (2018). Microwave assisted extraction of phenolic compounds from four economic brown macroalgae species and evaluation of their antioxidant activities and inhibitory effects on α-amylase, α-glucosidase, pancreatic lipase and tyrosinase. *Food Research International* 113: 288–297. DOI: 10.1016/j.foodres.2018.07.021.

Chapter 10

Microalgae: a unique source of poultry feed protein

Sahil Kalia, Andrew D. Magnuson, Guanchen Liu and Xin Gen Lei, Cornell University, USA

1 Introduction

In an animal feed industry, corn and soybean meal (SBM) are the two most commonly used ingredients to meet energy and protein requirements of various food-producing species, in particular, simple-stomached animals, such as poultry and swine. In the United States, 60% of corn and 47% of soy are used to feed mainly chickens and pigs. Ultimately, this practice creates a direct competition between feed and food for these two staples (Olson, 2006). Consequently, the world may not afford to continue the feed use of corn and SBM at the current rate or scale without jeopardizing the global food security of humans with an increasing global population. Therefore, alternative feed ingredients, in particular, high protein feedstuff supplement, must be explored to sustain the current growth in animal production (Austic et al., 2013). One such alternative may be microalgae (Christaki et al., 2011). Microalgae have gained popularity recently as the third generation of feedstocks for biofuel production because many species of microalgae grow rapidly, synthesize high levels of lipids and sequester large amounts of CO_2 (Spolaore et al., 2006). Defatted algal biomass remained after the extraction of oils from the full-fatted microalgae is a rich source of proteins, amino acids and bioactive nutrients.

http://dx.doi.org/10.19103/AS.2021.0091.16

2 Nutrient composition of microalgae as alternative feed protein and bioactive nutrient sources

Both full-fatted and defatted microalgal biomass can be used as animal feed. Approximately, 30% of total microalgal biomass cultivated around the world is currently sold as animal feed (Vanthoor-Koopmans et al., 2013). Concentrations of proteins, vitamins and minerals found in microalgae biomass are usually nearly equivalent or higher than that found in most conventional feedstuffs such as corn, oil extracted dehulled SBM, wheat and fish meal (Lum et al., 2013). Indeed, the high protein contents and balanced amino acid profiles of various microalgae species are one of the main factors that favor their use in poultry diets (Kovac et al., 2013). Moreover, various types of defatted microalgae biomass generated after extraction of lipid during the biofuel production are manifested with higher protein concentrations than those of the full-fatted microalgae (Becker, 2007).

Table 1 lists concentrations of crude protein and essential amino acids in various sources of full- and de-fatted microalgae, along with comparisons with those in corn, SBM, wheat and fish meal. The crude protein concentrations in the listed microalgae species range from 14% in the full-fatted diatom *Staurosira* biomass to 76% in the blue-green *Spirulina* biomass, compared with 8.5%, 14%, 49% and 62% in corn, wheat, SBM and fish meal, respectively. The metabolizable energy in microalgae was reported as 1320 kcal/kg in the defatted *Staurosira* biomass, compared with 3560 kcal/kg, 3170 kcal/kg, 2425 kcal/kg and 2950 kcal/kg in corn, wheat, SBM and fish meal, respectively (Batal and Dale, 2010; Austic et al., 2013).

Concentrations of essential amino acids histidine (His), isoleucine (Ile), leucine (Leu), phenylalanine (Phy), threonine (Thr), tryptophan (Trp) and valine (Val) in various microalgae species range from 0.77%, 1.8%, 3.5%, 2.1%, 1.97%, 0.55% and 2.33% in the defatted *Nannochloropsis oceanica* to 2.0%, 3.8%, 8.8%, 5.0%, 4.8%, 3.4% and 5.5% in the green *Chlorella vulgaris*. In comparison, concentrations of these amino acids are 1.4%, 2.3%, 3.9%, 2.6%, 2.02%, 0.7% and 2.4% in SBM and 1.5%, 2.4%, 4.4%, 2.3%, 2.8%, 0.5% and 2.8% in fish meal, respectively. Concentrations of lysine (Lys) and methionine (Met), the two most limiting amino acids in corn-SBM-based diets for poultry, range from 2.4% and 0.67% in the defatted *N. oceanica* to 8.4% and 2.2% in *C. vulgaris*, compared with 3.0% and 0.76% in SBM and 4.7% and 1.7% in fish meal, respectively (Batal and Dale, 2010; NRC, 2012; Tao et al., 2018). Our laboratory has been evaluating nutrient compositions and feeding values of several full-fatted and defatted microalgae including *Staurosira* spp., *Desmodesmus* spp., *N. oceanica*, *Haematococcus pluvialis* and *Aurantiochytrium* biomass in the diets for broiler chickens and laying hens. Crude protein concentrations in those microalgae used in our laboratory range from 10% to 45%, which is 21–93%

of that in the SBM and 4.2-fold greater than that in the corn. Concentrations of ether extract (EE), acid detergent fiber, neutral detergent fiber and ash range from 14%, 0.70%, 14% and 45% in the defatted *Staurosira* biomass to 38%, 7.4%, 24% and 20% in the defatted *N. oceanica*. Comparatively, those concentrations are 1.0%, 10%, 16% and 6.0% in SBM and 1.9%, 4.2%, 9.7% and 1.2% in corn, respectively (NRC, 2012; Austic et al., 2013; Gatrell et al., 2015).

Table 2 shows fatty acid profiles of microalgae studied in our laboratory, in comparison with those of other oil ingredients (corn oil, soybean oil, fish oil, SBM). Concentrations of saturated fatty acids (SFA) were found to be highest in full-fatted *Staurosira* spp., intermediate in *N. oceanica* (CO 18 strain) and lowest in *Desmodesmus* spp. (CO 46 strain), respectively, compared with those (on basis of % total fatty acids) in corn oil, soybean oil, fish oil, and SBM. Concentrations of monounsaturated fatty acids (MUFA) were found to be highest in defatted *N. oceanica*, intermediate in *Aurantiochytrium* and lowest in defatted *Desmodesmus* spp. (CO46-LEA), respectively, compared those in fish oil, corn oil, soybean oil and SBM (Moir et al., 1995; Oladiji et al., 2009; Lee et al., 2015; Kim et al., 2016; Sun et al., 2020a,b). Concentrations of polyunsaturated fatty acid (PUFA) were found to be highest in defatted *Desmodesmus* spp., intermediate in defatted *N. oceanica* and lowest in full-fatted *Staurosira* spp., respectively, compared with those in SBM, corn oil, soybean oil and fish oil. Concentrations of total n-3 PUFA were found to be highest in *Aurantiochytrium*, intermediate in defatted *Desmodesmus* spp. (CO46-LEA) and lowest in full-fatted *Staurosira* spp., respectively, compared with those in fish oil, SBM, soybean oil and corn oil. Overall, n-6/n-3 PUFA ratios were much lower in various sources of the microalgal biomass than those in all feed oils except for fish oil (Moir et al., 1995; Oladiji et al., 2009; Lee et al., 2015; Kim et al., 2016).

Table 3 lists vitamin and mineral contents of various microalgae, in comparison with SBM and fish meal. Concentrations of riboflavin and niacin were found to be much higher in *Spirulina* spp. and higher in *Chlorella* spp. than in SBM. Concentrations of thiamine and α-tocopherol were much higher in *Spirulina* spp. and higher in *Chlorella* spp. than in SBM. Moreover, concentrations of folic acid were found to be highest in fish meal, intermediate in *Chlorella* and *Spirulina* spp. and lowest in SBM. Concentrations of calcium and phosphorus were found to be highest in fish meal, intermediate in *Staurosira* and lowest in *Desmodesmus*. Concentrations of sodium and potassium were found to be highest in *Staurosira*, intermediate in *Desmodesmus* and lowest in SBM and fish meal (Batal and Dale, 2010; Andrade et al., 2018). High concentrations of ash or sodium in the microalgal biomass may produce adverse effects on nutrient metabolism or animal health by causing feed refusal and excessive drinking water needs by birds.

Table 1 Protein and amino acid profile (%) of selected microalgae species and other ingredients used in poultry diets

Source	CP	Glu	Pro	Ile	Leu	Lys	Asp	Ala	Val	Arg	Gly	Phe	Thr	Ser	Tyr	His	Met	Trp	Cys	References
SBM	48.5	9.20	2.41	2.31	3.90	3.23	5.95	2.42	2.41	3.73	2.16	2.64	2.02	2.58	1.47	1.35	0.76	0.68	0.68	Sriperm et al., 2011
Corn	8.50	1.41	0.66	0.27	0.93	0.26	0.52	0.58	0.37	0.41	0.32	0.40	0.28	0.38	0.28	0.22	0.17	0.065	0.17	Sriperm et al., 2011
Wheat	13.5	-	-	0.69	1.0	0.40	-	-	0.69	0.60	-	0.78	0.35	-	-	0.17	0.25	0.18	0.30	Batal and Dale, 2010
Fish meal	62.0	-	-	2.40	4.40	4.70	-	-	2.80	3.65	-	2.28	2.75	-	-	1.52	1.70	0.50	0.50	Batal and Dale, 2010
Aurantiochytrium	10.3	1.04	0.28	0.35	0.59	0.47	1.12	0.54	0.47	0.42	0.41	0.36	0.42	0.42	0.27	0.20	2.28	0.13	0.12	Sun et al., 2020a
Chlorella pyrenoidosa	44.0	-	-	1.75	3.79	2.06	-	-	2.47	2.06	2.20	1.81	2.12	-	0.80	0.62	0.36	0.80	-	Combs, 1952
Chlorella vulgaris	58.0	11.6	4.8	3.8	8.8	8.4	9.0	7.9	5.5	6.4	5.8	5.0	4.8	4.1	3.4	2.0	2.2	2.1	1.4	Becker, 2007
Desmodesmus spp. defatted	31.2	2.93	2.73	1.10	2.29	1.61	2.69	2.27	1.59	1.45	1.72	1.34	1.26	1.10	1.01	0.50	0.48	0.43	0.33	Ekmay et al., 2014
Desmodesmus spp. (CO 46)	41.0	-	-	-	-	-	-	-	-	-	-	-	-	-	-	-	-	-		Sun et al., 2020b
Desmodesmus spp. (CO46-LEA)	39.0	3.44	3.84	1.51	2.97	2.20	3.08	2.94	2.00	1.84	2.06	1.83	1.44	1.12	1.27	0.65	0.65	0.24	0.55	Sun et al., 2020b
Haematococcus pluvialis defatted	28.5	-	2.20	0.13	2.70	2.49	-	4.63	1.31	1.87	-	2.06	-	-	2.03	0.56	0.78	0.21	0.07	Sun et al., 2018; Magnuson et al., 2018

H. pluvialis	26.8	-	1.59	1.[illegible]6	3.22	2.59	-	4.89	1.67	0.87	-	2.45	-	-	1.27	0.42	0.87	0.17	0.07	Sun et al., 2018; Magnuson et al., 2018
Isochrysis spp.	27–45	4.60	2.40	1.80	3.90	2.50	4.20	3.20	2.40	2.50	2.60	3.80	2.40	2.20	3.80	0.90	1.40	0.60	1.40	Bandarra et al., 2003
Nannochloropsis oceanica	38.2	3.34	4.00	1.50	2.90	2.27	2.80	2.22	2.13	1.99	1.92	1.57	1.54	1.21	1.20	0.64	0.57	1.20	0.30	Gatrell et al., 2015
N. oceanica	45.1	4.13	2.38	1.83	3.51	2.37	3.53	-	2.33	2.23	2.36	2.05	1.97	1.63	1.52	0.77	0.67	0.55	0.42	Tao et al., 2018
N. oceanica (CO18 strain)	45.0	4.01	2.84	1.82	3.33	2.65	3.49	2.76	2.38	2.35	2.15	2.04	1.93	1.54	1.43	0.77	0.85	0.45	0.43	Unpublished report
Scenedesmus obliquus	55.0	10.7	3.9	3.6	7.3	5.6	8.4	9.0	6.0	7.1	7.1	4.8	5.1	3.8	3.2	2.1	1.5	0.3	0.6	Becker, 2007
Spirulina platensis	51.0–60.5	7.17	2.18	2.71	4.40	2.10	5.06	4.14	2.61	3.92	2.89	2.20	1.72	2.61	2.66	0.70	0.80	-	1.31	Tavernari et al., 2018
Spirulina spp.	76.0	6.12	1.72	2.55	3.88	2.10	4.58	3.21	2.94	-	2.32	-	1.97	-	-	-	1.07	-	0.42	Evans et al., 2015
Staurosira spp. defatted	19.1	1.81	0.65	0.78	1.33	0.83	1.88	1.09	0.98	0.93	0.96	0.86	0.88	0.76	0.57	0.30	0.33	1.18	0.32	Austic et al., 2013
Staurosira spp.	13.9	1.29	0.45	0.55	0.94	0.57	1.31	0.76	0.70	0.61	0.67	0.80	0.63	0.53	0.40	0.18	0.26	0.12	0.19	Ekmay et al., 2015

Ala, alanine; Arg, arginine; Asp, aspartic acid; CP, crude protein; Cys, cystine; Glu, glutamic acid; Gly, glycine; His, histidine; Ile, isoleucine; Leu, leucine; Lys, lysine; Met, methionine; Phe, phenylalanine; Pro, proline; SBM, soybean meal; Ser, serine; Thr, threonine; Trp, tryptophan; Tyr, tyrosine; Val, valine.

Table 2 Fatty acid composition of different microalgae species compared with conventional oils and soybean meal

Fatty acids	Content (% of total fatty acids)										
	Staurosira spp.[1]	Defatted *Desmodesmus* spp.[1]	Defatted *Nannochloropsis oceanica*[1]	*Desmodesmus* spp. (CO 46 strain)[2]	*Desmodesmus* spp. (CO 46-LEA)[2]	*N. oceanica* (CO 18 strain)[3]	*Aurantiochytrium*[4]	Corn oil[5]	Fish oil[6]	Soybean oil[7]	Soybean meal[8,9,10]
C14:0	8.31	1.21	7.44	0.59	0.48	4.77	2.19	0.1	9.70	0.10	0.14
C14:1	0.11	ND	0.12	ND	0.02	0.07	ND	ND	ND	ND	ND
C16:0	51.8	35.6	29.2	15.9	17.5	24.00	34.1	9.60	21.4	10.3	13.1
C16:1	35.7	1.59	25.2	3.08	2.66	21.0	ND	ND	15.2	5.20	0.13
C18:0	0.84	2.24	0.50	0.90	1.13	0.68	-	1.90	5.30	3.80	4.08
C18:1n9	0.71	19.3	12.1	9.03	8.19	5.79	-	29.0	15.9	22.8	16.2
C18:2n6	1.04	10.8	2.13	7.98	5.98	2.33	-	56.4	3.20	51.0	55.2
C18:3n3	ND	22.1	0.11	24.9	14.8	1.24	0.08	2.70	3.50	6.80	10.1
C18:3n6	0.24	1.64	0.42	1.38	1.22	0.28	-	ND	ND	ND	ND
C20:2n6	0.12	2.85	0.14	ND	ND	0.03	-	ND	ND	ND	ND
C20:4n6	0.41	ND	5.84	0.08	0.03	4.26	-	ND	ND	ND	ND
C20:5n3	0.49	0.41	16.4	0.47	ND	19.8	0.12	0.4	14.1	ND	ND
C22:5n3	ND	ND	ND	ND	ND	0.03	4.77	ND	1.80	ND	ND
C22:6n3	ND	ND	ND	0.10	0.05	0.09	23.01	ND	6.90	ND	ND
SFA	61.1	40.2	37.2	17.4	19.1	29.5	363	11.5	36.4	14.2	17.3
MUFA	36.6	21.9	37.7	12.1	10.9	26.8	-	29.0	31.0	28.0	16.3
PUFA	2.27	37.9	25.1	34.9	22.1	28.1	28.0	60.0	32.6	57.8	65.3

n-3	0.51	22.6	16.5	25.5	14.8	21.2	28.0	3.10	26.3	6.80	10.1
n-6	1.78	15.3	8.6	9.44	7.23	6.89	-	56.4	3.20	51.0	55.2
n-6/n-3	3.49	0.68	0.52	0.37	0.49	0.32	-	18.2	0.12	7.50	5.47
Ether extract (%)	30.1	1.10	1.50	-	-	-	-	6.0	9.20	20.0	1.51

MUFA, monounsaturated fatty acid; n-3, omega 3 fatty acid; n-6, omega 6 fatty acid; ND, not detected; PUFA, polyunsaturated fatty acid; SFA, saturated fatty acid.
Source: [1]Kim et al., 2016; [2]Sun et al., 2020b; [3]Unpublished report; [4]Sun et al., 2020a; [5]Moir et al., 1995; [6]NRC, 2012; [7]Oladiji et al., 2009; [8]Lee et al., 2015; [9]Tang et al., 2012; [10]Batal and Dale, 2010.

Table 3 Vitamins and mineral concentrations of different microalgae species, soybean and fish meal

	Chlorella spp.[1]	*Spirulina* spp.[1]	*Desmodesmus*[2]	*Staurosira*[3]	Soybean meal[4]	Fish meal[4]
Thiamine (mg/kg)	17	24	-	-	1.7	0.20
Riboflavin (mg/kg)	43	37	-	-	2.6	4.8
Niacin (mg/kg)	238	128	-	-	20.9	55.0
Pantothenic acid (mg/kg)	11	0	-	-	13	8.8
Folic acid (µg/kg)	940	940	-	-	700	1000
Vitamin B12 (µg/kg)	1	0	-	-	0	150
α-Tocopherol (mg/kg)	15	50	-	-	3.3	5.7
Calcium (%)	-	-	0.33	2.8	0.31	4.8
Phosphorus (%)	-	-	0.65	0.76	0.72	3.0
Sodium (%)	-	-	3.2	3.9	0.04	0.68
Potassium (%)	-	-	0.89	1.7	2.1	0.96

Source: [1]Andrade et al., 2018; [2]Ekmay et al., 2014; [3]Austic et al., 2013; [4]Batal and Dale, 2010.

3 Effects and values of microalgae as a supplement in broiler diets

Table 4 summarizes the effects of supplemental microalgae on growth performance, health status and chicken meat quality in diets for broilers. Although we attempt to provide a comprehensive and representative summary, the list may not be exhaustive. In earlier studies, freshwater green microalgae *Chlorella* biomass was investigated as a potential source of high-quality protein and antioxidants for broiler chickens. In a 28-day experiment, Combs (1952) supplemented the broiler diet with *Chlorella pyrenoidosa* at 10% to replace SBM. The supplement improved body weight gain (BWG) and feed use efficiency. Grau and Klein (1957) demonstrated that the inclusion of protein-rich, sewage-grown *Chlorella* for 18 days up to 20% in broiler diets was well tolerated by the birds. Likewise, Lipstein and Hurwitz (1983) reported that the high crude protein content of *C. vulgaris* allowed the replacements of fish meal and SBM by 5% and 10% in the diet of broilers without any negative effects on their BWG and feed conversion ratio (FCR). In a 35-day experiment, Kang et al. (2013) recorded improved BWG and intestinal microbial population when broilers were fed with 1% *C. vulgaris*. Similar results were observed by An et al. (2016) that 0.5% of *Chlorella* supplementation for 35 days improved the BWG and FCR and elevated the level of serum immunoglobulin G (IgG) and immunoglobulin M (IgM). Further, Dlouha et al. (2008) observed an improved growth performance and higher selenium (Se) and glutathione peroxidase activity in the broiler breast meat after feeding the animals with Se-enriched *Chlorella* at 0.3 mg Se/kg of diet for 42 days.

After 41 days of feeding, Ross and Dominy (1990) found no difference in the growth performance of broilers supplemented with 1.5%, 3%, 6% and 12% of dehydrated green *Spirulina platensis*, compared with the controls. They concluded that *Spirulina* up to the level of 12% could be used to replace other protein sources. Evans et al. (2015) reported that replacement of SBM and corn with dried *Spirulina maxima* up to 16% in the broiler diets produced no negative effect on the performance. Feeding broilers 4% or 8% of dried *Spirulina* from 21 days to 36 days of age by Toyomizu et al. (2001) enhanced yellow pigmentation in the muscles, skin, fat and liver. A longer study of 12 weeks by Venkataraman et al. (1994) demonstrated no negative effects of dried *S. platensis* up to 14% in the diets on the growth performance, dressing percentage and histopathology of various organs. Moreover, another study conducted for 36 days by Shanmugapriya et al. (2015) showed an improved BWG, FCR and villus length in chickens fed a diet with 1% *S. platensis* biomass. Bonos et al. (2016) reported that dietary supplementation of *S. platensis* (5 g/kg or 10 g/kg) in broiler diets for 42 days enriched the meat with eicosapentaenoic acid (EPA) and docosahexaenoic acid (DHA), without negative impact on growth performance.

Table 4 Effects of dietary inclusions of different microalgae species on broiler performance, health and meat quality

Microalgae	Level in the diet	Main findings	References
Aurantiochytrium spp. Full-fatted	1.2, 2.4 and 4.9 g DHA/kg diet	Dose-dependent enrichments of DHA and decreased n-6/n-3 PUFA ratios and nonesterified fatty acid concentration in plasma, liver, muscle and adipose tissue. Upregulation of ACCα, SCD1, FASN, elongases 2 and 5, desaturases 1 and 2, ACS and CPT2 expression in adipose tissues	Tolba et al., 2019
Chlorella pyrenoidosa	10% as replacement of SBM	Increased growth performance and improved feed use efficiency	Combs, 1952
Chlorella spp.	10–30%	Level up to 20% improved growth rate and was well tolerated by broilers	Grau and Klein, 1957
Chlorella spp.	0.3 mg Se/kg of diet	Improved growth performance with increased glutathione peroxidase and decreased lipid peroxidation in breast meat	Dlouha et al., 2008
Chlorella vulgaris	5–10%	No adverse effects on the growth performance, FCR, carcass fat or liver size	Lipstein and Hurwitz, 1983
C. vulgaris	0.5%	No effects on BWG but increased the phygocytic activity of leucocytes	Kotrbacek et al., 1994
C. vulgaris	1% as replacement of antibiotic growth promoters	Positive effects on BWG and IgG, IgM and *Lactobacillus* count in intestine	Kang et al., 2013
C. vulgaris	0.05%, 0.5% and 0.15%	Improved BWG, FCR and humoral immune response. No change in the meat quality, color, or pH	An et al., 2016
Desmodesmus spp. (DGM) defatted	15% as replacement of SBM and corn mixture	Improved FCR and altered the hepatic and muscle eIF4E, mTOR, S6 and pS6 protein levels	Ekmay et al., 2014
Haematococcus pluvalis	7, 36 and 179 mg astaxanthin/kg feed	No negative effects on BWG, feed intake, or FCR; increased astaxanthin and carotenoid concentrations in kidney, intestine and breast muscle.	Waldenstedt et al., 2003
H. pluvalis	10, 20, 40 and 80 mg/kg diet	Dose-dependent enrichments of astaxanthin, carotenoids, antioxidant capacity in plasma, liver, breast, thigh muscles; and improved meat quality	Sun et al., 2018
Nannochloropsis oceanica (DGA) defatted	2%, 4%, 8% and 16% as replacement of SBM and corn mixture	Enriched n-3 PUFA, EPA and DHA in plasma, liver, breast and thigh tissues. Decreased n-6/n-3 PUFA ratios and elevated the mRNA levels of fatty acid synthase and Δ-6 and Δ-9 desaturases. DGA inclusion at 2–8% had positive effects on BWG, FCR, relative organ weights and soluble inorganic phosphorus retention and excretion	Gatrell et al., 2015, 2017

N. oceanica defatted	10% as replacement of SBM and corn mixture	Increased EPA, DHA and antioxidant status in liver and muscles; decreased lipid peroxidation and cytochrome P450 (CYP2C23b, CYP2D6, CYP3A5 and CYF4V2) gene expression in liver and muscles	Tao et al., 2018
Schizochytrium JB5	0.1% and 0.2%	No adverse effects on growth performance or relative organ weights; improved meat quality with increased n-3 fatty acid and DHA content in breast meat	Yan and Kim, 2013
Schizochytrium limacinum	1% and 2% as replacement of SBM	Improvements in growth performance, carcass traits, serum composition, antioxidant capacity and EPA and DHA deposition in breast and thigh muscles	Long et al., 2018
Spirulina maxima	6%, 11%, 16% and 21%	Inclusion up to 16% had no adverse effects on broiler performance; improved digestibility of methionine	Evans et al., 2015
Spirulina platensis	1.5–12%	No adverse effects on BWG and FCR	Ross and Dominy, 1990
S. platensis	14–17% replacement of fish meal and groundnut cake	No negative effects on the BWG, meat quality, or histopathology of liver, heart and kidney	Venkataraman et al., 1994
S. platensis	0.5%, 1% and 1.5%	Improved BWG and FCR; increased villi height at inclusion level of 1%	Shanmugapriya et al., 2015
S. platensis	0.5% and 1%	No effects on growth performance or meat lipid peroxidation status; increased EPA and DHA content in thigh meat	Bonos et al., 2016
Spirulina spp.	4% and 8%	No negative effects on growth performance or relative weight of internal organs; increased yellow pigmentation in broiler flesh and liver	Toyomizu et al., 2001
Staurosira spp. (DFA) defatted	7.5% replacement of SBM or 7.5% and 10% replacements of SBM and corn mixture	Inclusion up to 7.5% for replacing a mixture of SBM and corn had no adverse effects on growth or plasma biochemistry; inclusion of 7.5% for replacing SBM improved performance in broilers with additions of essential amino acids (methionine, lysine, isoleucine, valine, tryptophan and threonine) to the diets	Austic et al., 2013

ACCα, acetyl-CoA carboxylase; ACS, acetyl-CoA synthetase; BWG, body weight gain; CPT2, carnitine palmitoyltransferase; DHA, docosahexaenoic acid; EPA, eicosapentaenoic acid; FASN, fatty acid synthase; FCR, feed conversion ratio; IgG, immunoglobulin G; IgM, immunoglobulin M; mRNA, messenger ribonucleic acid; mTOR, mammalian target of rapamycin; n-3, omega 3 fatty acid; n-6, omega 6 fatty acid; PUFA, polyunsaturated fatty acid; SBM, soybean meal; SCD1, stearoyl-CoA desaturase-1; Se, selenium.

In a 32-day experiment, Waldenstedt et al. (2003) evaluated the effects of supplemental astaxanthin-rich *Haematococcus pluvalis* microalgae meal (7 mg, 36 mg and 179 mg astaxanthin/kg feed) in broiler chickens infected with *Campylobacter jejuni*. They showed no difference in growth performance between different levels of supplementation, but tissue astaxanthin and carotenoid concentrations were significantly increased by higher levels of dietary microalgae. Yan and Kim (2013) reported increased DHA concentration and improved fatty acid profiles in broiler breast upon supplementation of 0.1–0.2% *Schizochytrium JB5*. Likewise, dietary inclusion of 1–2% of *Schizochytrium limacinum* for 42 days in broiler diets improved growth performance, decreased concentrations of serum cholesterol and low-density lipoprotein cholesterol (LDL-C) and increased concentrations of high-density lipoprotein cholesterol (HDL-C), antioxidants and fatty acids in breast and thigh muscle (Long et al., 2018).

Our laboratory has been testing the potential of full-fatted and/or defatted microalgal biomass of *Staurosira* spp., *Desmodesmus* spp., *N. oceanica* and *H. pluvialis* to replace conventional ingredients of corn and SBM (Table 4). Initially, Austic et al. (2013) reported that defatted diatom microalgae *Staurosira* biomass (DFA) could substitute for a mixture of 7.5% SBM and corn (1 part of SBM:3 parts of corn in a mixture) in a broiler diet. That level of replacement did not produce adverse effects on growth performance or plasma biomarkers. In contrast, the inclusion of DFA for replacing a mixture of 10% SBM and corn (1 part of SBM:4 parts of corn in a mixture) or 7.5% SBM impaired the growth performance of broilers. However, the lost performance associated with the 7.5% SBM replacement was partially restored by adding extra essential amino acids (methionine, lysine, isoleucine, valine, tryptophan and threonine) into the diet. A simple estimate is that the inclusion of 7.5% DFA in broiler diets to replace SBM could potentially spare over 2.4 million metric tons of soybean for human consumption annually (Austic et al., 2013).

Effects of feed ingredients on protein digestion and utilization in birds are often assessed by responses of plasma uric acid and amino acid concentrations (Milles and Featherston, 1974; Salter et al., 1974; Donsbough et al., 2010). The DFA supplementation in broiler diets decreased plasma uric acid concentrations, indicating an improved nitrogen utilization. In another study, Ekmay et al. (2014) evaluated the nutritional impacts of defatted green microalgae (*Desmodesmus* spp.) biomass (DGM) along with additions of protease and nonstarch polysaccharide degrading enzymes (NSPase) in diets for broilers for 42 days. Broilers fed with 15% DGM replacing the mixture of SBM and corn had improved growth performance and better gain/feed efficiency as compared with the control group. Intriguingly, supplementing NSPase and protease to the DGM-containing diets resulted in negative effects on growth performance. The DGM inclusion increased circulating plasma amino acid concentrations from

0.40 μmol/mL on day 21 to 2.20 μmol/mL on day 42 but did not alter plasma uric acid concentrations. The DGM inclusion further altered the hepatic and muscle levels of key proteins involved in the mammalian target of rapamycin (mTOR) pathway that regulates cell growth and nutrient metabolism.

Gatrell et al. (2015) investigated the effects of incorporating broiler diets with defatted green microalgae (*N. oceanica*) (DGA) for 42 days. Supplementing the diets for broilers with 0%, 2%, 4%, 8% and 16% DGA in replacing mixtures of corn and SBM resulted in linear increases in the concentrations of n-3 fatty acids, EPA and DHA, in the plasma, liver, breast and thigh. The 16% incorporation enriched the breast and thigh with EPA up to 0.54 and 0.88 mg/100 g tissue, DHA up to 0.79 and 1.04 mg/100 g tissue and n-3 PUFA up to 2.18 and 2.45 mg/100 g tissue, respectively. Broiler fed with 4% and 8% DGA displayed higher expression of Δ-6 and Δ-9 desaturase (genes involved in the biosynthesis of EPA and DHA) along with higher gene expression of fatty acid synthase than the control. The inclusions of DGA at 2%, 4% and 8% in the broiler diets did not create any adverse effect on BWG or FCR. In contrast, the highest level of inclusion (16%) led to adverse effects on the growth performance. The impairment was probably due to relative deficiency in the sulfur-containing amino acids methionine and cysteine, low digestibility of microalgal protein, or perhaps high nucleic acid content (ileal DNA retention was increased by DGA supplementation) (Gatrell et al., 2014; Gatrell et al., 2017). Meanwhile, a high sodium content of DGA also led to increased water intake of birds. However, the increasing DGA inclusions led to elevated soluble inorganic phosphorus retention and decreased inorganic phosphorus excretion. The decrease in the phosphorus excretion by the DGA inclusion may help minimize the environmental pollution of manure phosphorus excretion by broilers (Gatrell et al., 2017).

Tao et al. (2018) investigated the enrichments of n-3 PUFA in broilers by incorporating 10% defatted *N. oceanica* under various dietary corn oil, vitamin E and selenium concentrations. The incorporation of microalgae enriched the broiler breast and thigh with DHA up to 9.25 and 14.2 mg/100 g tissue, respectively. There was a significant downregulation in the hepatic expression of cytochrome P450 enzyme genes (CYP2C23b, CYP2D6, CYP3A5 and CYP4V2) involved in PUFA oxidation by the supplementation of defatted microalgae. In addition, supplementing 10% defatted *N. oceanica* in a corn-SBM basal diet increased the apparent retentions of dry matter and EE by 3.3% and 3.8%, respectively, compared with the controls (Sun et al., 2016). The supplementation decreased apparent ileal digestibilities of eight essential amino acids and six nonessential amino acids, ranging from 32% for isoleucine to 7% for glutamic acid (Sun et al., 2016).

Tolba et al. (2019) supplemented DHA-rich microalgae *Aurantiochytrium* to a corn-SBM basal diet for broilers for 42 days at 0, 1.2, 2.4 and 4.9 g DHA/ kg of diet, respectively, and found dose-dependent enrichments of DHA in the plasma, liver and muscle tissues. The DHA enrichments in the breast and thigh reached 82 and 96 mg/100 g tissue, respectively. As the United States Department of Agriculture (USDA) 2015–2020 Dietary Guidelines for Americans recommended daily intake of DHA is 250 mg of EPA + DHA (USDA, 2015; Cortinas et al., 2004), two servings (approximately 200 g) of the microalgae-supplementation derived chicken may meet a good portion of the suggested intake. Meanwhile, the microalgae incorporation upregulated the mRNA abundances of key genes involved in fatty acid de novo synthesis (ACCα, SCD1, FASN, elongase 2, elongase 5, desaturases 1 and desaturases 2) and oxidation (ACS and CPT2) in the adipose tissues but downregulated the same set of genes in the liver.

Sun et al. (2018) investigated the nutritional and metabolic effect of microalgal astaxanthin (AST) in diets for broilers under heat stress. Supplementation of AST from *H. pluvialis* at 80 mg/kg of diets enriched AST and carotenoids up to 2.16 mg/kg and 12.0 mg/kg, respectively, in the broiler breast muscle. The AST supplementation altered tissue redox status and intrinsic antioxidant enzyme activity and GSH concentrations. An increased meat color score and a decreased water holding capacity (WHC) were recorded in the breast and thighs. The AST supplementation decreased hepatic SFA, MUFA and PUFA concentrations, but did not affect the growth performance of chicks exposed to high room temperatures.

4 Effects and values of microalgae as a supplement in laying hen diets

Microalgal biomass incorporation into diets for laying hens is of particular interest due to the abundance of fat and the lipophilic nature of eggs for uptaking and depositing health-promoting compounds of microalgae. Table 5 outlines the effects of different microalgae supplementations in diets for laying hens on egg production and quality, along with related biochemical responses. Mariey et al. (2012) demonstrated that supplementing *S. platensis* at 0.10%, 0.15% and 0.20% for 24 weeks improved the egg production rate, egg mass, egg yolk color and egg yolk score and decreased yolk triglyceride and nonesterified free fatty acid concentrations. Selim et al. (2018) obtained similar results that *S. platensis* supplementation at 0.1%, 0.2% and 0.3% in hen diets from 38 weeks to 46 weeks of age enhanced egg production rate, egg mass, shell thickness and egg yolk score and decreased serum triglyceride concentrations and aspartate aminotransferase activities. Likewise, Zahroojian et al. (2013) reported significant increases in egg yolk color by supplementations of *S. platensis* at 1.5%, 2.0% and 2.5% of hen diets.

Walker et al. (2012) fed laying hens with AST-rich *H. pluvialis* at 0.49%, 1.47% and 2.94% in combination with tocomin for 8 weeks and noticed increased redness in color and concentration of tocopherol in the egg yolk. However, there was no change in egg production or egg yolk weight by the microalgae inclusion. The inclusions of *Chlorella* spp. at 30–120 g/kg of feed as a substitution of SBM in laying hen diets increased yellow pigmentation in the egg yolk whereas no changes occurred in egg production rate, egg weight, or shell quality (Lipstein et al., 1980). Grigorova (2005) reported that inclusions of 2% and 10% of dried biomass of freshwater *Chlorella* spp. in laying hen diets improved egg yolk pigmentation and egg production. Herber-McNeill and Elswyk (1998) observed increased redness and n-3 PUFA concentration in egg yolk by inclusions of *Schizochytrium* biomass at 2.4% and 4.8% in laying hen diets.

Inclusion of brown algae *Undaria pinnatifida* in laying hen diet at 0.5% enhanced egg production, organ weight and concentrations of blood albumin, triglycerides and total cholesterol (Choi et al., 2018). Carillo et al. (2012) fed birds with 2% of sardine oil in combination with 10% *Sargassum sinicola*, 10% *Enteromorpha* spp. or 10% *Macrocystis pyrifera* for 8 weeks. They found that supplementation of *S. sinicola* and *M. pyrifera* increased EPA whereas that of *Enteromorpha* spp. increased DHA concentrations in the egg yolk. Al-Harthi and El-Deek (2012) supplemented 3% and 6% of brown marine microalgae *Sargassum dentifebium* in hen diets from 23 to 42 weeks of age. The supplementations decreased cholesterol and triglyceride concentrations, but elevated high-density lipoproteins (HDL), carotenes and lutein plus zeaxanthin concentrations in egg yolk.

In our laboratory, Leng et al. (2014) supplemented 7.5% or 15% defatted *Staurosira* biomass (DFA) to replace SBM or a mixture of corn and SBM in laying hen diets for 8 weeks. Feeding DFA at 7.5% did not affect feed intake, egg weight, yolk weight, shell weight, shell thickness, egg specific gravity, egg-shell breaking strength or concentrations of yolk cholesterol and plasma uric acid. However, feeding DFA at 15% decreased feed intake, egg production and egg quality parameters as compared with the control. Inclusion of DFA increased redness and decreased lightness and yellowness of yolk color. This study revealed that 7.5% DFA could replace 7.5% SBM alone or a mixture of corn and SBM in diet of laying hens without adverse effects on health and egg quality. Ekmay et al. (2015) incorporated 25% defatted *Desmodesmus* and 11.7% full-fatted *Staurosira* in hen diets with or without added protease. Feeding hens with both microalgae for 8 weeks did not alter egg production, plasma concentrations of insulin, glutamine and uric acid, or plasma alkaline phosphatase activity, but reduced the plasma tartrate-resistant acid phosphatase activity (indicating status of bone health). The microalgae feeding also reduced plasma 3-methylhistidine concentration, improved the ileal amino acid digestibility, down-regulated duodenal *Lat1* and *Pept1* mRNA levels and up-regulated hepatic phospho-S6 ribosomal protein (PS6) production. This

Table 5 Effects of dietary inclusions of different microalgae on egg production and quality, health status and biochemical responses of laying hens

Microalgae	Level in the diet	Main findings	References
Chlorella spp.	3–12%	Egg yolks became a deep yellow color with no other effects.	Lipstein et al., 1980
Chlorella spp.	2.0–10%	Improved overall performance of layers; egg yolks became darker with algae inclusion.	Grigorova, 2005
Desmodesmus spp.	7.5–15%	No negative effects by 7.5% inclusion; reduced egg production and feed use efficiency by 15% inclusion; increased redness and decreased lightness and yellowness of yolk by all levels of inclusion.	Leng et al., 2014
Desmodesmus spp.	11.7%	Reduced plasma 3-methylhistidine concentration; improved ileal amino acid digestibility; decreased duodenal *Lat1* and *Pept1* mRNA levels; increased hepatic phospho-S6 ribosomal protein (PS6) levels.	Ekmay et al., 2015
Enteromorpha spp.	10%	Increased egg yolk content of n-3 fatty acid DHA.	Carillo et al., 2012
Haematococcus pluvialis	0.49–2.94%	Egg yolks becoming more orange-red in color with an increased tocopherol content.	Walker et al., 2012
H. pluvialis	0.2–0.8%	Microalgal astaxanthin improved the antioxidant capacity of eggs and the overall redox status of tissues.	Magnuson et al., 2018
Macrocystis pyrifera	10%	Increased total egg yolk n-3 fatty acid contents, particularly EPA.	Carillo et al., 2012
Nannochloropsis oceanica	0.1–1.0%	Enrichments of tissues and eggs with n-3 fatty acids EPA and DHA; darkening of egg yolk color.	Nitsan et al., 1999
N. oceanica	2.86–23%	Dose-dependent enrichments of EPA, DHA and n-3 fatty acids; altered mRNA expression of malic enzyme, fatty acid synthase, acetyl-CoA carboxylase, elongases 4 and 5, fatty acid desaturases 5, 6 and 9 in liver and muscles.	Manor et al., 2019
Sargassum dentifebium	3–6%	Elevation of plasma high-density lipoprotein; increased yolk palmitic acid, total carotene and lutein contents.	Al-Harthi and El-Deek, 2012
Sargassum sinicola	10%	Enriched egg yolks with EPA.	Carillo et al., 2012
Schizochytrium spp.	2.4–4.8%	Increased total egg yolk n-3 PUFA content; darken egg yolk color; no impact on taste by sensory panel.	Herber-McNeil and Elswyk, 1998

Spirulina platensis	0.1–0.2%	Enhanced egg production, egg size and FCR.	Mariey et al., 2012
S. platensis	1.5–2.5%	Darken yolk color.	Zahroojian et al., 2013
S. platensis	0.1–0.3%	Improved egg production and egg quality; showing hepatoprotective activity.	Selim et al., 2018
Staurosira spp.	3.0–15%	Incremental enrichments of n-3 fatty acids in yolks; combined supplementations of flaxseed oil and microalgae elevated egg ALA content by 13- to 15-fold over the control (15.4 mg/egg), DHA content by twofold over the control (38 mg/egg) and EPA from undetectable to 15.4 mg/egg.	Kim et al., 2016
Undaria pinnatifida	0.5%	Increased liver and cecum weights; elevated blood concentrations of albumen, triglycerides and total cholesterol.	Choi et al., 2018

ALA, α-Linolenic acid; DHA, docosahexaenoic acid; EPA, eicosapentaenoic acid; FCR, feed conversion ratio; Lat 1, L-type amino-acid transporter; mRNA, messenger ribonucleic acid; n-3, omega 3 fatty acid; Pept 1, peptide transporter; PUFA, polyunsaturated fatty acid.

study revealed the feasibility of the inclusion of microalgae biomass at a high level as a source of protein in laying hens diet.

Kim et al. (2016) evaluated the potential of combining three different microalgae with flaxseed oil in enriching eggs with n-3 PUFA. The combined supplementations of 7.5% full-fatted *Staurosira*, 7.5% defatted *Desmodesmus*, 7.5 or 15% defatted *N. oceanica* and 3% flaxseed oil did not affect daily feed intake, plasma uric acid and inorganic phosphorus concentrations and alkaline phosphatase activity, egg production, egg weight, egg yolk weight or egg-shell thickness. The combined supplementations elevated egg yolk α-linolenic acid (ALA) content by 13- to 15-fold over the control (15.4 mg/egg), DHA content by twofold over the control (38 mg/egg) and EPA from undetectable to 15.4 mg/egg. Furthermore, the combined supplementations of flaxseed oil and the three microalgae enhanced egg yolk contents of PUFA by 23% and decreased egg yolk contents of SFA up to 29% and MUFA up to 18%, respectively. Subsequently, Magnuson et al. (2018) supplemented hen diet with AST-rich *H. pluvialis* at 10, 20, 40 and 80 mg of AST/kg of diet. The highest level of supplementation enriched AST and total carotenoids to 36.2 μg/g and 114 μg/g of egg yolk, respectively. The microalgal AST was absorbed and deposited into the tissues and eggs of layers, and consequently improved their redox status. The decreases in total malondialdehyde, an indicator of lipid peroxidation, and increases in oxygen radical absorbance capacity in both eggs and tissues of layers were AST dose-dependent. Total glutathione concentrations and activities of glutathione peroxidase and glutathione-S transferase in the liver of hens were down-regulated by the *H. pluvialis* supplementation, which might reflect a metabolic coordination between the intrinsic antioxidant defense and the enrichment of extrinsic phytochemical.

Manor et al. (2019) supplemented laying hen diets with defatted green microalgae (*N. oceanica*) (DGA) for 42 days at 0%, 2.86%, 5.75%, 11.5% and 23% to replace corn and SBM. They observed DGA dose-dependent enrichments of n-3 PUFA, EPA and DHA in the egg yolk and the liver, breast and thigh tissues. The 23% inclusion enriched the egg yolk and liver with EPA up to 0.62 mg/g and 0.05 mg/g, DHA up to 4.8 mg/g and 1.2 mg/g, n-3 PUFA up to 5.9 mg/g and 1.3 mg/g, respectively. Furthermore, the DGA inclusion altered the mRNA expression of malic enzyme, fatty acid synthase, acetyl-CoA carboxylase, elongases 4 and 5 and fatty acid desaturases 5, 6 and 9 in the liver and muscles of hens.

5 Potential of microalgae as a supplement in diets for other poultry species

A limited number of studies have tested the effects of microalgae inclusions in diets for other poultry species. Overall, the inclusions have produced positive effects on growth performance and product quality (Table 6). In a 35-day

Table 6 Effects of dietary inclusions of different microalgae in diets for poultry species other than broilers and layers

Microalgae	Level in the diet	Species	Main findings	References
Chlorella vulgaris	0%, 0.1%, 0.2% replacement of soybean meal	Pekin ducks	Increased final body weight, BWG, feed intake and yellowness, pH and water holding capacity of the duck meat.	Oh et al., 2015
Crypthecodinium cohnii	1.5% replacement of soybean meal	Muscovy ducks	Increased DHA concentrations in the breast.	Schiavone et al., 2007
Schizochytrium spp.	0.5%	Japanese quail	Increased DHA concentrations; decreased n-6/n-3 PUFA ratios and cholesterol concentrations in the egg yolk.	Gladkowski et al., 2014
Spirulina platensis	0.5%, 1.0%, 1.5%, 2.0%	Japanese quail	Increased yolk color and fertility of quail.	Ross and Dominy, 1990
S. platensis	0.1% or 0.2%	Japanese quail	Increased weight gain and improved FCR.	Yusuf et al., 2016
S. platensis	0%, 0.25%, 1%	Japanese quail	Increased BWG and feed intake; improved FCR and fertility; lowered serum cholesterol and free fatty acid concentrations.	Abouelezz, 2017
S. platensis	0%, 5%, 10% and 15%	Japanese quail	Increased yolk color, MUFA levels and total antioxidant capacity; decreased yolk lipid peroxidation.	Boiago et al., 2019

BWG, body weight gain; DHA, docosahexaenoic acid; FCR, feed conversion ratio; MUFA, monounsaturated fatty acid; n-3, omega 3 fatty acid; n-6, omega 6 fatty acid; PUFA, polyunsaturated fatty acid.

experiment, Ross and Dominy (1990) fed 1-week-old Japanese quail with 0%, 1.5%, 3.0%, 6.0% and 12.0% of *S. platensis*. The supplementations showed no effects on growth performance, egg production, or hatchability, but improved fertility and increased yolk color. In a 42-day experiment on Japanese quail, Boiago et al. (2019) obtained improvements in the egg yolk color, MUFA level and total antioxidant capacity by inclusions of *S. platensis* at 5%, 10% and 15% of diets. Similarly, Abouelezz (2017) observed improvements in BWG, FCR, fertility rate and reduced serum cholesterol concentrations by feeding Japanese quail with *S. platensis* at 0.25% and 1% in their diets.

Gladkowski et al. (2014) reported that supplementation of *Schizochytrium* (0.5%) in combination with flaxseed and *Saccharomyces cerevisiae* (0.03%) in diets for Japanese quail for 19 weeks enriched their eggs with n-3 PUFA, ALA and DHA up to 7.33%, 4.57% and 4.35% of total fatty acid, respectively. Oh et al. (2015) supplemented Pekin ducks with fermented *C. vulgaris* at 0%, 0.1% and 0.2% for 42 days and observed improved BWG, feed intake and meat quality parameters including meat color, shear force, pH and water holding capacity. Schiavone et al. (2007) enhanced DHA concentrations in breast meat of Muscovy ducks by a supplementation of *Crypthecodinium cohnii* at 5 g/kg of diet.

6 Conclusion and future trends

Past research has generated sufficient evidences to qualify selected types of microalgal biomass as viable and human health value-added alternative feed proteins and bioactive nutrient sources for poultry feeding. Future research is needed to improve nutrient digestibility and utilization of the biomass and the biomass-containing diets by additions of exogenous hydrolytic enzymes, limiting essential amino acids and deficient trace minerals in diets to complete and complement the nutritional values of microalgae. Minimizing sodium concentrations in defatted marine microalgae will prevent resultant side effects of high salinity on feed and water intakes by birds. Special native or engineered microalgae strains should be developed to make the biomass a dual source of high-quality protein and bioactive nutrients including n-3 PUFA to various types of poultry. The threshold of microalgae supplementation in various poultry diets requires systematic evaluation. It remains unclear to us if the above described adverse effects of high microalgae inclusions such as the 16% DGA in broiler diets were attributed to the high salt content of seawater grown algae, high nucleic acid content in the unicellular diatom, or high resistance of microalgal cell wall to digestion. Identifying toxic and harmful microalgae strains and excluding their usage in poultry and other animal feeds should not be neglected.

The United States broiler and egg industries consume approximately 22 million tons of SBM annually (United Soybean Board, 2012). An estimated replacement of 5–20% of corn and SBM by microalgae in poultry diets would spare 575–2300 thousand tons of corn and SBM annually for human consumption (Bruinsma, 2003). Climate change, energy sustainability and food security are among the most important global challenges during the twenty-first century (Greene et al., 2016). The abundances of proteins, essential amino acids, n-3 PUFA and other essential micronutrients in microalgae will enable the development of not only alternative feed protein but also massive algal food industry toward commercialization of healthy and functional foods (Koyande et al., 2019). A dual production of algal food and fuel products at large scales would help the world to cope with the global challenges of food, energy, and water demands and climate changes (Efroymson et al., 2016).

7 Where to look for further information

Department of Animal Science, Cornell University (https://cals.cornell.edu/animal-science/research). The Lei Lab has been performing basic and applied research with a large group of world leading scientists to develop microalgae-based alternative feed and food protein and bioactive supplements to improve animal production, human health, and environmental sustainability to mitigate climate change.

- Marine AlGae Industrialization Consortium (MAGIC) (https://www.algaeconsortium.com/magic/).
- Department of Poultry Science at University of Arkansas (https://poultry-science.uark.edu/).
- United State Department of Agriculture (https://www.usda.gov/).

8 References

Abouelezz, F. M. K. (2017). Evaluation of *Spirulina* algae (*Spirulina platensis*) as a feed supplement for Japanese quail: nutritional effects on growth performance, egg production, egg quality, blood metabolites, sperm-egg penetration and fertility, *Poult. Sci. J.* 37(3): 707–719.

Al-Harthi, M. A. and El-Deek, A. A. (2012). Effect of different dietary concentrations of brown Marine algae (*Sargassum dentifebium*) prepared by different methods on plasma and yolk lipid profiles, yolk total carotene and lutein plus zeaxanthin of laying hens, *Ital. J. Anim. Sci.* 11(4). doi: 10.4081/ijas.2012.e64.

An, B.-K., Kim, K.-E., Jeon, J. Y. and Lee, K. W. (2016). Effect of dried *Chlorella vulgaris* and *Chlorella* growth factor on growth performance, meat qualities and humoral immune responses in broiler chickens, *Springerplus* 5(1): 718.

Andrade, L. M., Andrade, C. J., Dias, M., Nascimento, C. A. and Mendes, M. A. (2018). *Chlorella* and *Spirulina* microalgae as sources of functional foods, nutraceuticals, and food supplements; an overview, *MOJ Food Process. Technol.* 6(1): 5–58.

Austic, R. E., Mustafa, A., Jung, B., Gatrell, S. and Lei, X. G. (2013). Potential and limitation of a new defatted diatom microalgal biomass in replacing soybean meal and corn in diets for broiler chickens, *J. Agric. Food Chem.* 61(30): 7341–7348.

Bandarra, N. M., Pereira, P. A., Batista, I. and Vilela, M. H. (2003). Fatty acids, sterols and α-tocopherol in *Isochrysis galbana*, *J. Food Lipids* 10(1): 25–34.
Batal, A. and Dale, N. (2010). Ingredient analysis table: 2011 addition, *Feedstuffs* 2010: 16–17.
Becker, E. W. (2007). Micro-algae as a source of protein, *Biotechnol. Adv.* 25(2): 207–210.
Boiago, M. M., Dilkin, J. D., Kolm, M. A., Barreta, M., Souza, C. F., Baldissera, M. D., Dos Santos, I. D., Wagner, R., Tavernari, F. C., da Silva, M. L. B., Zampar, A., Stivanin, T. E. and Da Silva, A. S. (2019). *Spirulina platensis* in Japanese quail feeding alters fatty acid profiles and improves egg quality: benefits to consumers, *J. Food Biochem.* 43(7): e12860 doi: 10.1111/jfbc.12860.
Bonos, E., Kasapidou, E., Kargopoulos, A., Karampampas, A., Christaki, E., Florou-Paneri, P. and Nikolakakis, I. (2016). *Spirulina* as a functional ingredient in broiler chicken diets, *S. Afr. J. Anim. Sci.* 46(1): 94–102.
Bruinsma, J. (2003). *World Agriculture: Towards 2015/2030; an FAO Perspective*, Earthscan Publications, London, UK.
Carillo, S., Rios, V. H., Calvo, C., Carranco, M. E., Casas, M. and Pérez-Gil, F. (2012). n-3 fatty acid content in eggs laid by hens fed with marine algae and sardine oil and stored at different times and temperatures, *J. Appl. Phycol.* 24(3): 593–599.
Choi, Y., Lee, E. C., Na, Y. and Lee, S. R. (2018). Effects of dietary supplementation with fermented and non-fermented brown algae by-products on laying performance, egg quality, and blood profile in laying hens, *Asian-Australas. J. Anim. Sci.* 31(10): 1654–1659.
Christaki, E., Florou-Paneri, P. and Bonos, E. (2011). Microalgae: a novel ingredient in nutrition, *Int. J. Food Sci. Nutr.* 62(8): 794–799.
Combs, G. F. (1952). Algae (*Chlorella*) as a source of nutrients for the chick, *Science* 116(3017): 453–454.
Cortinas, L., Villaverde, C., Galobart, J., Baucells, M. D., Codony, R. and Barroeta, A. C. (2004). Fatty acid content in chicken thigh and breast as affected by dietary polyunsaturation level, *Poult. Sci.* 83(7): 1155–1164.
Dlouha, G., Sevcikova, S., Dokoupilova, A., Zita, L., Heindl, J. and Skřivan, M. (2008). Effect of dietary selenium sources on growth performance, breast muscle selenium, glutathione peroxidase activity and oxidative stability in broilers, *Czech J. Anim. Sci.* 53(6): 265–269.
Donsbough, A. L., Powell, S., Waguespac, A., Bidner, T. D. and Southern, L. L. (2010). Uric acid, urea, and ammonia concentrations in serum and uric acid concentration in excreta as indicators of amino acid utilization in diets for broilers, *Poult. Sci.* 89(2): 287–294.
Efroymson, R. A., Dale, V. H. and Langholtz, M. H. (2016). Socioeconomic indicators for sustainable design and commercial development of algal biofuel systems, *Glob. Change. Biol. Bioenergy* 9(6): 1005–1023.
Ekmay, R., Gatrell, S., Lum, K., Kim, J. and Lei, X. G. (2014). Nutritional and metabolic impacts of a defatted green marine microalgal (*Desmodesmus* sp.) biomass in diets for weanling pigs and broiler chickens, *J. Agric. Food Chem.* 62(40): 9783–9791.
Ekmay, R. D., Chou, K., Magnuson, A. and Lei, X. G. (2015). Continual feeding of two types of microalgal biomass affected protein digestion and metabolism in laying hens, *J. Anim. Sci.* 93(1): 287–297.
Evans, A. M., Smith, D. L. and Moritz, J. S. (2015). Effects of algae incorporation into broiler starter diet formulations on nutrient digestibility and 3 to 21 d bird performance, *J. Appl. Poult. Res.* 24(2): 206–214.

Gatrell, S., Lum, K., Kim, J. and Lei, X. G. (2014). Nonruminant nutrition symposium: potential of defatted microalgae from the biofuel industry as an ingredient to replace corn and soybean meal in swine and poultry diets, *J. Anim. Sci.* 92(4): 1306–1314.

Gatrell, S. K., Derksen, T. J., O'Neil, E. V. and Lei, X. G. (2017). A new type of defatted green microalgae exerts dose-dependent nutritional, metabolic, and environmental impacts in broiler chicks, *J. Appl. Poult. Res.* 26(3): 358–366.

Gatrell, S. K., Kim, J., Derksen, T. J., O'Neil, E. V. and Lei, X. G. (2015). Creating omega-3 fatty-acid-enriched chicken using defatted green microalgal biomass, *J. Agric. Food Chem.* 63(42): 9315–9322.

Gładkowski, W., Kiełbowicz, G., Chojnacka, A., Bobak, Ł., Spychaj, R., Dobrzański, Z., Trziszka, T. and Wawrzeńczyk, C. (2014). The effect of feed supplementation with dietary sources of n-3 polyunsaturated fatty acids, flaxseed and algae *Schizochytrium* sp., on their incorporation into lipid fractions of Japanese quail eggs, *Int. J. Food Sci. Technol.* 49(8): 1876–1885.

Grau, C. R. and Klein, N. W. (1957). Sewage-grown algae as a feedstuff for chicks, *Poult. Sci.* 36(5): 1046–1051.

Greene, C. H., Huntley, M. E., Archibald, I., Gerber, L. N., Sills, D. L., Granados, J., Tester, J., Beal, C., Walsh, M., Bidigare, R., Brown, S., Cochlan, W., Johnson, Z., Lei, X. G., Machesky, S., Redalje, D., Richardon, R., Kiron, V. and Corless, V. (2016). Marine microalgae: climate, energy, and food security from the sea, *Oceanography* 29(4): 10–15.

Grigorova, S. (2005). Dry biomass of freshwater algae of chlorella genus in the combined forages for laying hens, *J. Cent. Eur. Agric.* 6(4): 625–630.

Herber-McNeill, S. M. and Elswyk, M. E. Van (1998). Dietary marine algae maintain egg consumer acceptability while enhancing yolk color, *Poult. Sci.* 77(3): 493–496.

Kang, H. K., Salim, H. M., Akter, N., Kim, D. W., Kim, J. H., Bang, H. T., Kim, M. J., Na, J. C., Hwangbo, J., Choi, H. C. and Suh, O. S. (2013). Effect of various forms of dietary *Chlorella* supplementation on growth performance, immune characteristics, and intestinal microflora population of broiler chickens, *J. Appl. Poult. Res.* 22(1): 100–108.

Kim, J., Magnuson, A., Tao, L., Barcus, M. and Lei, X. G. (2016). Potential of combining flaxseed oil and microalgal biomass in producing eggs-enriched with n-3 fatty acids for meeting human needs, *Algal Res.* 17: 31–37.

Kotrbacek, V., Halouzka, R., Jurajda, V., Knotkova, Z. and Filka, J. (1994). Increased immune response in broilers after administration of natural food supplements, *Vet. Med. (Praha)* 39(6): 321–328.

Kovac, D. J., Simeunovic, J. B., Babic, O. B., Misan, A. C. and Milovanovic, I. L. (2013). Algae in food and feed, *Food Feed Res.* 40: 21–31.

Koyande, A. K., Chew, K. W., Rambabu, K., Tao, Y., Chu, D. T. and Show, P. (2019). Microalgae: a potential alternative to health supplementation for humans, *Food Sci. Hum. Well* 8(1): 16–24.

Lee, J. W., Kil, D. Y., Keever, B. D., Killefer, J., McKeith, F. K., Sulabo, R. C. and Stein, H. H. (2015). Carcass fat quality of pigs is not improved by adding corn germ, beef tallow, palm kernel oil, or glycerol to finishing diets containing distillers dried grains with solubles, *J. Anim. Sci.* 91(5): 2426–2437.

Leng, X. J., Hsu, K. N., Austic, R. E. and Lei, X. G. (2014). Effect of dietary defatted diatom biomass on egg production and quality of laying hens, *J. Anim. Sci. Biotechnol.* 5(1): 3. doi: 10.1186/2049-1891-5-3:Artn310.1186/2049-

Lipstein, B. and Hurwitz, S. (1983). The nutritional value of sewage-grown samples of *Chlorella* and *Micractinium* in broiler diets, *Poult. Sci.* 62(7): 1254–1260.

Lipstein, B., Hurwitz, S. and Bornstein, S. (1980). The nutritional value of algae for poultry. Dried *Chlorella* in layer diets, *Br. Poult. Sci.* 21(1): 23–27.

Long, S. F., Kang, S., Wang, Q. Q., Xu, Y. T., Pan, L., Hu, J. X., Li, M. and Piao, X. S. (2018). Dietary supplementation with DHA-rich microalgae improves performance, serum composition, carcass trait, antioxidant status, and fatty acid profile of broilers, *Poult. Sci.* 97(6): 1881-1890.

Lum, K. K., Kim, K. and Lei, X. G. (2013). Dual potential of microalgae as a sustainable feedstock and animal feed, *J. Anim. Sci. Technol.* 4: 1-7.

Magnuson, A. D., Sun, T., Yin, R., Liu, G., Tolba, S., Shinde, S. and Lei, X. G. (2018). Supplemental microalgal astaxanthin produced coordinated changes in intrinsic antioxidant systems of layer hens exposed to heat stress, *Algal Res.* 33: 84–90.

Manor, M. L., Derksen, T. J., Magnuson, A. D., Raza, F. and Lei, X. G. (2019). Inclusion of dietary defatted microalgae dose-dependently enriches omega-3 fatty acids in egg yolk and tissues of laying hens, *J. Nutr.* 149(6): 942–950.

Mariey, Y. A., Samak, H. R. and Ibrahem, M. A. (2012). Effect of using *Spirulina platensis* algae as a feed additive for poultry diets. 1. Productive and reproductive performances of local laying hens. *Egypt, Poult. Sci.* 32: 201–215.

Milles, R. D. and Featherston, W. R. (1974). Uric acid excretion as an indicator of the amino acid requirement of chicks, *Proc. Soc. Exp. Biol. Med.* 145(2): 686–689.

Moir, A. M. B., Park, B. S. and Zammit, V. A. (1995). Quantification in vivo of the effects of different types of dietary fat on theloci of control involved in hepatic triacylglycerol secretion, *Biochem. J.* 308(2): 537–542.

National Research Council (2012). *Nutrient Requirements of Swine* (11th rev. edn.), The National Academies Press, Washington, DC.

Nitsan, Z., Mokady, S. and Sukenik, A. (1999). Enrichment of poultry products with omega-3 512 fatty acids by dietary supplementation with the alga *Nannochloropsis* and mantur oil, *J. Agric. Food Chem.* 47(12): 5127-5132.

Oh, S. T., Zheng, L., Kwon, H. J., Choo, Y. K., Lee, K. W., Kang, C. W. and An, B. K. (2015). Effects of dietary fermented chlorella vulgaris (cbt(®)) on growth performance, relative organ weights, cecal microflora, tibia bone characteristics, and meat qualities in Pekin ducks, *Asian-Australas. J. Anim. Sci.* 28(1): 95–101.

Oladiji, A. T., Shoremekun, K. L. and Yakubu, M. T. (2009). Physicochemical properties of the oil from the fruit of *Blighia sapida* and toxicological evaluation of the oil-based diet in Wistar rats, *J. Med. Food* 12(5): 1127-1135.

Olson, R. D. (2006). *Below Cost Feed Crops: An Indirect Subsidy for Industrial Animal Factories*, Institute for Agriculture and Trade Policy, Minneapolis, MN.

Ross, E. and Dominy, W. (1990). The nutritional value of dehydrated, blue-green algae (*Spirulina plantensis*) for poultry, *Poult. Sci.* 69(5): 794–800.

Salter, D. N., Coates, M. E. and Hewitt, D. (1974). The utilization of protein and excretion of uric acid in germ-free and conventional chicks, *Br. J. Nutr.* 31(3): 307–318.

Schiavone, A., Chiarini, R., Marzoni, M., Castillo, A., Tassone, S. and Romboli, I. (2007). Breast meat traits of Muscovy ducks fed on a microalga (Crypthecodinium cohnii) meal supplemented diet, *Br. Poult. Sci.* 48(5): 573–579.

Selim, S., Hussein, E. and Abou-Elkhair, R. (2018). Effect of *Spirulina platensis* as a feed additive on laying performance, egg quality and hepatoprotective activity of laying hens, *Eur. Poult. Sci.* 82: 14–24.

Shanmugapriya, B., Babu, S. S., Hariharan, T., Sivaneswaran, S. and Anusha, M. B. (2015). Dietary administration of *Spirulina platensis* as probiotics on growth performance and histopathology in broiler chicks, *Int. J. Recent Sci. Res.* 6: 2650-2653.

Spolaore, P., Cassana, C. J., Duran, E. and Isamberta, A. (2006). Commercial applications of microalgae, *J. Biosci. Bioeng.* 101(2): 87-96.

Sriperm, N., Pesti, G. M. and Tillman, P. B. (2011). Evaluation of the fixed nitrogen-to-protein (N:P) conversion factor (6.25) versus ingredient specific N:P conversion factors in feedstuffs, *J. Sci. Food Agric.* 91(7): 1182-1186.

Sun, T., Magnuson, A. and Lei, X. G. (2016). New perspective of nutrient digestibility and retention in diets containing defatted microalgae, The 78th Cornell Nutrition Conference, East Syracuse, New York.

Sun, T., Tolba, S., Magnuson, A., Liu, G. and Lei, X. G. (2020a). Supplemental dietary DHA-rich microalgae affected growth performance, health status and meat quality of broiler chicks, Science Association Annual, Poultry Meeting, Virtual.

Sun, T., Wyman, B., Liu, G. and Lei, X. G. (2020b). Dietary supplemental full-fatted and defatted microalgae *Desmodesmus* sp. exerted similar impacts on growth performance and gut health of broiler chicks, Science Association Annual, Poultry Meeting, Virtual.

Sun, T., Yin, R., Magnuson, A. D., Tolba, S. A., Liu, G. and Lei, X. G. (2018). Dose-dependent enrichments and improved redox status in tissues of broiler chicks under heat stress by dietary supplemental microalgal astaxanthin, *J. Agric. Food Chem.* 66(22): 5521-5530.

Tang, J. W., Sun, H., Yao, X. H., Wu, Y. F., Wang, X. and Feng, J. (2012). Effects of replacement of soybean meal by fermented cottonseed meal on growth performance, serum biochemical parameters and immune function of yellow-feathered broilers, *Asian-Australas J. Anim. Sci.* 25(3): 393-400.

Tao, L., Sun, T., Magnuson, A. D., Qamar, T. R. and Lei, X. G. (2018). Defatted microalgae-mediated enrichment of n-3 polyunsaturated fatty acids in chicken muscle is not affected by dietary selenium, vitamin E, or corn oil, *J. Nutr.* 148(10): 1547-1555.

Tavernari, F. C., Roza, L. F., Surek, D., Sordi, C., Silva, M. L. B. D., Albino, L. F. T., Migliorini, M. J., Paiano, D. and Boiago, M. M. (2018). Apparent metabolisable energy and amino acid digestibility of microalgae *Spirulina platensis* as an ingredient in broiler chicken diets, *Br. Poult. Sci.* 59(5): 562-567.

Tolba, S. A., Sun, T., Magnuson, A. D., Liu, G. C., Abdel-Razik, W. M., El-Gamal, M. F. and Lei, X. G. (2019). Supplemental docosahexaenoic-acid-enriched microalgae affected fatty acid and metabolic profiles and related gene expression in several tissues of broiler chicks, *J. Agric. Food Chem.* 67(23): 6497-6507.

Toyomizu, M., Sato, K., Taroda, H., Kato, T. and Akiba, Y. (2001). Effects of dietary *Spirulina* on meat colour in muscle of broiler chickens, *Br. Poult. Sci.* 42(2): 197-202.

United Soybean Board (2012). *Animal Agriculture Economic Analysis: National, 2001-2011. A Report for United Soybean Board*, Agralytica, Inc., Alexandria, VA.

US Department of Agriculture (2015-2020). Dietary guidelines for Americans. https://www.fns.usda.gov/cnpp/dietary-guidelines-americans, accessed November 08, 2020.

Vanthoor-Koopmans, M., Wijffels, R. H., Barbosa, M. J. and Eppink, M. H. (2013). Biorefinery of microalgae for food and fuel, *Bioresour. Technol.* 135: 142-149.

Venkataraman, L. V., Somasekaran, T. and Becker, E. W. (1994). Replacement value of blue-green alga (*Spirulina platensis*) for fishmeal and a vitamin-mineral premix for broiler chicks, *Br. Poult. Sci.* 35(3): 373-381.

Waldenstedt, L., Inborr, J., Hansson, I. and Elwinger, K. (2003). Effects of astaxanthin-rich algal meal (*Haematococcus pluvalis*) on growth performance, caecal campylobacter and clostridial counts and tissue astaxanthin concentration of broiler chickens, *Anim. Feed Sci. Technol.* 108(1–4): 119–132.

Walker, L. A., Wang, T., Xin, H. and Dodle, D. (2012). Supplementation of laying-hen feed with palm tocos and algae astaxanthin for egg yolk nutrient enrichment, *J. Agric. Food Chem.* 60(8): 1989–1999.

Yan, L. and Kim, I. H. (2013). Effects of dietary ω-3 fatty acid-enriched microalgae supplementation on growth performance, blood profiles, meat quality, and fatty acid composition of meat in broilers, *J. Appl. Anim. Res.* 4: 392–397.

Yusuf, M. S., Hassan, M. A., Abdel-Daim, M. M., Nabtiti, A. S. E., Ahmed, A. M., Moawed, S. A., El-Sayed, A. K. and Cui, H. (2016). Value added by Spirulina platensis in two different diets on growth performance, gut microbiota, and meat quality of Japanese quails, *Vet. World* 9(11): 1287–1293.

Zahroojian, N., Moravej, H. and Shivazad, M. (2013). Effects of dietary marine algae (*Spirulina platensis*) on egg quality and production performance of laying hens, *J. Agr. Sci. Technol.* 15: 1353–1136.

Chapter 11

Developing macroalgae and microalgae as feed for fish

Mo Peng, Jiangxi Agricultural University, China; Chunxiang Ai, Xiamen University, China; Zhi Luo, Huazhong Agricultural University, China; and Qinghui Ai, Key Laboratory of Mariculture (Ministry of Education of China), Ocean University of China, China

1 Introduction

Aquaculture is a major global source of high-quality protein for human nutrition (Naylor et al., 2000; Froehlich et al., 2018). The rapid expansion of aquaculture has increased the demand for aquafeed ingredients such as fishmeal and fish oil (Olsen and Hasan, 2012; Fréon et al., 2014; FAO, 2020). Finding alternative, more sustainable feed ingredients is the key to maintaining the sustainable development of the aquaculture industry. In addition, the rapid expansion and intensification of the aquaculture industry has resulted in the occurrence of various fish diseases. The subsequent widespread use of antibiotics has increased the risk of resistant pathogenic bacteria and the accumulation of antibiotic residues in the aquatic food chain (Liu et al., 2017; Mo et al., 2017). Consequently, preventive strategies should be considered in feed formulation design to enhance the immunity of fish and thus reduce the reliance on antibiotics.

The substitution of fishmeal and fish oil in aquafeed by alternative ingredients has been a major area of aquaculture research in recent decades.

http://dx.doi.org/10.19103/AS.2021.0091.17

Among these alternatives, seaweed resources (macroalgae and microalgae) have significant potential since they contain all the essential amino acids and are rich in bioactive components such as essential fatty acids [for example, omega-3 long-chain polyunsaturated fatty acids (LC-PUFA)], vitamins, minerals, carotenes and other pigments. In addition, the development and utilization of seaweed resources could also potentially relieve the pressure on both scarce terrestrial and marine resources.

There have been numerous studies on the effects of various types of macroalgae and microalgae feed on reproduction, growth, body composition, physiology and immunity of different aquatic fish species (de Cruz et al., 2018; Nayak et al., 2018, 2020; Sánchez et al., 2000; Walker and Berlinsky, 2011; Xue et al., 2020). It has been found that macroalgae and microalgae can elevate omega-3 LC-PUFA content in fish and improve the quality and nutritional value of fish fillet (Byreddy et al., 2019; Cardinaletti et al., 2018; Perez-Velazquez et al., 2019). They could also improve fish immunity as a result of its rich content of bioactive natural compounds (Berri and Collen, 2016; dos Santos et al., 2019; Riccio and Lauritano, 2020; Seong et al., 2020).

In addition to the nutritional value of seaweed mentioned previously, algal-based integrated off-shore mariculture provides a basis for the expansion of integrated multi-trophic aquaculture (IMTA) for both fish and shrimp culture at different levels of intensification (Neori and Guttman, 2017). The IMTA has attracted interest given the concern over nutrient-loaded fish-cage effluents in traditional aquaculture which could create nutrient-rich surface plumes and thereby nitrify large areas of the ocean. Avoiding this problem by the use of macroalgae and microalgae could have a profound impact on aquaculture (Buck et al., 2018). This chapter discusses the potential impacts and challenges of microalgae and macroalgae utilization in aquafeed.

2 Effects of using macroalgae and microalgae in aquatic feed: growth performance and composition

Many algae, such as the macroalgae, *Ulva*, *Ascophyllum*, *Porphyra*, *Sargassum*, *Gracilaria* and *Laminaria,* and the microalgae, *Nannochloropsis* spp., *Phaeodactylum tricomutum*, *Chaetoceros curvisetus*, *Tetraselmis* and *Isochrysis galbana*, have relatively high protein levels, well-balanced amino acid profiles and relatively rich LC-PUFA content which contributes to the consideration of their use as suitable protein and lipid sources for aquafeeds (Dantagnan et al., 2009; de Oliveira et al., 2009; Huerlimann et al., 2010; Camacho-Rodríguez et al., 2013; Sørensen et al., 2016; Moutinho et al., 2018; Wan et al., 2019). Some seaweeds also contain phosphatidylethanolamine and phosphatidylcholine that act as an attractant and could enhance cultured fish feed intake and nutrient utilization (Sakata et al., 1991). As a result, there have been many studies on the effects of different forms of algae meals on growth performance and body composition of various fish species (Patterson and Gatlin III, 2013; Perez-Velazquez et al., 2018).

2.1 Growth performance

In the case of some commonly cultured fish, such as Atlantic salmon (*Salmo salar*), European sea bass (*Dicentrarchus labrax*), gilthead seabream (*Sparus aurata*), carp (*Cyprinus carpio*) and tilapia (*Oreochromis*), alternative algal sources have been used to replace fishmeal and fish oil with varying success rates, as shown in Table 1. Their inclusion in diets often depends on the fish species and the nutritional composition of the algae themselves (Davies et al., 1997; Dallaire et al., 2007). Compared to carnivorous fish, omnivorous and herbivorous fish seem to accept algae easily, and algae meals can be used to replace higher proportions of fishmeal. For example, up to 30% of dietary fishmeal replaced by *Spirulina* powder had no adverse effects on the growth of *Catla catla* (Nandeesha et al., 1998). Nile tilapia accepted diets with total replacement of fishmeal by *Nannochloropsis* sp. (Gbadamosi and Lupatsch, 2018). Shi et al. (2017b) found that *Chlorella* meal could totally replace fishmeal in diets for crucian carp. Cao et al. (2018) reported that 50% replacement of dietary fishmeal by *Spirulina* did not reduce the growth performance of gibel carp.

However, macroalgae and microalgae show a wide range of variation in nutritional composition between taxa and within genera (Ahlgren et al., 1992; Matos et al., 2016). Vizcaíno et al. (2016b) found that the incorporation of 25% *Gracilaria* in fishmeal significantly reduced the specific growth rate (SGR) in fish but the same level of *Ulva* had no negative effect on growth. Valente et al. (2006) suggested that adding 10% of *Gracilaria bursa-pastoris* and *Ulva rigida*, and up to 5% of *Gracilaria cornea* had no adverse effects on growth performance and feed utilization efficiency of European sea bass juveniles. The nutrient composition of algae meals is also affected by cultivation, harvesting, processing and drying conditions (Brown et al., 1997; Sánchez et al., 2000; Nielsen et al., 2016). For example, compared with algae reared in fresh seawater, algae produced in fishpond wastewater effluents had a high protein content (29–34%), as observed in *Ulva* spp., *Hypnea* spp. and *Gracilaria* spp. (Viera et al., 2005; Viera et al., 2011).

Generally speaking, reduced feed intake and lower apparent digestibility coefficients (ADCs) of nutrients are the main reasons limiting high inclusion levels of algae in fish diets (Tulli et al., 2012; Cardinaletti et al., 2018). The composition and thickness of algal cell walls can reduce nutrient bioavailability by resisting enzymatic digestion (Norambuena et al., 2015; Sørensen et al., 2016). Some algae also contain non-starch polysaccharides (NSPs) (e.g. cellulose, pectins

Table 1 Optimal levels of macroalgae and microalgae for fish species, evaluated by growth performance

Algae	Fish	Inclusion (%)	Evaluating parameters	Optimal level (%)	References
Macroalgae					
G. cornea	*Sparus aurata*	5, 15, 25	SGR	15	Vizcaíno et al., 2016b
U. rigida	*S. aurata*	5, 15, 25	SGR	25	Vizcaíno et al., 2016b
Gracilaria pulvinata	*Lates calcarifer*	3, 6, 9	WG and SGR	3	Morshedi et al., 2018
Gracilaria pygmaea	*Oncorhynchus mykiss*	3, 6, 9 and 12	SGR	6	Sotoudeh and Jafari, 2017
Palmaria palmata	*Salmo salar*	5, 10 and 15	Final weight, WG and SGR	15	Wan et al., 2016
Saccharina japonica	*Acanthopagrus schlegelii*	0.1, 0.2, 1.2, 2.5, 5 and 10	Final weight, and WG	2.5	Yu et al., 2020
G. lemaneiformis	*Lutjanus stellatus*	5, 10, 15 and 20	Weight gain	16.4	Zhu et al., 2017
Sargassum horneri	*L. stellatus*	5, 10, 15 and 20	Weight gain	15	Zhu et al., 2017
Microalgae					
Scenedesmus almeriensis	*S. aurata*	12, 20, 25 and 39	Final weight and SGR	39	Vizcaíno et al., 2014
Isochrysis sp.	*Dicentrarchus labrax*	7, 14	Total length	14	Tibaldi et al., 2015
Phaeodactylum tricornutum	*S. salar*	3, 6 and 12	Final weight, SGR and TGC	6	Sørensen et al., 2016
T. lutea + *Tetraselmis suecica* (2:1)	*D. labrax*	6, 12 and 18	Final weight and SGR	18	Cardinaletti et al., 2018
S. almeriensis	*O. mykiss*	1.3, 2.5, 5 and 10	Final weight, WG and SGR	<1.3	Tomas-Almenar et al., 2018
Schizochytrium sp. + *Arthrospira* sp. (1:4)	*M. chrysops* × *M. saxatilis*	6.9, 14, 21.2, 28.4 and 35.6	WG and TGC	21.2	Perez-Velazquez et al., 2019

Nannochloropsis ocenica	*Anarhichas minor*	7.5 and 15	Body weight, body length	15	Knutsen et al., 2019
Arthrospira platensis	*Carassius auratus gibelio*	3.38, 6.76, 13.52	Final weight, WG and SGR	3.38	Cao et al., 2018
Schizochytrium sp.	*Trachinotus ovatus*	3	WG and SGR	3	Xie et al., 2019
Spirulina platensis	*Salmo trutta caspius*	1.32, 2.64, 3.96 and 5.28	WG and SGR	5.28	Roohani et al., 2019
Amphora coffeaeformis	*Oreochromis mossambicus*	2.5, 5, 7.5 and 10	WG, DGI and SGR	10	Saleh et al., 2020
I. galbana	*T. ovatus*	2.9, 4.8, 6.7 and 8.6	Final weight, final length, WG and SGR	4.8	He et al., 2018
Heamatococcus pluvialis (defatted)	*Perca flavescens*	5, 10 and 15	Final weight, WG	5	Jiang et al., 2019
Chlorella spp.	*Ictalurus punctatus*	10, 20 and 40	WG and SGR	40	Kupchinsky et al., 2015
Nannochlorposis sp. (defatted)	*D. labrax*	5, 10, 15	DGI and Final weight	15	Valente et al., 2019
Tisochrysis lutea, Nannochloropsis gaditana or *S. almeriensis*	*Solea senegalensis*	15	DGI, final weight and SGR	15	Vizcaino et al., 2018

Notes: DGI: daily growth index; SGR: specific growth rate; TGC: thermal growth coefficient; WG, weight gain.

and hemicelluloses, xylans, agar, carrageenan, or alginates), which result in more rapid passage of food through the gastrointestinal tract and thus reduce feed utilization and growth rates of fish (Horie et al., 1995; de Oliveira et al., 2009; Rioux and Turgeon, 2015). The attraction and palatability of algae meals also affect feed intake and utilization and, ultimately, the growth performance of fish (Sánchez et al., 2000; Walker and Berlinsky, 2011; de Cruz et al., 2018). Ferreira et al. (2020) found that 4% sugar kelp in fishmeal influenced the Gh/igf-1 axis at the molecular level and reduced the growth of rainbow trout.

Some steps can be taken to address these problems and increase the proportions of algal meals in aquafeed. One approach is to mix algal proteins. For example, a blend of *Tisochrysis lutea* and *T. suecica* (in a 2:1 ratio) increased the proportion that could be added to fish diets (Tulli et al., 2012; Cardinaletti et al., 2018). A diet of 50% fishmeal, soy protein concentrate and fish oil was successfully replaced by the combination of *Arthrospira* sp. and *S. limacinum* meals for *S. ocellatus* (Perez-Velazquez et al., 2018) and hybrid striped bass (*Morone chrysops* × *M. saxatilis*) (Perez-Velazquez et al., 2019).

The second approach is the addition of NSP-degrading enzymes in the diets since NSP-degrading enzymes can enhance nutrient availability (Shi et al., 2017b; Shi et al., 2017c). Xie et al. (2019) found that the inclusion of 0.05% NSP-degrading enzymes in the diets increased the replacement levels of *Ulva prolifera*, *Gracilaria lemaneiformis* or *Ulva pertusa* to 12% for *Siganus canaliculatus*. The third approach is the addition of crystalline amino acids which could also increase nutrient availability. Gong et al. (2020) found that pre-extruded *N. oceanica* supplemented with essential amino acids significantly increased its incorporation level in fish diets as a substitute for fishmeal. The approach step is the use of processing steps such as extrusion, enzymatic hydrolysis and high-pressure homogenization which could break cell walls and thus increase the nutrient digestibility of algae (Tibbetts et al., 2017; Wang et al., 2018; Gong et al., 2020).

2.2 Composition: protein, lipid content and fatty acid content

Some studies suggest that the partial substitution of fishmeal with algae meal modifies fish body composition, though results are inconsistent (Dallaire et al., 2007; Kim et al., 2013). Several studies have found that algal meals reduced protein content in muscles and whole body of fish (Krogdahl and Bakke-McKellep, 2005; Teuling et al., 2017; Knutsen et al., 2019). These studies suggest that the complex carbohydrates in the cell walls of algae, which are difficult to be digested by the fish, reduce the absorption and utilization of algal protein. In contrast, other studies have indicated that some algae addition in diets increased muscle protein content in fish such as red seabream (Mustafa et al., 1995; Nakagawa, 1997), whilst other studies found that protein content in fish remained relatively constant after the addition of algae in the meal (Vizcaino et al., 2014; Tibaldi et al., 2015; Sørensen et al., 2016; Valente et al., 2019). Perez-Velazquez et al. (2019) found that the crude protein content of the whole body of fish showed no statistical differences after algal meal replacement of fishmeal and fish oil. Shi et al. (2017a) found that complete fish meal replacement by a mixture of rapeseed meal and *Chlorella* meal (at a 1:1 ratio) did not significantly influence protein contents in the liver and muscles of crucian carp.

Some studies have found reduced body lipid deposition after algal addition to the diets of ayu (Nematipour, 1990), red and black seabream (Nakagawa et al., 2000; Xuan et al., 2013), tilapia (Azaza et al., 2008), Atlantic salmon (Kiron et al., 2012), rainbow trout (*O. mykiss*) (Güroy et al., 2013), gilthead seabream (Vizcaíno et al., 2016a) and European sea bass (Cardinaletti et al., 2018). However, increased body lipid levels by the use of algae in the diet were reported in rainbow trout (Dantagnan et al., 2009), Atlantic salmon (Sørensen et al., 2016) and longfin yellowtail (*Seriola rivoliana*) (Kissinger et al., 2016). Other studies have indicated that fish body lipid content is not affected by dietary algal addition (Olvera-Novoa et al., 1998; Wassef et al., 2005; Valente et al., 2006; Sotoudeh and Jafari, 2017; Cortegano et al., 2019).

Some algae are primary producers of LC-PUFAs, including eicosapentaenoic acid (EPA) and docosahexaenoic acid (DHA). For example, *Schizochytrium* sp. contains both high levels of lipids (55–75% in dry matter) and up to 49–60% of DHA in relation to the total lipid content (Nakahara et al., 1996; Jiang et al., 2004; Ren et al., 2010). The contents of n-3 PUFAs of *I. galbana* reach up to 5.4% of dry biomass (Batista et al., 2013; Ryckebosch et al., 2014). The contents of EPA and/or DHA include 2.2–3.9% of dry matter in *P. tricornutum* (Veloso et al., 1991; Jiang and Gao, 2004; Cerón-García et al., 2013), 3% in *Monodus subterraneus* (Lu et al., 2002) and 2.8–4.3% in *Nannochloropsis* spp. (Huerlimann et al., 2010; Camacho-Rodríguez et al., 2013). The incorporation of these PUFAs into fish muscle offers potential health benefits for consumers (Mourente et al., 2005).

Many studies have pointed out that the addition of algae products in fish diets has increased PUFA content in various fish species (Kalla et al., 2008; Dantagnan et al., 2009; Norambuena et al., 2015; Ragaza et al., 2015). Dietary algae increased the deposition of EPA and DHA in Atlantic salmon (Kousoulaki et al., 2016), DHA deposition in hybrid striped bass (Perez-Velazquez et al., 2019), EPA, DHA and EPA+DHA contents in a range of species (Chen et al., 2019) and DHA content and n-3:n-6 ratio in tambaqui (Cortegano et al., 2019). Similar results have been reported in Nile tilapia (Sarker et al., 2016) and Jade perch (*Scortum barcoo*) (Van Hoestenberghe et al., 2016) when fed with the microalgae *Schizochytrium* sp., which is rich in DHA. He et al. (2018) reported that a moderate inclusion of *I. galbana* biomass in fish diets enhanced the total n-3 fatty acid, DHA and EPA contents in neutral and polar lipids of fish muscle and liver of *T. ovatus*. These results suggest that algae meal can influence fatty acid profiles and improve the quality of fish.

Looking more generally at body composition, Olvera-Novoa et al. (1998) found that the inclusion of *Spirulina* from 20% to 100% in experimental diets did not significantly influence the body composition of Nile tilapia. Xu et al. (2011) found that 33% of *G. lemaneiformis* in diets did not modify the chemical composition of *S. canaliculatus*. Morshedi et al. (2018) found that ash contents of fish carcasses were not significantly altered by adding 9% of *G. pulvinata* to fish diets. Vizcaino et al. (2014) noted that the muscle proximate composition of gilthead seabream did not show significant changes when fish were fed diets supplemented with *Scenedesmus almeriensis*. Peixoto et al. (2016) found that dietary addition of *Gracilaria* spp., *Ulva* spp. or *Fucus* spp. at levels of 2.5% or 7.5%, both separately and in combination, did not affect the whole-body composition of European seabass juveniles. The whole-body composition of moisture and ash showed no statistical differences in hybrid striped bass fed the algal diets (Perez-Velazquez et al., 2019).

The impact of dietary algae on the body composition of fish depends on the algae species, diet incorporation levels and the fish species tested. It is important to note that the nutritional composition of algae can be different among its species and can be markedly affected by its geographical location and environmental conditions (Colombo et al., 2006). For example, Fleurence (1999) reported that crude protein levels in red macroalgae *P. palmata* were much higher during the winter and spring months, compared to the summer and autumn seasons, while the total lipid contents of *S. horneri* and *Cystoseira hakodatensis* were 15% higher during the winter months compared to other seasons (Nomura et al., 2013).

3 Effects of using macroalgae and microalgae in aquatic feed: disease prevention and fish health

In the last few decades, the rapid expansion and intensification of the aquaculture industry has resulted in the occurrence of various diseases affecting fish which threaten further development of aquaculture (Lafferty et al., 2015; Sahoo et al., 2016). To maintain sustainable aquaculture production, various chemicals including disinfectants, antibiotics and anthelmintics have been used for infectious diseases treatment and prevention. However, these treatments have increased the risk of microbial dysbiosis, resistant pathogenic bacteria and accumulation of antibiotic residues in aquaculture food products that pose a risk to human health (Liu et al., 2017; Mo et al., 2017). There is a growing global acceptance of the need for antibiotic-free aquaculture (Schmidt et al., 2017; Wang et al., 2019). The use of marine plants (macro and microalgae) as dietary ingredients could potentially reduce the use of antibiotics as well as improve the health of fish due to the nutritional quality and health-promoting properties of algae.

3.1 Antibacterial properties

Macroalgae and microalgae possess antibacterial properties capable of inhibiting pathogens such as *Aeromonas* and *Vibrio* genera reported in a variety of economically important fish species (Cavallo et al., 2013; Pradhan et al., 2012; Rizzo et al., 2017; Velichkova et al., 2019). The antibacterial properties of macroalgae and microalgae are mediated by bioactive substances such as polyunsaturated aldehydes, glycosides, terpenes and chlorophyll-a derivatives (Natrah et al., 2014; Øverland et al., 2019). As shown in Table 2, *Chlorella vulgaris* supplementation can decrease the mortality of grey mullet (*Mugil cephalus*) after challenge with *Photobacterium damselae* (Akbary and Aminikhoei, 2019). Ethanol extract of the brown macroalga (*Sargassum vulgare*) is a promising active agent against multiple drug-resistant *A. hydrophila*, a potential cause of disease in Nile tilapia fish (Ali et al., 2016). However, a higher percentage of algae supplementation does not further improve performance against pathogens. Andrews et al. (2009) have revealed that 1% *Chlorella* sp. supplementation in fish diets significantly enhanced survival of Nile tilapia after challenge with *Aeromonas hydrophila*, whereas no significant effect was observed in fish fed with higher levels of *Chlorella* sp. (2% or 4% diet).

3.2 Fish immunity

Macroalgae and microalgae have been identified as rich sources of bioactive natural compounds that can improve animal immune function and health (Berri and Collen, 2016). Macroalgae and microalgae can exert immunostimulant effects due to their rich content of different immunostimulatory substances, including polysaccharides, PUFAs, astaxanthin or sulfolipids (Riccio and Lauritano, 2020). *In vitro* studies suggest that different algal extracts can stimulate immune cells and improve their immunity (Castro et al., 2004; Díaz-Rosales et al., 2007; Leiro et al., 2007). *In vivo* administration of algae or its extracts also show immunostimulating properties of various algae in fish (Table 2). Dietary supplementation of polysaccharides from *U. rigida* at a dose of 10 mg kg^{-1} was found to improve immune responses of grey mullet (Akbary and Aminikhoei, 2018). Polysaccharides from macroalgae such as *Padina gymnospora* (Rajendran et al., 2016) and *Caulerpa scalpelliformis* (Yengkhom et al., 2018) have been shown to improve immune response.

A major drawback of plant-derived products is the lack of omega-3 LC-PUFAs. In contrast, marine algae are rich sources of omega-3 and omega-6 LC-PUFAs which have a positive effect on fish immune function (Nayak et al.,

Table 2 Effects of macroalgae and microalgae on anti-infection capacity and health status of fish

Fish species	Algae species	Inclusion (%)	Response	Reference
Carnivorous fish				
S. aurata L.	*N. gaditana* or *Tetraselmis chuii*	5, 10	Anti-infection capacity (++)	Cerezuela et al., 2012
S. aurata L.	*P. tricornutum*	5, 10	Immunity (++)	Cerezuela et al., 2012
Salmo trutta caspius	*C. vulgaris*	1×10^8, 2×10^7, 3×10^6 /450 g diet	Immunity (++)	Saberi et al., 2017
S. senegalensis (Kaup)	*Porphyridium cruentum*	1	Immunity (++)	Díaz-Rosales et al., 2008
D. labrax	*S. platensis* and *C. vulgaris* (AM, 1:1)	3, 5, 7	Antioxidant capacity (++)	Hasanein et al., 2018
Clarias gariepinus	*S. platensis* or *C. vulgaris*	50, 75	Antioxidant capacity (++)	Raji et al., 2018
letalurus Punetaus	*S. platensis*	2.7	Immunity (++); Anti-infection capacity (++)	Duncan and Klesius, 1996
S. salar	*P. palmata*	5, 10, 15	Immunity (+)	Wan et al., 2016
O. mykiss	*Gracilariopsis persica, Sargassum bovianum*	5, 10	Immunity (++)	Kazemi et al., 2018
Omnivorous fish				
M. cephalus	*C. vulgaris*	0.5 to 1.5	Immunity (++), Antioxidant capacity (++), Anti-infection capacity (++)	Akbary and Aminikhoei, 2019
Carassius auratus gibelio	*Chlorella* sp.	0.4, 0.8, 1.2, 1.6, 2.0	Immunity (++) at 0.4 to 1.2% level; Immunity (+) at the level up to 1.6%	Zhang et al., 2014
Labeo rohita	*Chlorella* sp.	1, 2, 4	Anti-infection capacity (++) at 1% level; Anti-infection capacity (+) at 2, 4% level	Andrews et al., 2009
Carassius auratus gibelio	*Chlorella* sp.	0.4, 0.8, 1.2, 1.6, 2.0	Immunity (++), Antioxidant capacity (++)	Xu et al., 2014
C. carpio	sulfated polysaccharide from *P. gymnospora*	0.01, 0.1, 1	Immunity (++), Anti-infection capacity (++)	Rajendran et al., 2016
C. carpio	*Chlorella* sp.	2, 5, 7, 10	Immunity (++)	Khain et al., 2017
O. niloticus	*C. vulgaris*	5, 10	Immunity (++), Antioxidant capacity (++)	Zahran and Risha, 2014
O. niloticus	*S. platensis*	0.5, 0.75, 1, 1.5, 2	Immunity (++); Anti-infection capacity (++)	Ibrahem et al., 2013

++ Indicates a significant increase effect ($P < 0.05$); + indicates no significant effect ($P > 0.05$).

2018; Nayak et al., 2020; Xue et al., 2020). *Lobosphaera incisa* supplementation was found to modulate zebrafish immune function by elevating specific immune-related gene expression related to lysozyme and NF-kB, regulating the mucosal immune response (Nayak et al., 2020) and improving resistance to *Streptococcal* infection (Nayak et al., 2018). Astaxanthin was shown to improve the immune response of fish such as large yellow croaker (*Pseudosciaena crocea*) (Li et al., 2014), puffer fish (*Takifugu obscurus*) (Cheng et al., 2018) and Asian seabass (*L. calcarifer*) (Lim et al., 2019).

3.3 Antioxidant capacity

Redox homeostasis provides a microenvironment that is essential for many key physiological processes, including digestion and absorption, cell proliferation and apoptosis, as well as immune response (Circu and Aw, 2011; Sánchez-Nuño et al., 2019). Algae have been reported to exert antioxidant effects due to the interactions of several biomolecules with various antioxidant capabilities (Dell'Anno et al., 2020; Guedes et al., 2015). Bioactive compounds of phenolic origin and carotenoids are possible biomolecules that account for antioxidant and radical scavenging activities of microalgae (Safafar et al., 2015). As shown in Table 2, an investigation of 32 different microalgae species revealed that *T. suecica, Botryococcus braunii, Neochloris oleoabundans*, *Isochrysis* sp., *C. vulgaris and P. tricornutum* possessed high antioxidant capacities (Goiris et al., 2012). Dietary supplementation of *S. platensis* and *C. vulgaris* was shown to improve the antioxidant capacity of sea bass, seen by a significant increase in antioxidant parameters such as superoxide dismutase (SOD), catalase (CAT) and glutathione peroxidase (Gpx) (Hasanein et al., 2018). *S. platensis* supplementation was reported to elevate activities of CAT, GST and SOD of African catfish (Raji et al., 2018) and Nile tilapia (Ibrahem et al., 2013). Dietary *C. vulgaris* (5–10% diet) has been shown to protect Nile tilapia against immunosuppression and oxidative stresses induced by arsenic (Zahran and Risha, 2014).

4 Effects of using macroalgae and microalgae in aquatic feed: reproductive performance and meat quality

4.1 Reproductive performance

Healthy broodstock rearing and larval nutrition (with live food or artificial feed) are key factors in achieving a sustainable aquaculture industry. The age and nutritional status of broodstock are major factors that affect the reproductive capacity and quality of fish eggs (Izquierdo et al., 2001). Essential fatty acids, alfa tocopherol, ascorbic acid and carotenoids in feeds can stimulate sexual maturation and enhance fertility (Scabini et al., 2011). As macroalgae and microalgae are rich in various bioactive molecules, including carotenoids and LC-PUFAs, the inclusion of these algae in aquafeed could play a vital role in improving the reproductive performance of aquatic species (Bilbao et al., 2012; Roussel et al., 2019; Khanzadeh et al., 2016; Carneiro et al., 2020).

Dietary supplementation with carotenoid sources has been shown to improve the reproductive performance of fish (Vílchez et al., 2011; Watanabe and Vassallo-Agius, 2003). A number of studies have shown that algae meal can promote the reproductive performance of aquatic species (Bilbao et al., 2012; Güroy et al., 2012; Khanzadeh et al., 2016; Carneiro et al., 2020). Macroalgae enriched with ammonium may promote reproduction in female adult abalone broodstock (Bilbao et al., 2012). As shown in Table 2, microalgae supplementation improves the reproduction performance of omnivorous fish, such as three-spot gourami (*Trichopodus trichopterus*) (Khanzadeh et al., 2016) and yellow-tail cichlid (*Pseudotropheus acei*) (Güroy et al., 2012) and carnivorous fish, such as zebrafish (Carneiro et al., 2020), in terms of greater gonadosomatic indices, higher absolute fecundity, higher hatching percentages, higher total egg production and better larval survival. Supplementation with *Spirulina* and *Chlorella* up to 20% of the total diet is effective in promoting reproductive performance. More work needs to be done to explore the underlying mechanisms involved (Table 3).

4.2 Meat quality

It is well known that feed ingredients can affect the fillet composition and quality (Gatlin III et al., 2007). Fillets are the most nutritional and economically important part of fish, and fillet quality attributes such as fat content, color and texture are the primary determinants of consumer acceptability (Rasmussen, 2001). Dietary microalgae and macroalgae have been observed to improve the fillet composition by promoting triglyceride and protein deposition, increasing

Table 3 Effects of macroalgae and microalgae on the reproduction performance of fish

Fish species	Algae species	Inclusion (%)	Response	Reference
Omnivorous fish				
T. trichopterus	*S. platensis*	2.5, 5, 10, 20	Reproduction performance (++)	Khanzadeh et al., 2016
P. acei (Regan 1922)	*Spirulina*	2.5, 5, 10	Reproduction performance (++)	Güroy et al., 2012
Carnivorous fish				
Danio rerio	*Chlorella* sp.	1, 2, 3, 4, 5	Reproduction performance (++)	Carneiro et al., 2020

++ Indicates a significant increase effect ($P < 0.05$).

omega-3 fatty acid content and the carcass quality (Becker, 2004; Byreddy et al., 2019; Stoneham et al., 2018). As studies listed in Table 4 show, macroalgae and microalgae supplementation decreased muscle lipid content in Senegalese sole (*S. senegalensis*) (Sáez et al., 2020) and European sea bass (Cardinaletti et al., 2018). Macroalgae and microalgae in fish diets can increase LC-PUFAs (especially, DHA) within fillets of zebrafish (Byreddy et al., 2019), Nile tilapia (dos Santos et al., 2019), red seabream (*Pagrus major*) (Seong et al., 2020) and hybrid striped bass (Perez-Velazquez et al., 2019). *Ulva ohnoi* supplementation (at 5%) impacted favorably on the fillet texture and color attributes, increasing lightness and yellowness whilst reducing redness (Sáez et al., 2020). Dietary *Chlorella pyrenoidosa* at levels of 0.25%, 0.5% and 0.75% of the total diet was observed to enhance meat color, juiciness and overall acceptability of common

Table 4 Effects of macroalgae and microalgae on fillet quality and color of fish

Fish species	Algae species	Inclusion (%)	Response	Reference
Carnivorous fish				
S. senegalensis	*U. ohnoi*	5	Fillet quality (++); Pigment (++)	Sáez et al., 2020
D. rerio	*Schizochytrium* sp.	50, 100	Fillet quality (++)	Byreddy et al., 2019
P. major	*Schizochytrium*	11	Fillet quality (++)	Seong et al., 2020
D. labrax	*T. lutea* and *T. suecica* (2:1)	15, 30, and 45% fishmeal protein	Fillet quality (+); Pigment (++)	Cardinaletti et al., 2018
M. chrysops × *M. saxatilis*	*Arthrospira* sp. and *S. limacinum* (1:4)	6.9, 14, 21.2, 28.4 and 35.6	Fillet quality (++)	Perez-Velazquez et al., 2019
O. mykiss	autoclaved *U. rigida*	5, 10%	Fillet quality (++)	Güroy et al., 2013
Omnivorous fish				
C. carpio L.	*C. pyrenoidosa*	0.25, 0.5, 0.75	Fillet quality (+); Pigment (++)	Abdulrahman et al., 2019
P. acei (Regan 1922)	*Spirulina*	2.5, 5, 10	Pigment (++)	Güroy et al., 2012
Oreochromis	*Schizochytrium* sp.	1.75, 5. 26, 8.77	Fillet quality (++)	Stoneham et al., 2018
Oreochromis niloticus	*Schizochytrium* sp.	1, 2, 3, 4%	Fillet quality (++)	dos Santos et al., 2019

++ Indicates a significant increase effect ($P < 0.05$).

carp (Abdulrahman et al., 2019). Dietary macroalgae extract improved the rainbow trout fillet fatty acid (FA) composition as well as the activity of enzymes related to lipid metabolism (Bruni et al., 2020).

Macroalgae and microalgae can affect muscle as well as body color, which are particularly important characteristics for ornamental fish. The ornamental fish industry is a growing, multi-million-dollar global industry that provides an important source of revenue for exporting countries, particularly developing nations in Southeast Asia (Zhang et al., 2020). Pigmentation is a major determinant of the quality and price of ornamental fish. Carotenoid is the basis of pigmentation for ornamental fish (Storebakken et al., 1987). It cannot be directly synthesized by the fish (Goodwin, 1986). Hence, they derive their primary sources of color from carotenoids in microalgae, phytoplankton and higher plants (Güroy et al., 2012). Skin color changes can be seen in fish fed with a diet supplemented with microalgae (Table 4). *Spirulina* meal in fish diets can increase the yellow and blue coloration of yellow-tail cichlid, with the total concentration of carotenoids within skin positively correlating with the *spirulina* inclusion levels (Zhang et al., 2020). A greenish pigmentation together with some tendency toward redness was observed in European sea bass fed with a diet supplemented with *T. lutea* and *T. suecica* (Cardinaletti et al., 2018).

5 Uses of macroalgae and microalgae in fish feed

The cost of fish feed can be as high as 30–60% of a fish farm's operating costs and is predicted to keep on rising since worldwide fishmeal production has been flat for many years (Tacon and Metian, 2008). Due to the limited supply and the increasing prices of fishmeal and fish oil, as well as the demand for sustainable fish aquaculture, reducing dependence on fishmeal and fish oil is an important and urgent issue. Future fish feedstuffs need to be derived from more sustainable and scalable sources.

The use of marine plants (macro and microalgae) as dietary ingredients could potentially overcome the limitation of current feed ingredients used in aquafeed as well as improve the health condition of fish due to the nutritional quality and health-promoting properties of these dietary sources. As noted earlier, macroalgae and microalgae contain significant amounts of protein, omega-3 LC-PUFA-enriched lipids, essential amino acids and different bioactive functional compounds (Dawczynski et al., 2007; Fleurence, 1999; Garcia-Vaquero and Hayes, 2016; Kaparapu, 2017). What's more, as algae could be collected from coastal regions or farmed off-shore, its production would not rely on a limited stock of expensive arable land (Gupta and Abu-Ghannam, 2011; Roohinejad et al., 2017; Wan et al., 2019).

Many studies have been carried out on the use of algae as a possible ingredient for aquafeed for different fish species (Ahmad et al., 2020;

Garcia-Vaquero and Hayes, 2016; Hemaiswarya et al., 2011; Mosha, 2019; Norambuena et al., 2015; Ragaza et al., 2020; Rosas et al., 2019; Shah et al., 2018; Sirakov et al., 2015; Van Vo et al., 2020; Wan et al., 2019; Zhang et al., 2020). In general, small amounts (2.5–15% of the diet) of algae in fish diets resulted in positive effects, including an increase in growth performance, feed utilization efficiency, carcass quality, physiological activity, intestinal microbiota, disease-resistance, improved stress response, modulation of the lipid metabolism, quality of fillets and improved protein retention during the winter period of reduced feed intake. It has also been noted that the use of algae in aquafeed in high proportions might have a negative effect on fish growth and feed efficiency.

5.1 Use of macroalgae in fish feed

Macroalgae, commonly called seaweeds, are rich in nutraceutical or functional substances which improve feed nutritional quality and ultimately contribute to more sustainable aquaculture. The main species of macroalgae used in fish feed are *U. lactuca*, *Enteromorpha prolifera*, *Laminaria japonica*, *Sargassum* spp., *Porphyra* spp. and their extracts (Wan et al., 2019; Xie et al., 2019). Typical ratios of macroalgae in carnivorous fish feed are about 5–10% (Sotoudeh and Jafari, 2017; Morshedi et al., 2018; Zeynali et al., 2020). Ratios in feed for omnivorous and herbivorous fish are about 10–25% (Vizcaíno et al., 2016b; Wan et al., 2019; Xie et al., 2019). Ratios for use of macroalgae extracts in fish feed are mostly 0.5–1.5% (Kiadaliri et al., 2020).

Higher ratios for use of macroalgae in feed can be harmful. For example, dietary supplementation of *Ulva* spp. meal at levels of 15–20% in feed attenuated growth and feed utilization in black seabream (*A. schlegelii*) (Nakagawa et al., 1987), gilthead seabream (Wassef et al., 2005), common carp (Diler et al., 2007), Nile tilapia (Azaza et al., 2008) and African catfish (*C. gariepinus*) (Abdel-Warith et al., 2016), respectively. It is possible to have ratios of 10–15% of algae meal in feed for fishes that exhibit more amylase activity (herbivorous and/or omnivorous species) (Hidalgo et al., 1999), such as Nile tilapia (Azaza et al., 2008; Ergün et al., 2009), common carp (Diler et al., 2007) gilthead seabream (Wassef et al., 2005). Amylase activity allowed more efficient digestion of algal products (Montgomery and Gerking, 1980; Stanley and Jones, 1976). It is possible to add *G. lemaneiformis* up to 15% for black sea bream juveniles (Xuan et al., 2013). *G. lemaneiformis* supplementation at 15% in the diet of juvenile red sea bream had no negative influence on growth performance (Xuan et al., 2019).

5.2 Use of microalgae in fish feed

Microalgae (such as *Schizochytrium* sp. powder) can be used as an ingredient in fish meals (Seong et al., 2020). Other microalgae, such as *Spirulina, Chlorella* and *Schizochytrium* can also be used (Abdel-Tawwab et al., 2020; Ahmad et al., 2020; Samuelsen et al., 2018; Samuelsen et al., 2018; Sprague et al., 2015; Vizcaíno et al., 2014; Øverland et al., 2019; Zhang et al., 2020). The proportion of different microalgae used in fish feed varies widely between 1.3% and 40%. The common application ratio ranges from 5% to 25% (Vizcaíno et al., 2014; Tibaldi et al., 2015; Kupchinsky et al., 2015; Sørensen et al., 2016; Cardinaletti et al., 2018; Tomas-Almenar et al., 2018; He et al., 2018; Cao et al., 2018; Roohani et al., 2019; Perez-Velazquez et al., 2019; Knutsen et al., 2019; Xie et al., 2019; Jiang et al., 2019; Valente et al., 2019; Saleh et al., 2020).

An application ratio of 20% of *S. almeriensis* did not cause negative effects on the growth or nutrient utilization efficiency of gilthead seabream juveniles (Vizcaíno et al., 2014). Defatted microalgae (*Haematococcus pluvialis*) meal blended with soy protein isolate can be used to replace 25% of fishmeal in the diet of the yellow perch without compromising performance (*Perca flavscens*) (Jiang et al., 2019). Defatted microalgal *Nannochloropsis* sp. meal can replace 15% of fishmeal in European sea bass diets (Valente et al., 2019). Some studies show that the effects of microalgae fed to carnivorous fish on growth, feed utilization and nutrient digestibility depend on microalgal types as well as the amount used (Gong et al., 2018). Therefore, it is important to conduct feeding and digestibility trials with candidate microalgae.

Different fish species do not seem to affect the nutrient apparent digestibility coefficient (ADC) of microalgal sources. *A. maxima* had the highest protein ADC in fish species (81.4–82.5%), followed by *C. vulgaris* (80.7–80.9%), *N. gaditana* (72.4–74.7%) and *S. dimorphus* (67.0–68.3%). Ingredient fat ADC ranged between 65.1% and 89.1%. Unicellular NSPs, comprising the unicellular cell-wall fraction, are not inert in both fish species (ADC > 46.0%), probably due to fermentation. The digestibility data suggest that the differences in nutrient accessibility of unicellular sources are more important than differences in digestive systems between herbivorous and omnivorous fish (Teuling et al., 2019).

6 Challenges in using macroalgae and microalgae in fish feed

6.1 Digestibility and palatability of macro and microalgae for fish

Some algae contain anti-nutritional factors which would affect the digestibility and palatability of algae for fish (Teuling et al., 2017, 2019; Gong et al., 2018). In addition, the digestibility and palatability of different algae vary between fish species. This makes it important to assess and, where necessary, improve algal digestibility and palatability using methods such as fermentation (Bairagi et al., 2002; Marrion et al., 2003; Shobharani et al., 2013), enzymatic hydrolysis (Dumay et al., 2013), ultrasonication (Teuling et al., 2017, 2019) and bead milling (Teuling et al., 2017, 2019). Algae fermentation can reduce the number of indigestible complex carbohydrates, which have been shown to reduce nutritional digestibility; fermentation can also yield a range of beneficial by-products.

6.2 Product quality of macro and microalgae

The presence of anti-nutritional factors such as NSPs in macroalgae (such as *G. lemaneiformis*) limits their application in feed. Like terrestrial plants, the nutritional content of macroalgae and microalgae can vary greatly, depending on factors such as species, season, location and processing conditions (Griffiths et al., 2016; Yarnold et al., 2019; Wan et al., 2019). Processing methods (such as drying, extraction, fermentation, enzymatic hydrolysis and chemical hydrolysis) affect the nutrient value and presence of bioactives (such as the pigment astaxanthin, omega-3 PUFA and polysaccharide). Teuling et al. (2019) found that bead milling *N. gaditana* increased *in vivo* digestibility of dry matter, energy, protein, fat, ash and calcium in feed for Nile tilapia, compared to untreated algae biomass (including an increase in ADC of protein and fat from 62% to 78% and from 50% to 82%, respectively), while not affecting the ADC of total carbohydrates and phosphorus. Above all, it is crucial to test and establish standardized processing methods and product quality standards to ensure the quality of algae products (Wan et al., 2019).

6.3 Safety of macro and microalgae products

Algae live in a freshwater or seawater environment from where they take their nutrients. Algae may also accumulate toxic heavy metals (e.g., arsenic, lead, mercury and cadmium) (Ryan et al., 2012; Malea and Kevrekidis, 2014) as well as persistent organic pollutants (such as dichlorodiphenyltrichloroethane [DDT], dioxins and other polychlorinated biphenyls [PCBs]). Some algae also secrete algal toxins which could affect product safety (Wan et al., 2019). Algal toxins are organic molecules produced by a variety of algae in marine, brackish and fresh water. They are a problem in aquaculture when they are produced in sufficient quantities and with sufficient potency to decrease fish feeding and growth rates. These problems make it very important to ensure the safety of algal products. As a general rule, it is safer to breed algae in controlled conditions rather than harvest them in the wild.

6.4 Macroalgae in integrated multi-trophic aquaculture (IMTA)

Integrated multi-trophic aquaculture (IMTA) offers a sustainable and eco-friendly approach for aquaculture (Upling, 2020). IMTA involves the co-cultivation of fish species (such as finfish) together with plants and other species (such as algae).

By closing nutrient cycles, reducing waste and optimizing resources, it has the potential of reducing the environmental impact of open-cage fish farming while introducing new value chains (Buck et al., 2018; Carras et al., 2020; Shpigel et al., 2017). Algal-based integrated off-shore mariculture provides a way of developing IMTA (Park et al., 2015; Neori and Guttman, 2017). Macroalgae is being used as an extractive species in salmon-based IMTA (Fossberg et al., 2018; Handå et al., 2013; Ratcliff et al., 2016) and IMTA using gilthead seabream (Shpigel et al., 2018).

As an example, protein-rich *U. lactuca* replaced 100% of fishmeal in diets in an IMTA system using juvenile marine fish culture industry (*S. aurata*), effectively improving waste management and sustainability (Shpigel et al., 2018). The combination of Atlantic salmon, blue mussel (*Mytilus edulis*) and sugar macroalgae (*S. latissima*) in an IMTA operation resulted in a substantially higher net present value than a salmon monoculture (Carras et al., 2020). Seaweed can be integral to IMTA in providing inorganic nutrient extraction (Ratcliff et al., 2016). Its value depends on factors such as species and environmental conditions which need to be taken into account in system design and operation (Buschmann et al., 2001).

7 Conclusion and future trends

Many studies have highlighted the potential role of macroalgae and microalgae in fish nutrition and health. These include improvements in growth performance, feed utilization, immunity, stress tolerance, reproduction and fillet quality, making macroalgae and microalgae promising substitutes for fishmeal and fish oil.

It is necessary to carry out further research on the digestibility of macroalgae and microalgae at different levels of supplementation to optimize their use in different fish species. Other areas of research include the right form of algal feed such as the effects of algal powder and its extracts on growth, health, reproduction and meat quality of fish. Future research should also focus on the improvement of production conditions and processing methods and the genetic modification of strains. It is essential to establish standardized processing methods for an algal meal to ensure consistent product safety and quality. Understanding the effects of macro and microalgae on the gut microbiome of fish using novel, high-throughput techniques would give an interesting insight into how macro and microalgae affect fish health and nutrition. Next-generation aquafeed should be of better quality to ensure superior growth and improved health.

8 References

Abdel-Tawwab, M., Abdulrahman, N. M., Baiz, A. I., Nader, P. J. and Al-Refaiee, I. H. 2020. The using of *Chlorella pyrenoidosa* and *Daphnia magna* as feed supplements for common carp, *Cyprinus carpio*: growth performance, somatic indices, and hemato-biochemical biomarkers. *Journal of Applied Aquaculture*, 1–15.

Abdel-Warith, A. W., Younis, el-S. M. and Al-Asgah, N. A. 2016. Potential use of green macroalgae *Ulva lactuca* as a feed supplement in diets on growth performance, feed utilization and body composition of the African catfish, *Clarias gariepinus*. *Saudi Journal of Biological Sciences* 23(3), 404–409.

Abdulrahman, N. M., Abid, S. H. and Koyun, M. 2019. Effect of microalgae *Chlorella pyrenoidosa* as feed additive on common carp (*Cyprinus carpio* L.) nutritional performance, proximate composition and organoleptic evaluation. *Biological and Applies Environmental Research* 3, 118–126.

Ahlgren, G., Gustafsson, I. B. and Boberg, M. 1992. Fatty acid contnt and chemical of freshwater microalage. *Journal of Phycology* 28(1), 37–50.

Ahmad, M. T., Shariff, M., Md. Yusoff, F., Goh, Y. M. and Banerjee, S. 2020. Applications of microalga *Chlorella vulgaris* in aquaculture. *Reviews in Aquaculture* 12(1), 328–346.

Akbary, P. and Aminikhoei, Z. 2018. Effect of water-soluble polysaccharide extract from the green alga *Ulva rigida* on growth performance, antioxidant enzyme activity, and immune stimulation of grey mullet *Mugil cephalus*. *Journal of Applied Phycology* 30(2), 1345–1353.

Akbary, P. and Aminikhoei, Z. 2019. Impact of dietary supplementation of *Chlorella vulgaris* (Beijerinck, 1890) on the growth, antioxidant defense and immune status of the grey mullet, *Mugil cephalus* (Linnaeus, 1758). *Iranian Journal of Aquatic Animal Health* 5(1), 57–70.

Ali, S. S., Shaaban, M. T., Abomohra, A. E.-F. and El-Safity, K. 2016. Macroalgal activity against multiple drug resistant *Aeromonas hydrophila*: a novel treatment study towards enhancement of fish growth performance. *Microbial Pathogenesis* 101, 89–95.

Andrews, S. R., Sahu, N. P., Pal, A. K. and Kumar, S. 2009. Haematological modulation and growth of *Labeo rohita* fingerlings: effect of dietary mannan oligosaccharide, yeast extract, protein hydrolysate and *chlorella*. *Aquaculture Research* 41(1), 61–69.

Azaza, M. S., Mensi, F., Ksouri, J., Dhraief, M. N., Brini, B., Abdelmouleh, A. and Kraiem, M. M. 2008. Growth of Nile tilapia (*Oreochromis niloticus* L.) fed with diets containing graded levels of green algae *Ulva* meal (*Ulva riqida*) reared in geothermal waters of southern Tunisia. *Journal of Applied Ichthyology* 24(2), 202–207.

Bairagi, A., Sarkar Ghosh, K., Sen, S. K. and Ray, A. K. 2002. Duckweed (*Lemna polyrhiza*) leaf meal as a source of feedstuff in formulated diets for rohu (*Labeo rohita* Ham.) fingerlings after fermentation with a fish intestinal bacterium. *Bioresource Technology* 85(1), 17–24.

Batista, A. P., Gouveia, L., Bandarra, N. M., Franco, J. M. and Raymundo, A. 2013. Comparison of microalgal biomass profiles as novel functional ingredient for food products. *Algal Research* 2(2), 164–173.

Becker, W. 2004. *18 Microalgae in Human and Animal Nutrition, Handbook of Microalgal Culture: Biotechnology and Applied Phycology*. Wiley Online Library.

Berri, M. and Collen, P. N. 2016. Green algal sulfated polysaccharides: a natural alternative to antibiotics via modulation of the intestinal immune response. International Symposium on Alternatives to Antibiotics, Paris, France.

Bilbao, A., Uriarte, I., del Pino Viera, M., Sosa, B., Fernández-Palacios, H. and Hernández-Cruz, C. M. 2012. Effect of macroalgae protein levels on some reproductive aspects and physiological parameters for the abalone, *Haliotis tuberculata* coccinea (Reeve 1846). *Journal of the World Aquaculture Society* 43(6), 764–777.

Brown, M. R., Jeffrey, S. W., Volkman, J. K. and Dunstan, G. A. 1997. Nutritional properties of microalgae for mariculture. *Aquaculture* 151(1–4), 315–331.

Bruni, L., Secci, G., Mancini, S., Faccenda, F. and Parisi, G. 2020. A commercial macroalgae extract in a plant-protein rich diet diminished saturated fatty acids of *Oncorhynchus mykiss* walbaum fillets. *Italian Journal of Animal Science* 19(1), 373–382.

Buck, B. H., Troell, M. F., Krause, G., Angel, D. L., Grote, B. and Chopin, T. 2018. State of the art and challenges for offshore integrated multi-trophic aquaculture (IMTA). *Frontiers in Marine Science* 5, 165.

Buschmann, A. H., Troell, M. and Kautsky, N. 2001. Integrated algal farming: a review. *Cahiers de Biologie marine* 42, 83–90.

Byreddy, A. R., Yoganantharjah, P., Gupta, A., Gibert, Y. and Puri, M. 2019. Suitability of novel algal biomass as fish feed: accumulation and distribution of omega-3 long-chain polyunsaturated fatty acid in zebrafish. *Applied Biochemistry and Biotechnology* 188(1), 112–123.

Camacho-Rodríguez, J., Cerón-García, M. C., González-López, C. V., Fernández-Sevilla, J. M., Contreras-Gómez, A. and Molina-Grima, E. 2013. A low-cost culture medium for the production of *Nannochloropsis gaditana* biomass optimized for aquaculture. *Bioresource Technology* 144, 57–66.

Cao, S. P., Zou, T., Zhang, P. Y., Han, D., Jin, J. Y., Liu, H. K., Yang, Y. X., Zhu, X. M. and Xie, S. Q. 2018. Effects of dietary fishmeal replacement with *Spirulina platensis* on the growth, feed utilization, digestion and physiological parameters in juvenile gibel carp (*Carassis auratus* gibelio var. CAS III). *Aquaculture Research* 49(3), 1320 –1328.

Cardinaletti, G., Messina, M., Bruno, M., Tulli, F., Poli, B. M., Giorgi, G., Chini-Zittelli, G., Tredici, M. and Tibaldi, E. 2018. Effects of graded levels of a blend of *Tisochrysis lutea* and *Tetraselmis suecica* dried biomass on growth and muscle tissue composition of European sea bass (*Dicentrarchus labrax*) fed diets low in fish meal and oil. *Aquaculture* 485, 173–182.

Carneiro, W. F., Castro, T. F. D., Orlando, T. M., Meurer, F., Paula, D. AdJ., Virote, BdC. R., Vianna, A. RdC. B. and Murgas, L. D. S. 2020. Replacing fish meal by *Chlorella* sp. meal: effects on zebrafish growth, reproductive performance, biochemical parameters and digestive enzymes. *Aquaculture* 528, 735612.

Carras, M. A., Knowler, D., Pearce, C. M., Hamer, A., Chopin, T. and Weaire, T. 2020. A discounted cash-flow analysis of salmon monoculture and Integrated Multi-Trophic Aquaculture in eastern Canada. *Aquaculture Economics and Management* 24(1), 43–63.

Castro, R., Zarra, I. and Lamas, J. 2004. Water-soluble seaweed extracts modulate the respiratory burst activity of turbot phagocytes. *Aquaculture* 229(1–4), 67–78.

Cavallo, R., Acquaviva, M., Stabili, L., Cecere, E., Petrocelli, A. and Narracci, M. 2013. Antibacterial activity of marine macroalgae against fish pathogenic *Vibrio species*. *Open Life Sciences* 8(7), 646–653.

Cerezuela, R., Guardiola, F. A., Meseguer, J. and Esteban, M. Á. 2012. Enrichment of gilthead seabream (*Sparus aurata* L.) diet with microalgae: effects on the immune system. *Fish Physiology and Biochemistry* 38(6), 1729–1739.

Cerón-García, M. C., Fernández-Sevilla, J. M., Sánchez-Mirón, A., García-Camacho, F., Contreras-Gómez, A. and Molina-Grima, E. 2013. Mixotrophic growth of

Phaeodactylum tricornutum on fructose and glycerol in fed-batch and semi-continuous modes. *Bioresource Technology* 147, 569–576.

Chen, W., Wang, Y., Han, D., Zhu, X., Xie, S., Han, D. and Hu, Q. 2019. Two filamentous microalgae as feed ingredients improved flesh quality and enhanced antioxidant capacity and immunity of the gibel carp (*Carassius auratus gibelio*). *Aquaculture Nutrition* 25(5), 1145–1155.

Cheng, C. H., Guo, Z. X., Ye, C. X. and Wang, A. L. 2018. Effect of dietary astaxanthin on the growth performance, non-specific immunity, and antioxidant capacity of pufferfish (*Takifugu obscurus*) under high temperature stress. *Fish Physiology and Biochemistry* 44(1), 209–218.

Circu, M. L. and Aw, T. Y. 2011. Redox biology of the intestine. *Free Radical Research* 45(11–12), 1245–1266.

Colombo, M. L., Rise, P., Giavarini, F., De Angelis, L., Galli, C. and Bolis, C. L. 2006. Marine macroalgae as sources of polyunsaturated fatty acids. *Plant Foods for Human Nutrition* 61(2), 67–72.

Cortegano, C. A. A., de Alcântara, A. M., da Silva, A. F., Epifânio, C. M. F., Bentes, S. P. C., dos Santos, V. J., Visentainer, J. V. and Gonçalves, L. U. 2019. Finishing plant diet supplemented with microalgae meal increases the docosahexaenoic acid content in *Colossoma macropomum* flesh. *Aquaculture Research* 50(4), 1291–1299.

Díaz-Rosales, P., Chabrillón, M., Abdala, R. T., Figueroa, F. L., Balebona, M. C. and Moriñigo, M. A. 2008. Effect of dietary administration of *Porphyridium cruentum* on the respiratory burst activity of sole, *Solea senegalensis* (Kaup), phagocytes. *Journal of Fish Diseases* 31(7), 489–495.

Díaz-Rosales, P., Felices, C., Abdala, R., Figueroa, F. L., Gómez Pinchetti, J. L., Moriñigo, M. A. and Balebona, M. C. 2007. In vitro effect of the red alga *Hydropuntia cornea* (J. Agardh) on the respiratory burst activity of sole (*Solea senegalensis*, Kaup 1858) phagocytes. *Aquaculture Research* 38(13), 1411–1418.

Dallaire, V., Lessard, P., Vandenberg, G. and de la Noüe, J. 2007. Effect of algal incorporation on growth, survival and carcass composition of rainbow trout (*Oncorhynchus mykiss*) fry. *Bioresource Technology* 98(7), 1433–1439.

Dantagnan, P., Hernández, A., Borquez, A. and Mansilla, A. 2009. Inclusion of macroalgae meal (*Macrocystis pyrifera*) as feed ingredient for rainbow trout (*Oncorhynchus mykiss*): effect on flesh fatty acid composition. *Aquaculture Research* 41(1), 87–94.

Davies, S. J., Brown, M. T. and Camilleri, M. 1997. Preliminary assessment of the seaweed *Porphyra purpurea* in artificial diets for thick-lipped grey mullet (*Chelon labrosus*). *Aquaculture* 152(1–4), 249–258.

Dawczynski, C., Schubert, R. and Jahreis, G. 2007. Amino acids, fatty acids, and dietary fibre in edible seaweed products. *Food Chemistry* 103(3), 891–899.

de Cruz, C. R., Lubrano, A. and Gatlin III, D. M. 2018. Evaluation of microalgae concentrates as partial fishmeal replacements for hybrid striped bass *Morone sp. Aquaculture* 493, 130–136.

de Oliveira, M. N., Freitas, A. L. P., Carvalho, A. F. U., Sampaio, T. M. T., Farias, D. F., Alves Teixeira, D. I., Gouveia, S. T., Pereira, J. G. and Sena, M. MdC. Cd. 2009. Nutritive and non-nutritive attributes of washed-up seaweeds from the coast of Ceará, Brazil. *Food Chemistry* 115(1), 254–259.

Dell'Anno, M., Sotira, S., Rebucci, R., Reggi, S., Castiglioni, B. and Rossi, L. 2020. In vitro evaluation of antimicrobial and antioxidant activities of algal extracts. *Italian Journal of Animal Science* 19(1), 103–113.

Diler, I., Tekinay, A. A., Guroy, D., Guroy, B. K. and Soyuturk, M. 2007. Effects of *Ulva rigida* on the growth, feed intake and body composition of common carp, *Cyprinus carpio* L. *Journal of Biological Sciences* 7(2): 305–308.

dos Santos, S. K. A., Schorer, M., Moura, GdS., Lanna, E. A. T. and Pedreira, M. M. 2019. Evaluation of growth and fatty acid profile of Nile tilapia (*Oreochromis niloticus*) fed with *Schizochytrium* sp. *Aquaculture Research* 50(4), 1068–1074.

Dumay, J., Clément, N., Morançais, M. and Fleurence, J. 2013. Optimization of hydrolysis conditions of *Palmaria palmata* to enhance R-phycoerythrin extraction. *Bioresource Technology* 131, 21–27.

Duncan, P. L. and Klesius, P. H. 1996. Effects of feeding *Spirulina* on specific and nonspecific immune responses of channel catfish. *Journal of Aquatic Animal Health* 8(4), 308–313.

Ergün, S., Soyutürk, M., Güroy, B., Güroy, D. and Merrifield, D. 2009. Influence of *Ulva* meal on growth, feed utilization, and body composition of juvenile Nile tilapia (*Oreochromis niloticus*) at two levels of dietary lipid. *Aquaculture International* 17(4), 355–361.

FAO 2020. *The State of World Fisheries and Aquaculture 2020*. Sustainability in action. FAO, Rome. https://doi.org/10.4060/ca9229en.

Ferreira, M., Larsen, B. K., Granby, K., Cunha, S. C., Monteiro, C., Fernandes, J. O., Nunes, M. L., Marques, A., Dias, J., Cunha, I., Castro, L. F. C. and Valente, L. M. P. 2020. Diets supplemented with *Saccharina latissima* influence the expression of genes related to lipid metabolism and oxidative stress modulating rainbow trout (*Oncorhynchus mykiss*) fillet composition. *Food and Chemical Toxicology* 140, 111332.

Fleurence, J. 1999. Seaweed proteins: biochemical, nutritional aspects and potential uses. *Trends in Food Science and Technology* 10(1), 25–28.

Fossberg, J., Forbord, S., Broch, O. J., Malzahn, A. M., Jansen, H., Handå, A., Førde, H., Bergvik, M., Fleddum, A. L., Skjermo, J. and Olsen, Y. 2018. The potential for upscaling kelp (*Saccharina latissima*) cultivation in salmon-driven integrated multi-trophic aquaculture (IMTA). *Frontiers in Marine Science* 5, 418.

Fréon, P., Sueiro, J. C., Iriarte, F., Evar, O. F. M., Landa, Y., Mittaine, J.-F. and Bouchon, M. 2014. Harvesting for food versus feed: a review of Peruvian fisheries in a global context. *Reviews in Fish Biology and Fisheries* 24(1), 381–398.

Froehlich, H. E., Runge, C. A., Gentry, R. R., Gaines, S. D. and Halpern, B. S. 2018. Comparative terrestrial feed and land use of an aquaculture-dominant world. *Proceedings of the National Academy of Sciences of the United States of America* 115(20), 5295–5300.

Garcia-Vaquero, M. and Hayes, M. 2016. Red and green macroalgae for fish and animal feed and human functional food development. *Food Reviews International* 32(1), 15–45.

Gatlin III, D. M., Barrows, F. T., Brown, P., Dabrowski, K., Gaylord, T. G., Hardy, R. W., Herman, E., Hu, G., Krogdahl, Å., Nelson, R., Overturf, K., Rust, M., Sealey, W., Skonberg, D., J Souza, E., Stone, D., Wilson, R. and Wurtele, E. 2007. Expanding the utilization of sustainable plant products in aquafeeds: a review. *Aquaculture Research* 38(6), 551–579.

Gbadamosi, O. K. and Lupatsch, I. 2018. Effects of dietary *Nannochloropsis salina* on the nutritional performance and fatty acid profile of Nile tilapia, *Oreochromis niloticus*. *Algal Research* 33, 48–54.

Goiris, K., Muylaert, K., Fraeye, I., Foubert, I., De Brabanter, J. and De Cooman, L. 2012. Antioxidant potential of microalgae in relation to their phenolic and carotenoid content. *Journal of Applied Phycology* 24(6), 1477–1486.

Gong, Y., Guterres, H. A. D. S., Huntley, M., Sørensen, M. and Kiron, V. 2018. Digestibility of the defatted microalgae *Nannochloropsis sp.* and *Desmodesmus sp.* when fed to Atlantic salmon, *Salmo salar*. *Aquaculture Nutrition* 24(1), 56-64.

Gong, Y., Sørensen, S. L., Dahle, D., Nadanasabesan, N., Dias, J., Valente, L. M. P., Sørensen, M. and Kiron, V. 2020. Approaches to improve utilization of *Nannochloropsis oceanica* in plant-based feeds for Atlantic salmon. *Aquaculture* 522, 735122.

Goodwin, T. W. 1986. Metabolism, nutrition, and function of carotenoids. *Annual Review of Nutrition* 6, 273-297.

Griffiths, M., Harrison, S. T., Smit, M. and Maharajh, D. 2016. Major commercial products from micro- and macroalgae. In: Bux, F. and Chisti, Y. (Eds), *Algae Biotechnology*. Springer International Publishing, Geneva, Switzerland.

Guedes, A. C., Sousa-Pinto, I. and Malcata, F. X. 2015. *Application of Microalgae Protein to Aquafeed, Handbook of Marine Microalgae*. Elsevier, pp. 93-125.

Gupta, S. and Abu-Ghannam, N. 2011. Bioactive potential and possible health effects of edible brown seaweeds. *Trends in Food Science and Technology* 22(6), 315-326.

Güroy, B., Şahin, İ., Mantoğlu, S. and Kayalı, S. 2012. *Spirulina* as a natural carotenoid source on growth, pigmentation and reproductive performance of yellow tail cichlid *Pseudotropheus acei*. *Aquaculture International* 20(5), 869-878.

Güroy, B., Ergün, S., Merrifield, D. L. and Güroy, D. 2013. Effect of autoclaved *Ulva* meal on growth performance, nutrient utilization and fatty acid profile of rainbow trout, *Oncorhynchus mykiss*. *Aquaculture International* 21(3), 605-615.

Handå, A., Forbord, S., Wang, X., Broch, O. J., Dahle, S. W., Størseth, T. R., Reitan, K. I., Olsen, Y. and Skjermo, J. 2013. Seasonal-and depth-dependent growth of cultivated kelp (*Saccharina latissima*) in close proximity to salmon (*Salmo salar*) aquaculture in Norway. *Aquaculture* 414-415, 191-201.

Hasanein, S., Norhan E. Saleh, N. E. S., S. El-Sayed, H. and M. Helal, A. 2018. The Effect of dietary supplementation of *Spirulina platensis* and *Chlorella vulgaris* algae on the growth and disease resistance of the sea bass (*Dicentrarchus labrax*). *Egyptian Journal of Aquatic Biology and Fisheries* 22(4), 249-262.

He, Y., Lin, G., Rao, X., Chen, L., Jian, H., Wang, M., Guo, Z. and Chen, B. 2018. Microalga *Isochrysis galbana* in feed for *Trachinotus ovatus*: effect on growth performance and fatty acid composition of fish fillet and liver. *Aquaculture International* 26(5), 1261-1280.

Hemaiswarya, S., Raja, R., Kumar, R. R., Ganesan, V. and Anbazhagan, C. 2011. Microalgae: a sustainable feed source for aquaculture. *World Journal of Microbiology and Biotechnology* 27(8), 1737-1746.

Hidalgo, M. C., Urea, E. and Sanz, A. 1999. Comparative study of digestive enzymes in fish with different nutritional habits. Proteolytic and amylase activities. *Aquaculture* 170(3-4), 267-283.

Horie, Y., Sugase, K. and Horie, K. 1995. Physiological differences of soluble and insoluble dietary fibre fractions of brown algae and mushrooms in pepsin activity in vitro and protein digestibility. *Asia Pacific Journal of Clinical Nutrition* 4(2), 251-255.

Huerlimann, R., De Nys, R. and Heimann, K. 2010. Growth, lipid content, productivity, and fatty acid composition of tropical microalgae for scale-up production. *Biotechnology and Bioengineering* 107(2), 245-257.

Ibrahem, M. D., Mohamed, M. F. and Ibrahim, M. A. 2013. The role of *Spirulina platensis* (*Arthrospira platensis*) in growth and immunity of Nile tilapia (*Oreochromis niloticus*) and its resistance to bacterial infection. *Journal of Agricultural Science* 5(6).

Izquierdo, M. S., Fernandez-Palacios, H. and Tacon, A. G. J. 2001. Effect of broodstock nutrition on reproductive performance of fish. *Aquaculture* 197(1–4), 25–42.

Jiang, H. and Gao, K. 2004. Effects of lowering temperature during culture on the production of polyunsaturated fatty acids in the marine diatom *Phaeodactylum tricornutum* (bacillariophyceae) 1. *Journal of Phycology* 40(4), 651–654.

Jiang, Y., Fan, K. W., Tsz-Yeung Wong, R. T. and Chen, F. 2004. Fatty acid composition and squalene content of the marine microalga *Schizochytrium mangrovei*. *Journal of Agricultural and Food Chemistry* 52(5), 1196–1200.

Jiang, M., Zhao, H. H., Zai, S. W., Shepherd, B., Wen, H. and Deng, D. F. 2019. A defatted microalgae meal (*Haematococcus pluvialis*) as a partial protein source to replace fishmeal for feeding juvenile yellow perch *Perca flavescens*. *Journal of Applied Phycology* 31(2), 1197–1205.

Kalla, A., Yoshimatsu, T., Araki, T., Zhang, D.-M., Yamamoto, T. and Sakamoto, S. 2008. Use of *Porphyra spheroplasts* as feed additive for red sea bream. *Fisheries Science* 74(1), 104–108.

Kaparapu, J. 2017. Micro algal Immobilization Techniques. *J. Algal Biomass Utln* 8, 64–70.

Kazemi, M., Abediankenari, A. and Rabiei, R. 2018. Effect of marine macroalgae on growth performance and immune response in rainbow trout fingerlings. *Fisheries Science and Technology* 7, 9–16.

Khain, M., Soltani, M., Mehejan, M. S., Foroudi, F. and Ghaeni, M. 2017. The effect of *Chlorella vulgaris* (Chlorophyta, Volvocales) microalga on some hematological and immune system parameters of Koi carp (*Cyprinus carpio*). *Iranian Journal of Ichthyology* 4, 62–68.

Khanzadeh, M., Esmaeili Fereidouni, A. and Seifi Berenjestanaki, S. 2016. Effects of partial replacement of fish meal with *Spirulina platensis* meal in practical diets on growth, survival, body composition, and reproductive performance of three-spot gourami (*Trichopodus trichopterus*) (Pallas, 1770). *Aquaculture International* 24(1), 69–84.

Kiadaliri, M., Firouzbakhsh, F. and Deldar, H. 2020. Effects of feeding with red algae (*Laurencia caspica*) hydroalcoholic extract on antioxidant defense, immune responses, and immune gene expression of kidney in rainbow trout (*Oncorhynchus mykiss*) infected with *Aeromonas hydrophila*. *Aquaculture* 526, 735361.

Kim, S. S., Ly, H. V., Kim, J., Choi, J. H. and Woo, H. C. 2013. Thermogravimetric characteristics and pyrolysis kinetics of Alga *Sagarssum sp.* biomass. *Bioresource Technology* 139, 242–248.

Kiron, V., Phromkunthong, W., Huntley, M., Archibald, I. and De Scheemaker, G. 2012. Marine microalgae from biorefinery as a potential feed protein source for Atlantic salmon, common carp and whiteleg shrimp. *Aquaculture Nutrition* 18(5), 521–531.

Kissinger, K. R., García-Ortega, A. and Trushenski, J. T. 2016. Partial fish meal replacement by soy protein concentrate, squid and algal meals in low fish-oil diets containing *Schizochytrium limacinum* for longfin yellowtail *Seriola rivoliana*. *Aquaculture* 452, 37–44.

Knutsen, H. R., Johnsen, I. H., Keizer, S., Sørensen, M., Roques, J. A. C., Hedén, I., Sundell, K. and Hagen, Ø. 2019. Fish welfare, fast muscle cellularity, fatty acid and body-composition of juvenile spotted wolffish (*Anarhichas minor*) fed a combination of plant proteins and microalgae (*Nannochloropsis oceanica*). *Aquaculture* 506, 212–223.

Kousoulaki, K., Mørkøre, T., Nengas, I., Berge, R. K. and Sweetman, J. 2016. Microalgae and organic minerals enhance lipid retention efficiency and fillet quality in Atlantic salmon (*Salmo salar* L.). *Aquaculture* 451, 47–57.

Krogdahl, A. and Bakke-McKellep, A. M. 2005. Fasting and refeeding cause rapid changes in intestinal tissue mass and digestive enzyme capacities of Atlantic salmon (*Salmo salar* L.). *Comparative Biochemistry and Physiology. Part A, Molecular and Integrative Physiology* 141(4), 450–460.

Kupchinsky, Z. A., Coyle, S. D., Bright, L. A. and Tidwell, J. H. 2015. Evaluation of heterotrophic algae meal as a diet ingredient for channel catfish, *Ictalurus punctatus*. *Journal of the World Aquaculture Society* 46(4), 445–452.

Lafferty, K. D., Harvell, C. D., Conrad, J. M., Friedman, C. S., Kent, M. L., Kuris, A. M., Powell, E. N., Rondeau, D. and Saksida, S. M. 2015. Infectious diseases affect marine fisheries and aquaculture economics. *Annual Review of Marine Science* 7, 471–496.

Leiro, J. M., Castro, R., Arranz, J. A. and Lamas, J. 2007. Immunomodulating activities of acidic sulphated polysaccharides obtained from the seaweed *Ulva rigida* C. Agardh. *International Immunopharmacology* 7(7), 879–888.

Li, M., Wu, W., Zhou, P., Xie, F., Zhou, Q. and Mai, K. 2014. Comparison effect of dietary *astaxanthin* and *Haematococcus pluvialis* on growth performance, antioxidant status and immune response of large yellow croaker *Pseudosciaena crocea*. *Aquaculture* 434, 227–232.

Lim, K. C., Yusoff, F. M., Shariff, M., Kamarudin, M. S. and Nagao, N. 2019. Dietary supplementation of astaxanthin enhances hemato-biochemistry and innate immunity of Asian seabass, *Lates calcarifer* (Bloch, 1790). *Aquaculture* 512, 734339.

Liu, X., Steele, J. C. and Meng, X. Z. 2017. Usage, residue, and human health risk of antibiotics in Chinese aquaculture: a review. *Environmental Pollution* 223, 161–169.

Lu, C., Acién Fernández, F. G., Cañizares Guerrero, E., Hall, D. O. and Molina Grima, E. 2002. Overall assessment of *Monodus subterraneus* cultivation and EPA production in outdoor helical and bubble column reactors. *Journal of Applied Phycology* 14(5), 331–342.

Malea, P. and Kevrekidis, T. 2014. Trace element patterns in marine macroalgae. *Science of the Total Environment* 494–495, 144–157.

Marrion, O., Schwertz, A., Fleurence, J., Guéant, J. L. and Villaume, C. 2003. Improvement of the digestibility of the proteins of the red alga *Palmaria palmata* by physical processes and fermentation. *Nahrung* 47(5), 339–344.

Matos, Â. P., Feller, R., Moecke, E. H. S., de Oliveira, J. V., Junior, A. F., Derner, R. B. and Sant'Anna, E. S. 2016. Chemical characterization of six microalgae with potential utility for food application. *Journal of the American Oil Chemists' Society* 93(7), 963–972.

Mo, W. Y., Chen, Z., Leung, H. M. and Leung, A. O. W. 2017. Application of veterinary antibiotics in China's aquaculture industry and their potential human health risks. *Environmental Science and Pollution Research International* 24(10), 8978–8989.

Montgomery, W. L. and Gerking, S. D. 1980. Marine macroalgae as foods for fishes: an evaluation of potential food quality. *Environmental Biology of Fishes* 5(2), 143–153.

Morshedi, V., Bahabadi, M. N., Sotoudeh, E., Azodi, M. and Hafezieh, M. 2018. Nutritional evaluation of *Gracilaria pulvinata* as partial substitute with fish meal in practical diets of barramundi (*Lates calcarifer*). *Journal of Applied Phycology* 30(1), 619–628.

Mosha, S. S. 2019. The significance of *Spirulina* meal on fishmeal replacement in aquaculture: a review. *Journal of Fisheries and Aquaculture Development* 145. ISSN 2577-1493.

Mourente, G., Dick, J. R., Bell, J. G. and Tocher, D. R. 2005. Effect of partial substitution of dietary fish oil by vegetable oils on desaturation and β-oxidation of [1-^{14}C] 18: 3n − 3 (LNA) and [1-^{14}C] 20: 5n − 3 (EPA) in hepatocytes and enterocytes of European sea bass (*Dicentrarchus labrax* L.). *Aquaculture* 248(1–4), 173–186.

Moutinho, S., Linares, F., Rodríguez, J. L., Sousa, V. and Valente, L. M. P. 2018. Inclusion of 10% seaweed meal in diets for juvenile and on-growing life stages of Senegalese sole (*Solea senegalensis*). *Journal of Applied Phycology* 30(6), 3589–3601.

Mustafa, G., Wakamatsu, S., Takeda, T.-A., Umino, T. and Nakagawa, H. 1995. Effects of algae meal as feed additive on growth, feed efficiency, and body composition in red sea bream. *Fisheries Science* 61(1), 25–28.

Nakagawa, H. 1997. Effect of dietary algae on improvement of lipid metabolism in fish. *Biomedicine and Pharmacotherapy* 51(8), 345–348.

Nakagawa, H., Kasahara, S. and Sugiyama, T. 1987. Effect of *Ulva* meal supplementation on lipid metabolism of black sea bream, *Acanthopagrus schlegeli* (Bleeker). *Aquaculture* 62(2), 109–121.

Nakagawa, H. E., Mustafa, M. D. G. H., Takii, K. E., Umino, T. E. and Kumai, H. I. 2000. Effect of dietary catechin and *Spirulina* on vitamin C metabolism in red sea bream. *Fisheries Science* 66(2), 321–326.

Nakahara, T., Yokochi, T., Higashihara, T., Tanaka, S., Yaguchi, T. and Honda, D. 1996. Production of docosahexaenoic and docosapentaenoic acids by *Schizochytrium sp.* isolated from Yap Islands. *Journal of the American Oil Chemists' Society* 73(11), 1421–1426.

Nandeesha, M. C., Gangadhar, B., Varghese, T. J. and Keshavanath, P. 1998. Effect of feeding *Spirulina platensis* on the growth, proximate composition and organoleptic quality of common carp, *Cyprinus carpio* L. *Aquaculture Research* 29(5), 305–312.

Natrah, F. M. I., Bossier, P., Sorgeloos, P., Yusoff, F. M. and Defoirdt, T. 2014. Significance of microalgal-bacterial interactions for aquaculture. *Reviews in Aquaculture* 6(1), 48–61.

Nayak, S., Khozin-Goldberg, I., Cohen, G. and Zilberg, D. 2018. Dietary supplementation with ω6 LC-PUFA-rich algae modulates zebrafish immune function and improves resistance to *streptococcal* infection. *Frontiers in Immunology* 9, 1960.

Nayak, S., Al Ashhab, A., Zilberg, D. and Khozin-Goldberg, I. 2020. Dietary supplementation with omega-6 LC-PUFA-Rich microalgae regulates mucosal immune response and promotes microbial diversity in the zebrafish gut. *Biology* 9(6), 119.

Naylor, R. L., Goldburg, R. J., Primavera, J. H., Kautsky, N., Beveridge, M. C., Clay, J., Folke, C., Lubchenco, J., Mooney, H. and Troell, M. 2000. Effect of aquaculture on world fish supplies. *Nature* 405(6790), 1017–1024.

Nematipour, G. R. 1990. Effect of Chlorella-extract supplementation to diet on in vitro lipolysis in ayu. 日本水産学会誌 56, 777–782.

Neori, A. and Guttman, L. 2017. Thoughts on algae cultivation toward an expansion of aquaculture to the scale of agriculture. Proceedings of the 7th International Conference on Innovation in Chemical, Agricultural, Biological and Environmental Sciences (ICABES-2017), pp. 53–58.

Nielsen, M. M., Manns, D., D'Este, M., Krause-Jensen, D., Rasmussen, M. B., Larsen, M. M., Alvarado-Morales, M., Angelidaki, I. and Bruhn, A. 2016. Variation in biochemical composition of Saccharina latissima and Laminaria digitata along an estuarine salinity gradient in inner Danish waters. *Algal Research* 13, 235–245.

Nomura, M., Kamogawa, H., Susanto, E., Kawagoe, C., Yasui, H., Saga, N., Hosokawa, M. and Miyashita, K. 2013. Seasonal variations of total lipids, fatty acid composition, and fucoxanthin contents of *Sargassum horneri* (Turner) and *Cystoseira hakodatensis* (Yendo) from the northern seashore of Japan. *Journal of Applied Phycology* 25(4), 1159–1169.

Norambuena, F., Hermon, K., Skrzypczyk, V., Emery, J. A., Sharon, Y., Beard, A. and Turchini, G. M. 2015. Algae in fish feed: performances and fatty acid metabolism in juvenile Atlantic salmon. *PLoS ONE* 10(4), e0124042.

Olsen, R. L. and Hasan, M. R. 2012. A limited supply of fishmeal: impact on future increases in global aquaculture production. *Trends in Food Science and Technology* 27(2), 120–128.

Olvera-Novoa, M. A., Domínguez-Cen, L. J., Olivera-Castillo, L. and Martínez-Palacios, C. A. 1998. Effect of the use of the microalga *Spirulina maxima* as fish meal replacement in diets for tilapia, *Oreochromis mossambicus* (Peters), fry. *Aquaculture Research* 29(10), 709–715.

Øverland, M., Mydland, L. T. and Skrede, A. 2019. Marine macroalgae as sources of protein and bioactive compounds in feed for monogastric animals. *Journal of the Science of Food and Agriculture* 99(1), 13–24.

Park, H. J., Han, E., Lee, W. C., Kwak, J. H., Kim, H. C., Park, M. S. and Kang, C. K. 2015. Trophic structure in a pilot system for the integrated multi-trophic aquaculture off the east coast of Korean peninsula as determined by stable isotopes. *Marine Pollution Bulletin* 95(1), 207–214.

Patterson, D. and Gatlin III, D. M. 2013. Evaluation of whole and lipid-extracted algae meals in the diets of juvenile red drum (*Sciaenops ocellatus*). *Aquaculture* 416–417, 92–98.

Peixoto, M. J., Salas-Leitón, E., Pereira, L. F., Queiroz, A., Magalhães, F., Pereira, R., Abreu, H., Reis, P. A., Gonçalves, J. F. M. and de Almeida Ozório, R. O. 2016. Role of dietary seaweed supplementation on growth performance, digestive capacity and immune and stress responsiveness in European seabass (*Dicentrarchus labrax*). *Aquaculture Reports* 3, 189–197.

Perez-Velazquez, M., Gatlin III, D. M., González-Félix, M. L. and García-Ortega, A. 2018. Partial replacement of fishmeal and fish oil by algal meals in diets of red drum *Sciaenops ocellatus*. *Aquaculture* 487, 41–50.

Perez-Velazquez, M., Gatlin III, D. M., González-Félix, M. L., García-Ortega, A., de Cruz, C. R., Juárez-Gómez, M. L. and Chen, K. 2019. Effect of fishmeal and fish oil replacement by algal meals on biological performance and fatty acid profile of hybrid striped bass (*Morone crhysops*♀ × *M. saxatilis*♂). *Aquaculture* 507, 83–90.

Pradhan, J., Das, B. K., Sahu, S., Marhual, N. P., Swain, A. K., Mishra, B. K. and Eknath, A. E. 2012. Traditional antibacterial activity of freshwater microalga *Spirulina platensis* to aquatic pathogens. *Aquaculture Research* 43(9), 1287–1295.

Ragaza, J. A., Hossain, M. S., Meiler, K. A., Velasquez, S. F. and Kumar, V. 2020. A review on Spirulina: alternative media for cultivation and nutritive value as an aquafeed. *Reviews in Aquaculture* 12(4), 2371–2395.

Ragaza, J. A., Koshio, S., Mamauag, R. E., Ishikawa, M., Yokoyama, S. and Villamor, S. S. 2015. Dietary supplemental effects of red seaweed *Eucheuma denticulatum* on growth performance, carcass composition and blood chemistry of juvenile Japanese flounder, *Paralichthys olivaceus*. *Aquaculture Research* 46(3), 647–657.

Rajendran, P., Subramani, P. A. and Michael, D. 2016. Polysaccharides from marine macroalga, *Padina gymnospora* improve the nonspecific and specific immune responses of *Cyprinus carpio* and protect it from different pathogens. *Fish and Shellfish Immunology* 58, 220–228.

Raji, A. A., Alaba, P. A., Yusuf, H., Abu Bakar, N. H., Mohd Taufek, N., Muin, H., Alias, Z., Milow, P. and Abdul Razak, S. 2018. Fishmeal replacement with Spirulina Platensis and *Chlorella vulgaris* in African catfish (Clarias gariepinus) diet: effect on antioxidant enzyme activities and haematological parameters. *Research in Veterinary Science* 119, 67–75.

Rasmussen, R. S. 2001. Quality of farmed salmonids with emphasis on proximate composition, yield and sensory characteristics. *Aquaculture Research* 32(10), 767–786.

Ratcliff, J. J., Wan, A. H. L., Edwards, M. D., Soler-Vila, A., Johnson, M. P., Abreu, M. H. and Morrison, L. 2016. Metal content of kelp (*Laminaria digitata*) co-cultivated with Atlantic salmon in an Integrated Multi-Trophic Aquaculture system. *Aquaculture* 450, 234–243.

Ren, L. J., Ji, X. J., Huang, H., Qu, L., Feng, Y., Tong, Q. Q. and Ouyang, P. K. 2010. Development of a stepwise aeration control strategy for efficient docosahexaenoic acid production by *Schizochytrium* sp. *Applied Microbiology and Biotechnology* 87(5), 1649–1656.

Riccio, G. and Lauritano, C. 2020. Microalgae with immunomodulatory activities. *Marine Drugs* 18(1), 2.

Rioux, L.-E. and Turgeon, S. L. 2015. *Seaweed Carbohydrates, Seaweed Sustainability*. Elsevier, pp. 141–192.

Rizzo, C., Genovese, G., Morabito, M., Faggio, C., Pagano, M., Spanò, A., Zammuto, V., Minicante, S. A., Manghisi, A., Cigala, R. M., Crea, F., Marino, F. and Gugliandolo, C. 2017. Potential antibacterial activity of marine macroalgae against pathogens relevant for aquaculture and human health. *Journal of Pure and Applied Microbiology* 11(4), 1695–1706.

Roohani, A. M., Abedian Kenari, A., Fallahi Kapoorchali, M., Borani, M. S., Zoriezahra, S. J., Smiley, A. H., Esmaeili, M. and Rombenso, A. N. 2019. Effect of Spirulina *Spirulina platensis* as a complementary ingredient to reduce dietary fish meal on the growth performance, whole-body composition, fatty acid and amino acid profiles, and pigmentation of Caspian brown trout (*Salmo trutta* caspius) juveniles. *Aquaculture Nutrition* 25(3), 633–645.

Roohinejad, S., Koubaa, M., Barba, F. J., Saljoughian, S., Amid, M. and Greiner, R. 2017. Application of seaweeds to develop new food products with enhanced shelf-life, quality and health-related beneficial properties. *Food Research International* 99(3), 1066–1083.

Rosas, V. T., Poersch, L. H., Romano, L. A. and Tesser, M. B. 2019. Feasibility of the use of Spirulina in aquaculture diets. *Reviews in Aquaculture* 11(4), 1367–1378.

Roussel, S., Caralp, C., Leblanc, C., Le Grand, F., Stiger-Pouvreau, V., Coulombet, C., Le Goïc, N. and Huchette, S. 2019. Impact of nine macroalgal diets on growth and initial reproductive investment in juvenile abalone *Haliotis tuberculata*. *Aquaculture* 513, 734385.

Ryan, S., McLoughlin, P. and O'Donovan, O. 2012. A comprehensive study of metal distribution in three main classes of seaweed. *Environmental Pollution* 167, 171–177.

Ryckebosch, E., Bruneel, C., Termote-Verhalle, R., Goiris, K., Muylaert, K. and Foubert, I. 2014. Nutritional evaluation of microalgae oils rich in omega-3 long chain polyunsaturated fatty acids as an alternative for fish oil. *Food Chemistry* 160, 393–400.
Biochemical Engineering Journal 6(1), 13–18.
Saberi, A., Zorriehzahra, M., Emadi, H., Kakoolaki, S. and Fatemi, S. 2017. Effects of *Chlorella vulgaris* on blood and immunological parameters of Caspian Sea salmon (*Salmo trutta caspius*) fry exposed to Viral Nervous Necrosis (VNN) virus. مجله علوم شیلاتی ایران 16, 494–510.
Sáez, M. I., Vizcaíno, A., Galafat, A., Anguís, V., Fernández-Díaz, C., Balebona, M. C., Alarcón, F. J. and Martínez, T. F. 2020. Assessment of long-term effects of the macroalgae *Ulva ohnoi* included in diets on Senegalese sole (*Solea senegalensis*) fillet quality. *Algal Research* 47, 101885.
Safafar, H., Van Wagenen, J., Møller, P. and Jacobsen, C. 2015. Carotenoids, phenolic compounds and tocopherols contribute to the antioxidative properties of some microalgae species grown on industrial wastewater. *Marine Drugs* 13(12), 7339–7356.
Sahoo, T. K., Jena, P. K., Patel, A. K. and Seshadri, S. 2016. Bacteriocins and their applications for the treatment of bacterial diseases in aquaculture: a review. *Aquaculture Research* 47(4), 1013–1027.
Sakata, K., Kato, K., Iwase, Y., Okada, H., Ina, K. and Machiguchi, Y. 1991. Feeding-stimulant activity of algal glycerolipids for marine herbivorous gastropods. *Journal of Chemical Ecology* 17(1), 185–193.
Saleh, N. E., Ismail, R. F., Sayed, A. E. H., Zaghloul, E. H. and Saleh, H. 2020. Comprehensive assessment of benthic diatom (*Amphora coffeaeformis*) as a feed additive in Nile tilapia (*Oreochromis niloticus*) diet. *Aquaculture Research* 51(9), 3506 –3519.
Samuelsen, T. A., Oterhals, Å. and Kousoulaki, K. 2018. High lipid microalgae (*Schizochytrium sp.*) inclusion as a sustainable source of n-3 long-chain PUFA in fish feed–effects on the extrusion process and physical pellet quality. *Animal Feed Science and Technology* 236, 14–28.
Sánchez-Nuño, S., Carbonell, T. and Valls, A. I. 2019. *Redox Balance Affects Fish Welfare, Redox*. IntechOpen.
Sánchez, S., Martínez, M. E. and Espinola, F. 2000. Biomass production and biochemical variability of the marine microalga *Isochrysis galbana* in relation to culture medium.
Sarker, P. K., Kapuscinski, A. R., Lanois, A. J., Livesey, E. D., Bernhard, K. P. and Coley, M. L. 2016. Towards sustainable aquafeeds: complete substitution of fish oil with marine microalga *Schizochytrium sp.* improves growth and fatty acid deposition in juvenile Nile tilapia (*Oreochromis niloticus*). *PLoS ONE* 11(6), e0156684.
Scabini, V., Fernández-Palacios, H., Robaina, L., Kalinowski, T. and Izquierdo, M. S. 2011. Reproductive performance of gilthead seabream (*Sparus aurata* L., 1758) fed two combined levels of carotenoids from paprika oleoresin and essential fatty acids. *Aquaculture Nutrition* 17(3), 304–312.
Schmidt, V., Gomez-Chiarri, M., Roy, C., Smith, K. and Amaral-Zettler, L. 2017. Subtle microbiome manipulation using probiotics reduces antibiotic-associated mortality in fish. *mSystems* 2(6).
Seong, T., Kitagima, R., Haga, Y. and Satoh, S. 2020. Non-fish meal, non-fish oil diet development for red sea bream, *Pagrus major*, with plant protein and graded levels

of *Schizochytrium* sp.: effect on growth and fatty acid composition. *Aquaculture Nutrition* 26(4), 1173–1185.

Shah, M. R., Lutzu, G. A., Alam, A., Sarker, P., Kabir Chowdhury, M. A., Parsaeimehr, A., Liang, Y. and Daroch, M. 2018. Microalgae in aquafeeds for a sustainable aquaculture industry. *Journal of Applied Phycology* 30(1), 197–213.

Shi, X., Chen, F., Chen, G. H., Pan, Y. X., Zhu, X. M., Liu, X. and Luo, Z. 2017a. Fishmeal can be totally replaced by a mixture of rapeseed meal and *Chlorella* meal in diets for crucian carp (*Carassius auratus* gibelio). *Aquaculture Research* 48(11), 5481–5489.

Shi, X., Luo, Z., Chen, F., Huang, C., Zhu, X. -M. and Liu, X. 2017b. Effects of dietary cellulase addition on growth performance, nutrient digestibility and digestive enzyme activities of juvenile crucian carp *Carassius auratus*. *Aquaculture Nutrition* 23(3), 618–628.

Shi, X., Luo, Z., Chen, F., Wei, C. C., Wu, K., Zhu, X. M. and Liu, X. 2017c. Effect of fish meal replacement by *Chlorella* meal with dietary cellulase addition on growth performance, digestive enzymatic activities, histology and myogenic genes' expression for crucian carp *Carassius auratus*. *Aquaculture Research* 48(6), 3244–3256.

Shobharani, P., Halami, P. M. and Sachindra, N. M. 2013. Potential of marine lactic acid bacteria to ferment *Sargassum* sp. for enhanced anticoagulant and antioxidant properties. *Journal of Applied Microbiology* 114(1), 96–107.

Shpigel, M., Guttman, L., Shauli, L., Odintsov, V., Ben-Ezra, D. and Harpaz, S. 2017. *Ulva lactuca* from an integrated multi-trophic aquaculture (IMTA) biofilter system as a protein supplement in gilthead seabream (*Sparus aurata*) diet. *Aquaculture* 481, 112–118.

Shpigel, M., Shauli, L., Odintsov, V., Ben-Ezra, D., Neori, A. and Guttman, L. 2018. The sea urchin, *Paracentrotus lividus*, in an Integrated Multi-Trophic Aquaculture (IMTA) system with fish (*Sparus aurata*) and seaweed (*Ulva lactuca*): nitrogen partitioning and proportional configurations. *Aquaculture* 490, 260–269.

Sirakov, I., Velichkova, K., Stoyanova, S. and Staykov, Y. 2015. The importance of microalgae for aquaculture industry. Review. *International Journal of Fisheries and Aquatic Studies* 2, 81–84.

Sørensen, M., Berge, G. M., Reitan, K. I. and Ruyter, B. 2016. Microalga *Phaeodactylum tricornutum* in feed for Atlantic salmon (*Salmo salar*)–effect on nutrient digestibility, growth and utilization of feed. *Aquaculture* 460, 116–123.

Sotoudeh, E. and Jafari, M. 2017. Effects of dietary supplementation with red seaweed, *Gracilaria pygmaea*, on growth, carcass composition and hematology of juvenile rainbow trout, *Oncorhynchus mykiss*. *Aquaculture International* 25(5), 1857–1867.

Sprague, M., Walton, J., Campbell, P. J., Strachan, F., Dick, J. R. and Bell, J. G. 2015. Replacement of fish oil with a DHA-rich algal meal derived from *Schizochytrium* sp. on the fatty acid and persistent organic pollutant levels in diets and flesh of Atlantic salmon (*Salmo salar*, L.) post-smolts. *Food Chemistry* 185, 413–421.

Stanley, J. G. and Jones, J. B. 1976. Feeding algae to fish. *Aquaculture* 7(3), 219–223.

Stoneham, T. R., Kuhn, D. D., Taylor, D. P., Neilson, A. P., Smith, S. A., Gatlin, D. M., Chu, H. S. S. and O'Keefe, S. F. 2018. Production of omega-3 enriched tilapia through the dietary use of algae meal or fish oil: improved nutrient value of fillet and offal. *PLoS ONE* 13(4), e0194241.

Storebakken, T., Foss, P., Schiedt, K., Austreng, E., Liaaen-Jensen, S. and Manz, U. 1987. Carotenoids in diets for salmonids: IV. Pigmentation of Atlantic salmon with *astaxanthin, astaxanthin dipalmitate* and *canthaxanthin*. *Aquaculture* 65(3–4), 279–292.

Tacon, A. G. J. and Metian, M. 2008. Global overview on the use of fish meal and fish oil in industrially compounded aquafeeds: trends and future prospects. *Aquaculture* 285(1-4), 146-158.

Teuling, E., Schrama, J. W., Gruppen, H. and Wierenga, P. A. 2017. Effect of cell wall characteristics on algae nutrient digestibility in Nile tilapia (*Oreochromis niloticus*) and African catfish (*Clarus gariepinus*). *Aquaculture* 479, 490-500.

Teuling, E., Wierenga, P. A., Agboola, J. O., Gruppen, H. and Schrama, J. W. 2019. Cell wall disruption increases bioavailability of *Nannochloropsis gaditana* nutrients for juvenile Nile tilapia (*Oreochromis niloticus*). *Aquaculture* 499, 269-282.

Tibaldi, E., Chini Zittelli, G., Parisi, G., Bruno, M., Giorgi, G., Tulli, F., Venturini, S., Tredici, M. R. and Poli, B. M. 2015. Growth performance and quality traits of European sea bass (*D. labrax*) fed diets including increasing levels of freeze-dried *Isochrysis* sp. (T-ISO) biomass as a source of protein and n-3 long chain PUFA in partial substitution of fish derivatives. *Aquaculture* 440, 60-68.

Tibbetts, S. M., Mann, J. and Dumas, A. 2017. Apparent digestibility of nutrients, energy, essential amino acids and fatty acids of juvenile Atlantic salmon (*Salmo salar* L.) diets containing whole-cell or cell-ruptured *Chlorella vulgaris* meals at five dietary inclusion levels. *Aquaculture* 481, 25-39.

Tomás-Almenar, C., Larrán, A. M., de Mercado, E., Sanz-Calvo, M. A., Hernández, D., Riaño, B. and García-González, M. C. 2018. Scenedesmus almeriensis from an integrated system waste-nutrient, as sustainable protein source for feed to rainbow trout (*Oncorhynchus mykiss*). *Aquaculture* 497, 422-430.

Tulli, F., Chini Zittelli, G., Giorgi, G., Poli, B. M., Tibaldi, E. and Tredici, M. R. 2012. Effect of the inclusion of dried *Tetraselmis suecica* on growth, feed utilization, and fillet composition of European sea bass juveniles fed organic diets. *Journal of Aquatic Food Product Technology* 21(3), 188-197.

Upling, J. Y. 2020. A review on waste absorption efficiency of different extractive integrated multi-trophic aquaculture (IMTA) species: implications in coastal and offshore aquaculture waste management. *GSC Biological and Pharmaceutical Sciences* 11(2), 257-264.

Valente, L. M. P., Gouveia, A., Rema, P., Matos, J., Gomes, E. F. and Pinto, I. S. 2006. Evaluation of three seaweeds *Gracilaria bursa pastoris*, *Ulva rigida* and *Gracilaria cornea* as dietary ingredients in European sea bass (*Dicentrarchus labrax*) juveniles. *Aquaculture* 252(1), 85-91.

Valente, L. M. P., Custódio, M., Batista, S., Fernandes, H. and Kiron, V. 2019. Defatted microalgae (*Nannochloropsis* sp.) from biorefinery as a potential feed protein source to replace fishmeal in European sea bass diets. *Fish Physiology and Biochemistry* 45(3), 1067-1081.

Van Hoestenberghe, S., Fransman, C. A., Luyten, T., Vermeulen, D., Roelants, I., Buysens, S. and Goddeeris, B. M. 2016. *Schizochytrium* as a replacement for fish oil in a fishmeal free diet for jade perch, *Scortum barcoo* (McCulloch & Waite). *Aquaculture Research* 47(6), 1747-1760.

Van Vo, B., Siddik, M. A. B., Fotedar, R., Chaklader, M. R., Hanif, M. A., Foysal, M. J. and Nguyen, H. Q. 2020. Progressive replacement of fishmeal by raw and enzyme-treated alga, *Spirulina platensis* influences growth, intestinal micromorphology and stress response in juvenile barramundi, Lates calcarifer. *Aquaculture* 529, 735741.

Velichkova, K., Sirakov, I. and Denev, S. 2019. In vitro antibacterial effect of *Lemna minuta*, *Chlorella vulgaris* and *Spirulina sp.* extracts against fish pathogen *Aeromonas hydrophila*. *Aquaculture, Aquarium, Conservation & Legislation* 12, 936–940.

Veloso, V., Reis, A., Gouveia, L., Fernandes, H. L., Empis, J. A. and Novais, J. M. 1991. Lipid production by *Phaeodactylum tricornutum*. *Bioresource Technology* 38(2-3), 115–119.

Viera, M. P., De Vicose, G. C., Gómez-Pinchetti, J. L., Bilbao, A., Fernandez-Palacios, H. and Izquierdo, M. S. 2011. Comparative performances of juvenile abalone (*Haliotis tuberculata* coccinea Reeve) fed enriched vs non-enriched macroalgae: effect on growth and body composition. *Aquaculture* 319(3-4), 423–429.

Viera, M. P., Gómez Pinchetti, J. L., Courtois de Vicose, G., Bilbao, A., Suárez, S., Haroun, R. J. and Izquierdo, M. S. 2005. Suitability of three red macroalgae as a feed for the abalone *Haliotis tuberculata* coccinea Reeve. *Aquaculture* 248(1-4), 75–82.

Vílchez, C., Forján, E., Cuaresma, M., Bédmar, F., Garbayo, I. and Vega, J. M. 2011. Marine carotenoids: biological functions and commercial applications. *Marine Drugs* 9(3), 319–333.

Vizcaíno, A. J., López, G., Sáez, M. I., Jiménez, J. A., Barros, A., Hidalgo, L., Camacho-Rodríguez, J., Martínez, T. F., Cerón-García, M. C. and Alarcón, F. J. 2014. Effects of the microalga *Scenedesmus almeriensis* as fishmeal alternative in diets for gilthead sea bream, *Sparus aurata*, juveniles. *Aquaculture* 431, 34–43.

Vizcaíno, A. J., Rodiles, A., López, G., Sáez, M. I., Herrera, M., Hachero, I., Martínez, T. F., Cerón-García, M. C. and Alarcón, F. J. 2018. Growth performance, body composition, and digestive functionality of Senegalese sole (*Solea senegalensis* Kaup, 1858) juveniles fed diets including microalgae freeze-dried biomass. *Fish Physiology and Biochemistry* 44(2), 661–677.

Vizcaíno, A. J., Saéz, M. I., López, G., Arizcun, M., Abellán, E., Martínez, T. F., Cerón-García, M. C. and Alarcón, F. J. 2016a. *Tetraselmis suecia* and *Tisochrysis lutea* meal as dietary ingredients for gilthead sea bream (*Sparus aurata* L.) fry. *Journal of Applied Phycology* 28(5), 2843–2855.

Vizcaíno, A. J., Mendes, S. I., Varela, J. L., Ruiz-Jarabo, I., Rico, R., Figueroa, F. L., Abdala, R., Moriñigo, M. Á., Mancera, J. M. and Alarcón, F. J. 2016b. Growth, tissue metabolites and digestive functionality in *Sparus aurata* juveniles fed different levels of macroalgae, *Gracilaria cornea* and *Ulva rigida*. *Aquaculture Research* 47(10), 3224–3238.

Walker, A. B. and Berlinsky, D. L. 2011. Effects of partial replacement of fish meal protein by microalgae on growth, feed intake, and body composition of Atlantic cod. *North American Journal of Aquaculture* 73, 76–83.

Wan, A. H. L., Davies, S. J., Soler-Vila, A., Fitzgerald, R. and Johnson, M. P. 2019. Macroalgae as a sustainable aquafeed ingredient. *Reviews in Aquaculture* 11(3), 458–492.

Wan, A. H. L., Soler-Vila, A., O'Keeffe, D., Casburn, P., Fitzgerald, R. and Johnson, M. P. 2016. The inclusion of *Palmaria palmata* macroalgae in Atlantic salmon (*Salmo salar*) diets: effects on growth, haematology, immunity and liver function. *Journal of Applied Phycology* 28(5), 3091–3100.

Wang, A., Ran, C., Wang, Y., Zhang, Z., Ding, Q., Yang, Y., Olsen, R. E., Ringø, E., Bindelle, J. and Zhou, Z. 2019. Use of probiotics in aquaculture of China–a review of the past decade. *Fish and Shellfish Immunology* 86, 734–755.

Wang, M., Cheng, H., Chen, S., Wen, S., Wu, X., Zhang, D., Yuan, Q. and Cong, W. 2018. Microalgal cell disruption via extrusion for the production of intracellular valuables. *Energy* 142, 339-345.

Wassef, E. A., El-Sayed, A. F. M., Kandeel, K. M. and Sakr, E. M. 2005. Evaluation of Pterocla Dia (*Rhodophyta*) and Ulva (*Chlorophyta*) meals as additives to gilthead seabream *Sparus aurata* diets. *Egyptian Journal of Aquatic Research* 31, 321-332.

Watanabe, T. and Vassallo-Agius, R. 2003. Broodstock nutrition research on marine finfish in Japan. *Aquaculture* 227(1-4), 35-61.

Xie, D., Li, X., You, C., Wang, S. and Li, Y. 2019. Supplementation of macroalgae together with non-starch polysaccharide-degrading enzymes in diets enhanced growth performance, innate immune indexes, and disease resistance against Vibrio parahaemolyticus in rabbitfish *Siganus canaliculatus*. *Journal of Applied Phycology* 31(3), 2073-2083.

Xu, S., Zhang, L., Wu, Q., Liu, X., Wang, S., You, C. and Li, Y. 2011. Evaluation of dried seaweed *Gracilaria lemaneiformis* as an ingredient in diets for teleost fish *Siganus canaliculatus*. *Aquaculture International* 19(5), 1007-1018.

Xu, W., Gao, Z., Qi, Z., Qiu, M., Peng, J.-q. and Shao, R. 2014. Effect of dietary Chlorella on the growth performance and physiological parameters of gibel carp, *Carassius auratus* gibelio. *Turkish Journal of Fisheries and Aquatic Sciences* 14, 53-57.

Xuan, X., Li, W., Zhu, W. and Wang, S. 2019. Effects of different levels of macroalga *Gracilaria lemaneiformis* on growth performance and feed utilization on the red sea bream, *Pagrosomus major*. *Journal of Applied Phycology* 31(5), 3213-3222.

Xuan, X., Wen, X., Li, S., Zhu, D. and Li, Y. 2013. Potential use of macro-algae *Gracilaria lemaneiformis* in diets for the black sea bream, *Acanthopagrus schlegelii*, juvenile. *Aquaculture* 412-413, 167-172.

Xue, X., Hall, J. R., Caballero-Solares, A., Eslamloo, K., Taylor, R. G., Parrish, C. C. and Rise, M. L. 2020. Liver transcriptome profiling reveals that dietary DHA and EPA levels influence suites of genes involved in metabolism, redox homeostasis, and immune function in Atlantic salmon (*Salmo salar*). *Marine Biotechnology* 22(2), 263-284.

Yarnold, J., Karan, H., Oey, M. and Hankamer, B. 2019. Microalgal aquafeeds as part of a circular bioeconomy. *Trends in Plant Science* 24(10), 959-970.

Yengkhom, O., Shalini, K. S., Subramani, P. A. and Michael, R. D. 2018. Non-specific immunity and disease resistance are enhanced by the polysaccharide fraction of a marine chlorophycean macroalga in *Oreochromis niloticus* (Linnaeus, 1758). *Journal of Applied Ichthyology* 34(3), 556-567.

Yu, C., Liu, G., Yu, J., Lin, F. and Wen, X. 2020. Dietary *Saccharina japonica* is a natural and effective tool to fortify marine teleost black sea bream. *Journal of Applied Phytology* 32, 3447-3456.

Zahran, E. and Risha, E. 2014. Modulatory role of dietary *Chlorella vulgaris* powder against arsenic-induced immunotoxicity and oxidative stress in Nile tilapia (*Oreochromis niloticus*). *Fish and Shellfish Immunology* 41(2), 654-662.

Zeynali, M., Nafisi Bahabadi, M., Morshedi, V., Ghasemi, A. and Torfi Mozanzadeh, M. 2020. Replacement of dietary fishmeal with *Sargassum ilicifolium* meal on growth, innate immunity and immune gene mRNA transcript abundance in *Lates calcarifer* juveniles. *Aquaculture Nutrition* 26(5), 1657-1668.

Zhang, F., Man, Y. B., Mo, W. Y. and Wong, M. H. 2020. Application of *Spirulina* in aquaculture: a review on wastewater treatment and fish growth. *Reviews in Aquaculture* 12(2), 582-599.

Zhang, Q., Qiu, M., Xu, W., Gao, Z., Shao, R. and Qi, Z. 2014. Effects of dietary administration of *Chlorella* on the immune status of gibel carp, *Carassius auratus* gibelio. *Italian Journal of Animal Science* 13(3), 3168.

Zhu, D., Wen, X., Li, S., Xuan, X. and Li, Y. 2017. Evaluation of the red alga *Gracilaria lemaneiformis* and brown alga *Sargassum horneri* as ingredients in diets for white spotted snapper *Lutjanus stellatus* Akazaki juveniles. *Journal of Applied Phycology* 29(6), 3211–3219.

Index